Primality Testing and Integer Factorization in Public-Key Cryptography

Second Edition

Advances in Information Security

Sushil Jajodia

Consulting Editor
Center for Secure Information Systems
George Mason University
Fairfax, VA 22030-4444
email: jajodia@gmu.edu

The goals of the Springer International Series on ADVANCES IN INFORMATION SECURITY are, one, to establish the state of the art of, and set the course for future research in information security and, two, to serve as a central reference source for advanced and timely topics in information security research and development. The scope of this series includes all aspects of computer and network security and related areas such as fault tolerance and software assurance.

ADVANCES IN INFORMATION SECURITY aims to publish thorough and cohesive overviews of specific topics in information security, as well as works that are larger in scope or that contain more detailed background information than can be accommodated in shorter survey articles. The series also serves as a forum for topics that may not have reached a level of maturity to warrant a comprehensive textbook treatment.

Researchers, as well as developers, are encouraged to contact Professor Sushil Jajodia with ideas for books under this series.

Additional titles in the series:

INSIDER ATTACK AND CYBER SECURITY: Beyond the Hacker edited by Salvatore Stolfo, Steven M. Bellovin, Angelos D. Keromytis, Sara Sinclaire, Sean W. Smith; ISBN: 978-0-387-77321-6

INTRUSION DETECTION SYSTEMS edited by Robert Di Pietro and Luigi V. Mancini; ISBN: 978-0-387-77265-3

VULNERABILITY ANALYSIS AND DEFENSE FOR THE INTERNET edited by Abhishek Singh; ISBN: 978-0-387-74389-9

BOTNET DETECTION: Countering the Largest Security Threat edited by Wenke Lee, Cliff Wang and David Dagon; ISBN: 978-0-387-68766-7

PRIVACY-RESPECTING INTRUSION DETECTION by Ulrich Flegel; ISBN: 978-0-387-68254-9

SYNCHRONIZING INTERNET PROTOCOL SECURITY (SIPSec) by Charles A. Shoniregun; ISBN: 978-0-387-32724-2

SECURE DATA MANAGEMENT IN DECENTRALIZED SYSTEMS edited by Ting Yu and Sushil Jajodia; ISBN: 978-0-387-27694-6

NETWORK SECURITY POLICIES AND PROCEDURES by Douglas W. Frye; ISBN: 0-387-30937-3

DATA WAREHOUSING AND DATA MINING TECHNIQUES FOR CYBER SECURITY by Anoop Singhal; ISBN: 978-0-387-26409-7

SECURE LOCALIZATION AND TIME SYNCHRONIZATION FOR WIRELESS SENSOR AND AD HOC NETWORKS edited by Radha Poovendran, Cliff Wang, and Sumit Roy; ISBN: 0-387-32721-5

PRESERVING PRIVACY IN ON-LINE ANALYTICAL PROCESSING (OLAP) by Lingyu Wang, Sushil Jajodia and Duminda Wijesekera; ISBN: 978-0-387-46273-8

SECURITY FOR WIRELESS SENSOR NETWORKS by Donggang Liu and Peng Ning; ISBN: 978-0-387-32723-5

MALWARE DETECTION edited by Somesh Jha, Cliff Wang, Mihai Christodorescu, Dawn Song, and Douglas Maughan; ISBN: 978-0-387-32720-4

Additional information about this series can be obtained from http://www.springer.com

Primality Testing and Integer Factorization in Public-Key Cryptography

Second Edition

by

Song Y. Yan

Harvard University
and
Massachusetts Institute of Technology
USA

 Springer

Author:
Dr. Song Y. Yan
Visiting Professor
Department of Mathematics
Harvard University
One Oxford Street
Cambridge, MA 02138-2901
syan@math.harvard.edu

and

Department of Mathematics
Massachusetts Institute of Technology
77 Massachusetts Avenue
Cambridge, MA 02139-4307
syan@math.mit.edu

Library of Congress Control Number: 2008935407

ISBN-13: 978-0-387-77267-7   e-ISBN-13: 978-0-387-77268-4

Printed on acid-free paper

springer.com

In Memory of Prof Shiing-Shen Chern (1911–2004)

Founding Director, Mathematical Sciences Research Institute, Berkeley

Table of Contents

Preface to the Second Edition ix

Preface to the First Edition xi

1. Number-Theoretic Preliminaries 1
1.1 Problems in Number Theory 1
1.2 Groups, Rings and Fields 13
1.3 Divisibility Properties 23
1.4 Euclid's Algorithm and Continued Fractions 34
1.5 Arithmetic Functions $\sigma(n), \tau(n), \phi(n), \lambda(n), \mu(n)$ 50
1.6 Linear Congruences 63
1.7 Quadratic Congruences 85
1.8 Primitive Roots and Power Residues 103
1.9 Arithmetic of Elliptic Curves 113
1.10 Chapter Notes and Further Reading 124

2. Primality Testing and Prime Generation 127
2.1 Computing with Numbers and Curves 127
2.2 Riemann ζ and Dirichlet L Functions 139
2.3 Rigorous Primality Tests 149
2.4 Compositeness and Pseudoprimality Tests 157
2.5 Lucas Pseudoprimality Test 168
2.6 Elliptic Curve Primality Tests 172
2.7 Superpolynomial-Time Tests 177
2.8 Polynomial-Time Tests 182
2.9 Comparison of General Purpose Primality Tests 188
2.10 Primality Tests for Special Numbers 192
2.11 Prime Number Generation 201
2.12 Chapter Notes and Further Reading 207

3. Integer Factorization and Discrete Logarithms 209
3.1 Introduction 209
3.2 Simple Factoring Methods 212
3.3 Elliptic Curve Method (ECM) 221

3.4   General Factoring Congruence 226
3.5   Continued FRACtion Method (CFRAC) 230
3.6   Quadratic Sieve (QS) 234
3.7   Number Field Sieve (NFS) 239
3.8   Quantum Factoring Algorithm 251
3.9   Discrete Logarithms 257
3.10 kth Roots 270
3.11   Elliptic Curve Discrete Logarithms 278
3.12   Chapter Notes and Further Reading 285

4.   **Number-Theoretic Cryptography** 287
4.1   Public-Key Cryptography 287
4.2   RSA Cryptosystem 292
4.3   Security and Cryptanalysis of RSA 301
4.4   Rabin Cryptography 314
4.5   Quadratic Residuosity Cryptography 320
4.6   Discrete Logarithm Cryptography 326
4.7   Elliptic Curve Cryptography 331
4.8   Zero-Knowledge Techniques 338
4.9   Deniable Authentication 341
4.10   Non-Factoring Based Cryptography 346
4.11   Chapter Notes and Further Reading 351

Bibliography 353

Index 367

About the Author 373

Preface to the Second Edition

The mathematician's patterns, like the painter's or the poet's must be beautiful; the ideas, like the colours or the words must fit together in a harmonious way. Beauty is the first test: there is no permanent place in this world for ugly mathematics.

G. H. HARDY (1877–1947)

The success of the first edition of the book encourages me to prepare this second edition. I have taken this opportunity to try to make the book as updated and self-contained as possible by including new developments and results in the field. Notable features of this new edition are that several new sections and more than 100 new pages are added. These include a new section in Chapter 2 on the comparison of the Rabin-Miller probabilistic test in $\mathcal{RP}$, the Atkin-Morain elliptic curve test in $\mathcal{ZPP}$ and the AKS deterministic test in $\mathcal{P}$, a new section in Chapter 3 on recent work in quantum factoring, and a new section in Chapter 4 on post-quantum cryptography. To make the book suitable as an advanced undergraduate and/or postgraduate text/reference, about ten problems at various levels of difficulty are added at the end of each section, making about 300 problems in total contained in the book; most of problems are research-oriented and with prizes offered by individuals or organizations to a total amount over five million US dollars.

During the preparation of the book, I had a good opportunity to talk and to discuss with the later Prof Shiing-Shen Chern, the first Director of the Mathematical Sciences Research Institute in Berkeley, many times in his Home of Geometry at Nankai University, Tianjin, when I was a Specially Appointed Visiting Professor at Nankai. Although a geometer, Prof Chern had long been interested in number theory, partly because many of his close friends, e.g., André Weil (1906–1998), were working in both geometry and number theory. In contrast to G. H. Hardy (1877–1947) and L. E. Dickson (1874–1954), Chern always regarded number theory as a branch of applied mathematics as it is applicable to many other branches of mathematics and other science subjects. Special thanks must be given to Prof Michael Siper of the Massachusetts Institute of Technology, Prof Benedick Gross and Prof

Barry Mazue of Harvard University, Prof Glyn James of Coventry University, Prof Yuan Wang and Prof Zhexian Wan of Chinese Academy of Sciences, Prof Zikun Wang of Beijing Normal University, and the editors, Susan Lagerstrom-Fife and Sharon Palleschi, of Springer in Boston for their encouragements and help during the preparation of this second edition. Finally, I would like to thank Prof Glyn James, Prof Zuowen Tan and Prof Duanqiang Xie for reading the manuscripts of the book.

This work was financially supported by a Global Research Award from the Royal Academy of Engineering, London, UK to work at Harvard University and the Massachusetts Institute of Technology. The author is very grateful to Prof Ivor Smith, Prof John McWhirter and Dr Chris Coulter of the Royal Academy of Engineering for their support, and to Prof Cliff Taubes, Chairman of the Department of Mathematics at Harvard University and Prof Michael Sipser, Head of the Department of Mathematics at the Massachusetts Institute of Technology for appointing me to work at the two institutions.

Cambridge, Massachusetts, September 2008   S. Y. Y.

Preface to the First Edition

*The problem of distinguishing prime numbers from composite, and
of resolving composite numbers into their prime factors, is one of
the most important and useful in all arithmetic. ... The dignity of
science seems to demand that every aid to the solution of such an
elegant and celebrated problem be zealously cultivated.*

C. F. GAUSS (1777–1855)

Primality testing and integer factorization, as identified by Gauss in his *Disquisitiones Arithmeticae*, Article 329, in 1801, are the two most fundamental problems (as well as two most important research fields) in number theory. With the advent of modern computers, they have also been found unexpected applications in public-key cryptography and information security. In this book, we shall introduce various methods/algorithms for primality testing and integer factorization, and their applications in public-key cryptography and information security. More specifically, we shall first review some basic concepts and results in number theory in Chapter 1. Then in Chapter 2 we shall discuss various algorithms for primality testing and prime number generation, with an emphasis on the Miller-Rabin probabilistic test, the Goldwasser-Kilian and Atkin-Morain ellptic curve tests, and Agrawal-Kayal-Saxena deterministic test for primality. There is also an introduction to large prime number generation in Chapter 2. In Chapter 3 we shall introduce various algorithms, particularly the Elliptic Curve Method (ECM), the Quadratic Sieve (QS) and the Number Field Sieve (NFS) for integer factorization. Also in Chapter 3 we shall discuss some other computational problems that are related to factoring, such as the square root problem, the discrete logarithm problem and the quadratic residuosity problem. In Chapter 4, we shall discuss the most widely used cryptographic systems based on the intractability of the integer factorization, square roots, discrete logarithms, elliptic curve discrete logarithms and quadratic residuosity problems.

I have tried to make this book as self-contained as possible, so that it can be used either as a textbook suitable for a course for final-year undergraduate or first-year postgraduate students, or as a basic reference in the field.

Acknowledgments

I would like to thank the three anonymous referees for their very helpful comments and kind encouragements. I would also like to thank Susan Lagerstrom-Fife and Sharon Palleschi of Kluwer Academic Publishers for her encouragements. Special thanks must be given to Bob Newman for his support during the preparation of this book. Finally, I would like to thank Prof Shiing-Shen Chern, Director Emeritus of the Mathematical Sciences Research Institute in Berkeley for his kind encouragements and guidance.

Coventry, England, September 2003   S. Y. Y.

Notation

Notation	Explanation
$\mathbb{N}$	set of natural numbers: $\mathbb{N} = \{1, 2, 3, \ldots\}$
$\mathbb{Z}$	set of integers (whole numbers): $\mathbb{Z} = \{0, \pm n : n \in \mathbb{N}\}$
$\mathbb{Z}^+$	set of positive integers: $\mathbb{Z}^+ = \mathbb{N}$
$\mathbb{Z}_{>1}$	set of positive integers greater than 1: $\mathbb{Z}_{>1} = \{n : n \in \mathbb{Z} \text{ and } n > 1\}$
$\mathbb{Q}$	set of rational numbers: $\mathbb{Q} = \left\{\dfrac{a}{b} : a, b \in \mathbb{Z} \text{ and } b \neq 0\right\}$
$\mathbb{R}$	set of real numbers: $\mathbb{R} = \{n + 0.d_1 d_2 d_3 \cdots : n \in \mathbb{Z}, \ d_i \in \{0, 1, \ldots, 9\}$ and no infinite sequence of 9's appears$\}$
$\mathbb{C}$	set of complex numbers: $\mathbb{C} = \{a + bi : a, b \in \mathbb{R} \text{ and } i = \sqrt{-1}\}$
$\mathbb{Z}/n\mathbb{Z}$	also denoted by $\mathbb{Z}_n$, residue classes modulo n; ring of integers modulo n; field if n is prime
$(\mathbb{Z}/n\mathbb{Z})^*$	multiplicative group; the elements of this group are the elements in $\mathbb{Z}/n\mathbb{Z}$ that are relatively prime to n: $(\mathbb{Z}/n\mathbb{Z})^* = \{[a]_n \in \mathbb{Z}/n\mathbb{Z} : \ \gcd(a, n) = 1\}$
$\#((\mathbb{Z}/n\mathbb{Z})^*)$	also denoted by $|(\mathbb{Z}/n\mathbb{Z})^*|$, order of the group $(\mathbb{Z}/n\mathbb{Z})^*$, i.e., the number of elements in the group
$\mathbb{F}_p$	also denoted by $\mathbb{Z}/p\mathbb{Z}$, finite field with p elements, where p is a prime number

$\mathbb{F}_q$	finite field with $q = p^k$ a prime power
$\mathbb{Z}[x]$	set of polynomials with integer coefficients
$\mathbb{Z}_n[x]$	set of polynomials with coefficients from $\mathbb{Z}/n\mathbb{Z}$
$\mathbb{Z}[x]/h(x)$	set of polynomials modulo polynomial $h(x)$, with integer coefficients
$\mathbb{Z}_p[x]/h(x)$	also denoted by $\mathbb{F}_p[x]/h(x)$; set of polynomials modulo polynomial $h(x)$, with coefficients from $\mathbb{Z}_p$
G	group
$|G|$	also denoted by $\#(G)$, order of group G
R	ring
K	(arbitrary) field
E	elliptic curve $y^2 = x^3 + ax + b$
$E/\mathbb{Q}$	elliptic curve over $\mathbb{Q}$
$E/\mathbb{Z}/n\mathbb{Z}$	elliptic curve over $\mathbb{Z}/n\mathbb{Z}$
$E/\mathbb{F}_p$	elliptic curve over $\mathbb{F}_p$
$\mathcal{O}_E$	point at infinity on E
$E(\mathbb{Q})$	elliptic curve group formed by points on $E/\mathbb{Q}$
$|E(\mathbb{Q})|$	number of points in $E(\mathbb{Q})$
$\Delta(E)$	discriminant of E, $\Delta(E) = -16(4a^3 + 27b^2) \neq 0$
F_n	Fermat numbers: $F_n = 2^{2^n} + 1$, $n \geq 0$
$\mathcal{P}$	class of problems solvable in deterministic polynomial time
$\mathcal{NP}$	class of problems solvable in non-deterministic polynomial time
$\mathcal{RP}$	class of problems solvable in random polynomial time with one-sided errors
$\mathcal{BPP}$	class of problems solvable in random polynomial time with two-sided errors
$\mathcal{ZPP}$	class of problems solvable in random polynomial time with zero errors
IFP	Integer Factorization Problem
DLP	Discrete Logarithm Problem
ECDLP	Elliptic Curve Discrete Logarithm Problem
SQRTP	SQuare RooT Problem

QRP	Quadratic Residuosity Problem
CFRAC	Continued FRACtion method (for factoring)
ECM	Elliptic Curve Method
NFS	Number Field Sieve
QS/MPQS	Quadratic Sieve/Multiple Polynomial Quadratic Sieve
ECPP	Elliptic Curve Primality Proving
DHM	Diffie-Hellman-Merkle
RSA	Rivest-Shamir-Adleman
DSA/DSS	Digital Signature Algorithm/Digital Signature Standard
$a \mid b$	a divides b
$a \nmid b$	a does not divide b
$p^{\alpha} \| n$	$p^{\alpha} \mid n$ but $p^{\alpha+1} \nmid n$
$\gcd(a, b)$	greatest common divisor of (a, b)
$\text{lcm}(a, b)$	least common multiple of (a, b)
$\lfloor x \rfloor$	floor: also denoted by $[x]$; the greatest integer less than or equal to x
$\lceil x \rceil$	ceiling: the least integer greater than or equal to x
$x \bmod \mathrm{n}$	remainder: $x - n \left\lfloor \dfrac{x}{n} \right\rfloor$
$x = y \bmod n$	x is equal to y reduced to modulo n
$x \equiv y \pmod{\mathrm{n}}$	x is congruent to y modulo n
$x \not\equiv y \pmod{\mathrm{n}}$	x is not congruent to y modulo n
$f(x) \equiv g(x) \pmod{\mathrm{h(x)}, \mathrm{n}}$	$f(x)$ is congruent to $g(x)$ modulo $h(x)$, with coefficients modulo n
$[a]_n$	residue class of a modulo n
$+_n$	addition modulo n
$-_n$	subtraction modulo n
$\cdot_n$	multiplication modulo n
$\sqrt{x} \pmod{n}$	square root of x modulo n
$\sqrt[k]{x} \pmod{n}$	kth root of x modullo n
$x^k \bmod n$	x to the power k modulo n
$\log_x y \bmod n$	discrete logarithm of y to the base x modulo n

x^k	x to the power k
kP	$kP = \underbrace{P \oplus P \oplus \cdots \oplus P}_{k \text{ summands}}$, where P is a point (x, y) on elliptic curve $E : \ y^2 = x^3 + ax + b$
$kP \bmod n$	kP modulo n, where P is a point on E
$\log_P Q \bmod n$	elliptic curve discrete logarithm of Q to the base P modulo n, where P and Q are points on elliptic curve E
$\mathrm{ord}_n(a)$	order of an integer a modulo n; also denoted by $\mathrm{ord}(a, n)$
$\mathrm{ind}_{g,n} a$	index of a to the base g modulo n; also denoted by $\mathrm{ind}_g a$ whenever n is fixed
$\sim$	asymptotic equality
$\approx$	approximate equality
∞	infinity
$\Longrightarrow$	implication
$\Longleftrightarrow$	equivalence
$\square$	blank symbol; end of proof
$\sqcup$	space
Prob	probability measure
$|S|$	cardinality of set S
$\in$	member of
$\subset$	proper subset
$\subseteq$	subset
$\star, *$	binary operations
$\oplus$	binary operation (addition)
$\odot$	binary operation (multiplication)
$f(x) \sim g(x)$	$f(x)$ and $g(x)$ are asymptotically equal
$\perp$	undefined
$f(x)$	function of x
f^{-1}	inverse of f
$\binom{n}{i}$	binomial coefficient: $\binom{n}{i} = \dfrac{n!}{i!(n-i)!}$
$\int$	integration

$\mathrm{Li}(x)$ logarithmic integral: $\mathrm{Li}(x) = \displaystyle\int_2^x \frac{dt}{\ln t}$

$\displaystyle\sum_{i=1}^{n} x_i$ sum: $x_1 + x_2 + \cdots + x_n$

$\displaystyle\prod_{i=1}^{n} x_i$ product: $x_1 x_2 \cdots x_n$

$n!$ factorial: $n(n-1)(n-2)\cdots 3 \cdot 2 \cdot 1$

$\log_b x$ logarithm of x to the base b ($b \neq 1$): $x = b^{\log_b x}$

$\log x$ binary logarithm: $\log_2 x$

$\ln x$ natural logarithm: $\log_e x$, $e = \displaystyle\sum_{n \geq 0} \frac{1}{n!} \approx 2.7182818$

$\exp(x)$ exponential of x: $e^x = \displaystyle\sum_{n \geq 0} \frac{x^n}{n!}$

$\pi(x)$ number of primes less than or equal to x:
$$\pi(x) = \sum_{\substack{p \leq x \\ p \in \text{ Primes}}} 1$$

$\tau(n)$ number of positive divisors of n: $\tau(n) = \displaystyle\sum_{d|n} 1$

$\sigma(n)$ sum of positive divisors of n: $\sigma(n) = \displaystyle\sum_{d|n} d$

$\phi(n)$ Euler's totient function: $\phi(n) = \displaystyle\sum_{\substack{0 \leq k \leq n \\ \gcd(k,n)=1}} 1$

$\lambda(n)$ Carmichael's function:
$$\lambda(n) = \mathrm{lcm}\left(\lambda(p_1^{\alpha_1}), \lambda(p_2^{\alpha_2}), \ldots, \lambda(p_k^{\alpha_k})\right) \text{ if } n = \prod_{i=1}^{k} p_i^{\alpha_i}$$

$\mu(n)$ Möbius function

$\zeta(s)$ Riemann zeta-function: $\zeta(s) = \displaystyle\prod_{n=1}^{\infty} \frac{1}{n^s}$,
            where s is a complex variable

$\left(\dfrac{a}{p}\right)$ Legendre symbol, where p is prime

$\left(\dfrac{a}{n}\right)$ Jacobi symbol, where n is composite

Q_n set of all quadratic residues of n

$\overline{Q}_n$ set of all quadratic non-residues of n

J_n $J_n = \left\{ a \in (\mathbb{Z}/n\mathbb{Z})^* : \left(\dfrac{a}{n}\right) = 1 \right\}$

$\tilde{Q}_n$ set of all pseudo-squares of n: $\tilde{Q}_n = J_n - Q_n$

$K(k)_n$ set of all kth power residues of n, where $k \geq 2$

$\overline{K(k)}_n$	set of all kth power non-residues of n, where $k \geq 2$
$[q_0, q_1, q_2, \ldots, q_n]$	finite simple continued fraction
$C_k = \dfrac{P_k}{Q_k}$	kth convergent of a continued fraction
$[q_0, q_1, q_2, \ldots]$	infinite simple continued fraction
$[q_0, q_1, \ldots, q_k, \overline{q_{k+1}, q_{k+2}, \ldots, q_{k+m}}]$	periodic simple continued fraction
e_k	encryption key
d_k	decryption key
$E_{e_k}(M)$	encryption process $C = E_{e_k}(M)$, where M is the plain-text
$D_{d_k}(C)$	decryption process $M = D_{d_k}(C)$, where C is the cipher-text
$\mathcal{O}(\cdot)$	upper bound: $f(n) = \mathcal{O}(g(n))$ if there exists *some* constant $c > 0$ such that $f(n) \leq c \cdot g(n)$
$\mathcal{O}(n^k)$	polynomial-time complexity measured in terms of arithmetic operations, where $k > 0$ is a constant
$\mathcal{O}\left((\log n)^k\right)$	polynomial-time complexity measured in terms of bit operations, where $k > 0$ is a constant
$\mathcal{O}\left((\log n)^{c \log n}\right)$	superpolynomial complexity, where $c > 0$ is a constant
$\mathcal{O}\left(\exp\left(c\sqrt{\log n \log \log n}\,\right)\right)$	subexponential complexity, $\mathcal{O}\left(\exp\left(c\sqrt{\log n \log \log n}\,\right)\right) = \mathcal{O}\left(n^{c\sqrt{\log \log n / \log n}}\right)$
$\mathcal{O}\left(\exp(x)\right)$	exponential complexity, sometimes denoted by $\mathcal{O}\left(e^x\right)$
$\mathcal{O}\left(n^\epsilon\right)$	exponential complexity measured in terms of bit operations; $\mathcal{O}\left(n^\epsilon\right) = \mathcal{O}\left(2^{\epsilon \log n}\right)$, where $\epsilon > 0$ is a constant

1. Number Theoretic Preliminaries

Mathematics is the Queen of the sciences, and number theory is the Queen of mathematics.

C. F. GAUSS (1777–1855)

It will be another million years, at least, before we understand the primes.

PAUL ERDÖS (1913–1996)

1.1 Problems in Number Theory

The theory of numbers is primarily the theory of the *properties* of integers (whole numbers), such as parity, divisibility, primality, additivity, multiplicativity, and unique factorization, etc. One of the important features of number theory is that problems in number theory are generally easy to state but often very difficult to solve. The following are just some of the hard problems in number theory that are still open to this day:

(1) **Riemann Hypothesis:** Let s be a complex variable (we write $s = \sigma + it$ with σ and t real, where $\sigma = \mathrm{Re}(s)$ is the real part of s, whereas $t = \mathrm{Im}(s)$ is the imaginary part of s). Then the Riemann ζ-function, $\zeta(s)$, is defined to be the sum of the following series

$$\zeta(s) = \sum_{n=1}^{\infty} n^{-s}. \tag{1.1}$$

Bernhard Riemann (1826–1866) in 1859 calculated the first five complex zeros of the ζ-function above the σ-axis and found that they lie on the vertical line $\sigma = 1/2$. He then conjectured that *all the nontrivial (complex) zeros ρ of $\zeta(s)$ lying in the critical strip $0 < \mathrm{Re}(s) < 1$ must lie on the critical line $\mathrm{Re}(s) = 1/2$*. That is, $\rho = 1/2 + it$, where ρ denotes a

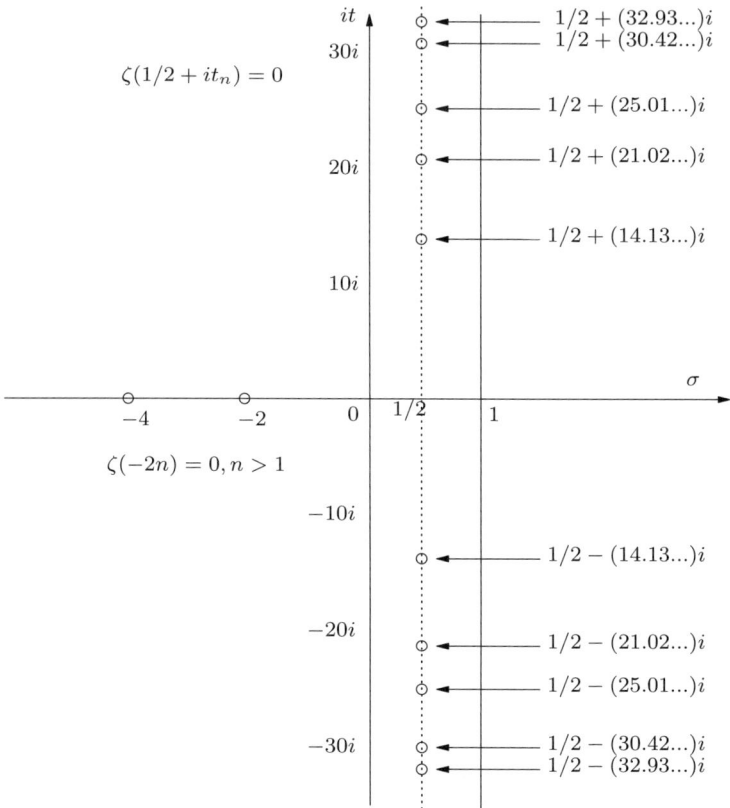

Figure 1.1. First five zeros of ζ-function above σ-axis

nontrivial zero of $\zeta(s)$. Note that the real zeros are the trivial zeros, as $\zeta(-2n) = 0$ when $n = 1, 2, 3, \ldots$. The Riemann hypothesis may be true, it may also be false. At present, no-one knows whether or not it is true. The Clay Mathematics Institute of Cambridge in Massachusetts has named the Riemann hypothesis as one of its seven millennium Prize Problems [27]; the Board of Directors of the institute designated on 24 May 2000 \$7 million prize fund for the solution to these problems, with \$1 million allocated to each. The Riemann hypothesis is intimately connected to the distribution of prime numbers over positive integers. Let

$$\pi(x) = \sum_{\substack{p \leq x \\ p \in \text{Primes}}} 1.$$

Then

$$\pi(x) \quad \sim \quad \infty \quad \text{(Euclid, 250 BC)}$$

$$\sim \quad \frac{x}{\ln x} \quad \text{(Hadamard and De la Vallée–Poussin, 1896)}$$

$$\sim \quad \int_2^x \frac{dt}{\ln t}$$

$$= \quad \int_2^x \frac{dt}{\ln t} + \mathcal{O}(xe^{-A\sqrt{\ln x}}) \quad \text{(De la Vallée–Poussin, 1899)}$$

In number theory, the logarithmic integral above is often denoted by Li(x). That is,

$$\mathrm{Li}(x) = \int_2^x \frac{dt}{\ln t}.$$

If the Riemann hypothesis is true, then

$$\pi(x) = \mathrm{Li}(x) + \mathcal{O}\left(\sqrt{x}\ln x\right).$$

(2) **The Goldbach Conjecture:** In a letter to Euler (1707–1783), dated 7 June 1742 (see Figure 1.2), the Prussian mathematician Christian Goldbach (1690–1764) proposed a conjecture concerning the additive properties of integers. After some rephrasing, this conjecture may be expressed as the following two conjectures:

- **Binary (or even, strong) Goldbach Conjecture** (BGC): Every even number greater than 4 is the sum of two primes. That is, $2n = p_1 + p_2$, for $n > 2$ and $p_1, p_2 \in$ Primes. For example, $6 = 3 + 3$, $8 = 3 + 5$, $10 = 3 + 7, 12 = 5 + 7, \cdots$.
- **Ternary (or odd, weak) Goldbach Conjecture** (TGC): Every odd number greater than 7 is the sum of three primes. That is, $2n + 1 = p_1 + p_2 + p_3$, for $n > 3$ and $p_1, p_2, p_3 \in$ Primes. For example, $9 = 3 + 3 + 3$, $11 = 3 + 3 + 5, 13 = 3 + 5 + 5, 15 = 3 + 5 + 7, \cdots$.

Clearly, the first implies the second. For BGC, J. R. Chen [45] proved in 1973 that every sufficiently large even number $\aleph$ can be represented as a sum of one prime and a product of at most two primes. That is, $\aleph = p_1 + p_2 \cdot p_3$, but not quantity value is given to $\aleph$. On the other hand, BGC is numerically verified to be true up to $4 \cdot 10^{14}$ by Richstein [196]), and later this bound was extended to 10^{15}. However, there is still a long way to go for proving/disproving BGC. Compared to BGC, TGC is essentially solved. Unconditionally, TGC is true for all odd numbers $\geq 10^{43000}$; this is a refinement of Chen and Wang [46] over Vinogradov's famous *three-prime theorem* [248]. Assuming the Generalized Riemann Hypothesis (the Generalized Riemann Hypothesis or the Extended Riemann Hypothesis is the same assertion of the Riemann Hypothesis for

Figure 1.2. Goldbach's letter to Euler

certain generalizations of the ζ-function, called the Dirichlet L functions; we shall discuss the Dirichlet L functions and the Generalized Riemann Hypothesis in Section 2.2), every odd number ≥ 7 can be represented as a sum of three prime numbers, due to Deshouillers, Effinger, Te Riele and Zinoviev [66]. Numerically, TGC was verified to be true up to 10^{20} by Saouter [213] in 1995. Figure 1.3 shows the progress of the research in both BGC and TGC.

(i) Binary Goldbach conjecture, with ℵ a large unknown number

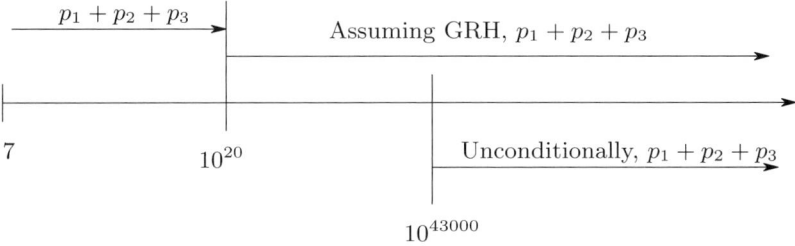

(ii) Ternary Goldback conjecture

Figure 1.3. Known results on Goldbach's conjecture to date

(3) **Twin Prime Conjecture:** Twin primes are defined to be pairs $(p,\ p+2)$, such that both p and $p+2$ are prime. For example, $(3,5)$, $(17,19)$ and $(1997,1999)$ are twin primes. Table 1.1 gives 20 large known twin primes. It is conjectured but not yet proven that: let $\pi_2(x)$ be the number of primes p such that if $p \le x$ is prime, and $p+2$ is also prime, then

(a) **Weak form of the Twin Prime Conjecture:** There are infinitely many twin primes. That is,

$$\lim_{x \to \infty} \pi_2(x) = \infty. \tag{1.2}$$

No.	Twin Primes	Digits	Year
1	$2003663613 \cdot 2^{195000} \pm 1$	58711	2007
2	$194772106074315 \cdot 2^{171960} \pm 1$	51780	2007
3	$100314512544015 \cdot 2^{171960} \pm 1$	51780	2006
4	$16869987339975 \cdot 2^{171960} \pm 1$	51779	2005
5	$33218925 \cdot 2^{169690} \pm 1$	51090	2002
6	$60194061 \cdot 2^{114689} \pm 1$	34533	2002
7	$1765199373 \cdot 2^{107520} \pm 1$	32376	2002
8	$318032361 \cdot 2^{107001} \pm 1$	32220	2001
9	$1046619117 \cdot 2^{100000} \pm 1$	30113	2007
10	$1807318575 \cdot 2^{98305} \pm 1$	29603	2001
11	$7473214125 \cdot 2^{83125} \pm 1$	25033	2006
12	$11694962547 \cdot 2^{83124} \pm 1$	25033	2006
13	$58950603 \cdot 2^{83130} \pm 1$	25033	2006
14	$5583295473 \cdot 2^{80828} \pm 1$	24342	2006
15	$134583 \cdot 2^{80828} \pm 1$	24337	2005
16	$665551035 \cdot 2^{80025} \pm 1$	24099	2000
17	$1046886225 \cdot 2^{70000} \pm 1$	21082	2004
18	$8544353655 \cdot 2^{66666} \pm 1$	20079	2005
19	$8179665447 \cdot 2^{66666} \pm 1$	20079	2006
20	$6968409117 \cdot 2^{66666} \pm 1$	20079	2005

Table 1.1. Twenty known large twin primes

(b) **Strong form of the Twin Prime Conjecture:** Let

$$\mathrm{Li}_2(x) \;=\; 2 \prod_{p \geq 3} \frac{p(p-2)}{(p-1)^2} \int_2^x \frac{dt}{\ln^2 t}$$

$$\approx \; 1.320323632 \int_2^x \frac{dt}{\ln^2 t} \tag{1.3}$$

then

$$\lim_{x \to \infty} \frac{\pi_2(x)}{\mathrm{Li}_2(x)} = 1. \tag{1.4}$$

Note that in practice the estimate seems to be very good; see Table 1.2. Using very complicated arguments based on sieve methods, similar to his proof on Goldbach's conjecture, the Chinese mathematician J. R. Chen showed that *there are infinitely many pairs of integers $(p,\ p+2)$, with p prime and $p+2$ a product of at most two primes.*

x	$\pi_2(x)$	$\mathrm{Li}_2(x)$	$\pi_2(x)/\mathrm{Li}_2(x)$
10^3	35	46	.7608695652
10^4	205	214	.9579439252
10^5	1224	1249	.9799839872
10^6	8169	8248	.9904219205
10^7	58980	58754	1.003846547
10^8	440312	440368	.9998728336
10^9	3424506	3425308	.9997658605
10^{10}	27412679	27411417	1.000046039
10^{11}	224376048	224368865	1.000032014
10^{12}	1870585220	1870559867	1.000013554
10^{13}	15834664872	15834598305	1.000004204
10^{14}	135780321665	135780264894	1.000000418
10^{15}	1177209242304	1177208491861	1.000000637

Table 1.2. Number of twin primes $\leq x$

A prime p is called a *Sophie Germain Primes*, named after the great French woman mathematician Marie-Sophie Germain (1776–1831), if $2p + 1$ is also a prime. The first few Sophie Germain primes are:

$$2, 3, 5, 11, 23, 29, 41, 53, 83, 89, 113, 131, 173, 179, 191, 233.$$

Table 1.3 gives 20 large known Sophie Germain primes. Let the number of Sophie Germain primes less than or equal to x be $\pi_s(x)$, and

$$\mathrm{Li}_s(x) \;=\; 2 \prod_{p \geq 3} \frac{p(p-2)}{(p-1)^2} \int_2^x \frac{\mathrm{d}x}{\log x \log(2x+1)}$$

$$\sim \;\; \frac{1.320323632x}{(\log x)^2}.$$

Then $\pi_s(x)$ is conjectured to be asympototic to $\mathrm{Li}_s(x)$. This estimate works surprisingly well (See Table 1.4 for more information).

No.	Prime p	Digits	Year of Discovery
1	$48047305725 \cdot 2^{172403} - 1$	51910	Jan 2007
2	$137211941292195 \cdot 2^{171960} - 1$	51780	May 2006
3	$7068555 \cdot 2^{121301} - 1$	36523	Jan 2005
4	$2540041185 \cdot 2^{114729} - 1$	34547	Jan 2003
5	$1124044292325 \cdot 2^{107999} - 1$	32523	Dec 2006
6	$112886032245 \cdot 2^{108000} - 1$	32523	Dec 2006
7	$18912879 \cdot 2^{98395} - 1$	29628	Nov 2002
8	$10495740081 \cdot 2^{83125} - 1$	25034	Feb 2006
9	$61078155 \cdot 2^{82002} - 1$	24693	Feb 2006
10	$1213822389 \cdot 2^{81131} - 1$	24432	Aug 2002
11	$2566851867 \cdot 2^{70001} - 1$	21082	Jun 2007
12	$1040131975 \cdot 2^{66458} + 1$	20015	Feb 2007
13	$109433307 \cdot 2^{66452} - 1$	20013	Feb 2001
14	$984798015 \cdot 2^{66444} - 1$	20011	Mar 2001
15	$3714089895285 \cdot 2^{60000} - 1$	18075	Mar 2000
16	$3379174665 \cdot 2^{58502} - 1$	17621	Jan 2008
17	$909004827 \cdot 2^{56789} - 1$	17105	Mar 2005
18	$1162665081 \cdot 2^{55649} - 1$	16762	May 2004
19	$790717071 \cdot 2^{54254} - 1$	16341	Jan 2007
20	$4127632557 \cdot 2^{50001} - 1$	15062	Jun 2007

Table 1.3. Twenty known larger sophie Germain primes

Conjecture 1.1.1 (Sophie Germain Prime Conjecture).

$$\lim_{x \to \infty} \frac{\pi_s(x)}{\mathrm{Li}_s(x)} = 1.$$

(4) **Mersenne Prime Conjecture:** Mersenne primes are primes of the form $2^p - 1$, where p is prime. The largest known Mersenne prime to date is $2^{43112609} - 1$, with 12978189 digits, and found in 23 August 2008. Forty-six Mersenne primes have been found to date (see Table 1.5; the first four appeared in Euclid's *Elements* 2000 year ago).

Conjecture 1.1.2 (Mersenne Prime Conjecture).   There are infinitely many Mersenne primes.

An interesting property of the Mersenne primes is that whenever $2^p - 1$ is a Mersenne prime, $2^{p-1}(2^p - 1)$ will be a perfect number (note that a positive integer n is a perfect number if $\sigma(n) = 2n$, where $\sigma(n)$ denotes the sum of all the positive divisors of n and is considered in Section 1.5; for example, 6,28,496,8128 are the first four perfect numbers since $\sigma(6) = 1 + 2 + 3 + 6 = 2 \cdot 6$, $\sigma(28) = 1 + 2 + 4 + 7 + 14 + 28 = 2 \cdot 28$, $\sigma(496) = 1 + 2 + 4 + 8 + 16 + 31 + 62 + 124 + 248 + 496 = 2 \cdot 492$, $\sigma(8128) = 1 + 2 + 4 + 8 + 16 + 32 + 64 + 127 + 254 + 508 + 1016 +$

x	$\pi_s(x)$	$\mathrm{Li}_s(x)$	$\pi_s(x)/\mathrm{Li}_s(x)$
10^3	37	39	.9487179487
10^4	190	195	.9743589744
10^5	1171	1166	1.004288165
10^6	7746	7811	.9916784023
10^7	56032	56128	.9982896237
10^8	423140	423295	.9996338251
10^9	3308859	3307888	1.000293541
10^{10}	26569515	26568824	1.000026008
10^{11}	218116524	218116102	1.000001935

Table 1.4. Number of Sophie Germain primes $\leq x$

$2032 + 4064 + 8128 = 2 \cdot 8128$). That is, n is an even perfect number if and only if $n = 2^{p-1}(2^p - 1)$, where $2^p - 1$ is a Mersenne prime. This is the famous Euclid-Euler theorem which took about 2000 years to prove; the sufficient condition was established by Euclid in his *Elements*, but the necessary condition was established by Euler in work published posthumously. Again, we do not know if there are infinitely many perfect numbers, particularly, we do not know if there is an *odd perfect number*; what we know is that there is no odd perfect numbers up to 10^{300}, due to Brent, et al [38]. Whether or not there is an odd perfect number is one of the most important unsolved problems in all of mathematics.

(5) **Arithmetic Progression of Consecutive Primes:** An arithmetic progression of primes is defined to be the sequence of primes satisying:

$$p, p + d, p + 2d, \ldots, p + (k - 1)d \qquad (1.5)$$

where p is the first term, d the common difference, and $p + (k - 1)d$ the last term of the sequence. For example, the following are some sequences of the arithmetic progression of primes:

2				
2	3			
3	5	7		
5	11	17	23	
5	11	17	23	29

The longest arithmetic progression of primes to date is the following sequence with 23 terms: $56211383760397 + k \cdot 44546738095860$ with $k = 0, 1, \ldots, 22$.

No.	p	Digits	Time	No.	p	Digits	Time
1	2	1	–	2	3	1	–
3	5	2	–	4	7	3	–
5	13	4	1456	6	17	6	1588
7	19	6	1588	8	31	10	1772
9	61	19	1883	10	89	27	1911
11	107	33	1914	12	127	39	1876
13	521	157	1952	14	607	183	1952
15	1279	386	1952	16	2203	664	1952
17	2281	687	1952	18	3217	969	1957
19	4253	1281	1961	20	4423	1332	1961
21	9689	2917	1963	22	9941	2993	1963
23	11213	3376	1963	24	19937	6002	1971
25	21701	6533	1978	26	23209	6987	1979
27	44497	13395	1979	28	86243	25962	1982
29	110503	33265	1988	30	132049	39751	1983
31	216091	65050	1985	32	756839	227832	1992
33	859433	258716	1994	34	1257787	378632	1996
35	1398269	420921	1996	36	2976221	895932	1997
37	3021377	909526	1998	38	6972593	2098960	1999
39	13466917	4053946	2001	40	20996011	6320430	2003
41	24036583	7235733	2004	42	25964951	7816230	2005
43	30402457	9152052	2005	44	32582657	9808358	2006
43	30402457	9152052	2005	44	32582657	9808358	2006
45	37156667	11185272	2008	46	43112609	12978189	2008

Table 1.5. Forty-six known Mersenne primes

56211383760397	100758121856257
145304859952117	189851598047977
234398336143837	278945074239697
323491812335557	368038550431417
412585288527277	457132026623137
501678764718997	546225502814857
590772240910717	635318979006577
679865717102437	724412455198297
768959193294157	813505931390017
858052669485877	902599407581737
947146145677597	991692883773457
1036239621869317	

Thanks to Green and Tao [92] who proved that *there are arbitrary long arithmetic progressions of primes* (i.e., k can be any arbitrary large natural number), which enabled, among others, Tao to receive a Field Prize in 2006, an equivalent Nobel Prize for Mathematics. However, their result cannot be extended to arithmetic progressions of consecutive primes; we still do not know if there are arbitrary long arithmetic progressions of consecutive primes, although Chowla [48] proved in 1944 that there exists an infinity of three consecutive primes of arithmetic progressions. Note that an arithmetic progression of consecutive primes is a sequence of consecutive primes in the progression. In 1967, Jones, Lal and Blundon [114] found an arithmetic progression of five consecutive primes $10^{10} + 24493 + 30k$ with $k = 0, 1, 2, 3, 4$:

10000024493	10000024523
10000024553	10000024583
10000024613	

In the same year, Lander and Parkin [127] discovered six in an arithmetic progression $121174811 + 30k$ with $k = 0, 1, 2, 3, 4, 5$.

121174811	121174841
121174871	121174901
121174931	121174961

The longest arithmetic progression of consecutive primes to date, discovered by Manfred Toplic in 1998, is $507618446770482 \cdot 193\# + x77 + 210k$, where $193\#$ is the product of all primes ≤ 193, i.e., $193\# = 2 \cdot 3 \cdot 5 \cdot 7 \cdots 193$, $x77$ is a 77-digit number:

54538241683887582668189703590110659057865934764604873840781 923513421103495579,

and $k = 0, 1, 2, \ldots, 9$. Table 1.6 lists some large examples of arithmetic progressions of consecutive primes.

In this book we shall be mainly concerned with those number theoretic problems such as primality testing and integer factorization that have connections to cryptography and information security. However before discussing these problems and their solutions, we shall provide a gentle but modern introduction to the basic concepts and results of number theory, including divisibility theory, congruence theory and the arithmetic of elliptic curves.

Problems for Section 1.1

Problem 1.1.1. Prove or disprove the Riemann hypothesis. (This problem has been opened since 1859, and is one of the seven Millennium Prize Problems, with one million US dollar prize attached to it.)

Problem 1.1.2. Prove or disprove the Goldbach conjecture. (This problem has been opened since 1742. The British publisher Tony Faber offered a one million US dollar prize for a proof of the conjecture in 2000, if a proof was submitted before April 2002. The prize was never claimed.)

Problem 1.1.3. Prove or disprove that there are infinitely many twin primes. (This problem has been opened for more than 2000 years.)

No.	p	Digits	Terms	Difference	Year
1	$197418203 \cdot 2^{25000} + 6089$	7535	3	6090	2005
2	$87 \cdot 2^{24582} + 2579$	7402	3	1290	2004
3	$4811 \cdot 2^{20219} + 16091$	6091	3	3738	1996
4	$(84055657369 \cdot 205881 \cdot 4001\#$ $\cdot(205881 \cdot 4001\# + 1) + 210) \cdot$ $(205881 \cdot 4001\# - 1)/35 + 13$	5132	3	6	2006
5	$(613103465292058814001\#$ $(2058814001\# + 1) + 210)$ $(2058814001\# - 1)/35 + 13$	5132	3	6	2005
6	$2^{5900} + 469721940591$	1777	4	2880	2007
7	$186728916584099\#$ $+1591789579$	1763	4	218	2003
8	$23^{963} + 1031392866$	1312	4	1500	2005
9	$49197618052999\# + 6763$	1284	4	38	2003
10	$11^{1008} + 998672782$	1050	4	1080	2005
11	$11^{1005} + 260495538$	1047	4	60	2005
12	$1426611576262411\# + 71427877$	1038	5	30	2002

Table 1.6. Twelve large consecutive primes in arithmetic progressions

Problem 1.1.4. Prove or disprove that there are infinitely many Sophie Germain primes.

Problem 1.1.5. Prove or disprove that there are infinitely many Mersenne primes (or perfect numbers). (This problem has been opened for more than 2000 years.)

Problem 1.1.6. Show that there exists at least one odd perfect numbers. (This problem has been opened for more than 2000 years.)

Problem 1.1.7. Show that there are arbitrary long arithmetic progressions of *consecutive* primes. (This problem has been opened for a very long time.)

Problem 1.1.8. (Green-Tao) Show that there are arbitrary long arithmetic progressions of primes. (By solving this problem, among others, Tao won the 2006 Fields Prize.)

Problem 1.1.9. Let $\sigma(x)$ be the sum of all positive divisors of x. A pair of positive integers (a, b) are amicable if $\sigma(a) = \sigma(b) = a + b$; for example, $(220, 284)$ is an amicable (in fact, the smallest) pair, since

$$\sigma(220) = 1 + 2 + 4 + 5 + 10 + 11 + 20 + 22 + 44 + 55 + 110 + 220 = 504$$
$$\sigma(284) = 1 + 2 + 4 + 71 + 142 + 284 = 504$$

Prove or disprove that there are infinitely many amicable pairs. (The first amicable pair $(220, 284)$ was found by the legendary Pythagras at least 2300 years ago.)

Problem 1.1.10. Show that k positive integers $a_1, a_2, \ldots, a_k$ are sociable if

$$\begin{cases} \sigma(a_1) &= a_1 + a_2 \\ \sigma(a_2) &= a_2 + a_3 \\ &\vdots \\ \sigma(a_k) &= a_k + a_1 \end{cases}$$

Problem 1.1.11. Fermat numbers are defined to be $F_n = 2^{2^n} + 1$, for $n = 0, 1, 2, 3, \ldots$. To date, only F_0, F_1, F_2, F_3, F_4 are known to be prime, the rest are either composite or their primality are not known. Prove or disprove that there are infinitely many Fermat composites (or primes).

Problem 1.1.12. In 1769 Euler conjectured that if $n \geq 3$, then fewer than n nth powers cannot sum to an nth power. This conjecture has been proved to be false when $n = 4, 5$ with the following counter-examples:

$$
\begin{aligned}
95800^4 + 217519^4 + 414560^4 &= 422481^4 \\
2682440^4 + 15365639^4 + 18796760^4 &= 20615673^4 \\
27^5 + 84^5 + 110^5 + 133^5 &= 144^5 \\
55^5 + 3183^5 + 28969^5 + 85282^5 &= 85359^5
\end{aligned}
$$

The conjecture is, however, still open for $n > 5$. Find integer solutions to $a^6 + b^6 + c^6 + d^6 + e^6 = f^6$. Prove or disprove Euler's conjecture for $n > 5$.

1.2 Groups, Rings and Fields

Number theory can be best studied from an algebraic point of view. Thus, in this section, a brief introduction to the basic concepts and results of groups, rings and fields is provided.

Definition 1.2.1. A *group*, denoted by G, is a nonempty set G of elements together with a binary operation $\star$ (e.g., the ordinary addition or multiplication), such that the following axioms are satisfied:

(1) Closure: $a \star b \in G, \quad \forall a, b \in G$.
(2) Associativity: $(a \star b) \star c = a \star (b \star c), \quad \forall a, b, c \in G$.
(3) Existence of identity: There is a unique element $e \in G$, called the identity, such that $e \star a = a \star e = a, \quad \forall a \in G$.
(4) Existence of inverse: For every $a \in G$ there is a unique element b such that $a \star b = b \star a = e$. This b is denoted by a^{-1} and called the inverse of a.

The group G is called a *commutative group* if it satisfies a further axiom:

(5) Commutativity: $a \star b = b \star a, \quad \forall a, \ b \in G$.

A commutative group is also called an *Abelian group*, in honor of the Norwegian mathematician N. H. Abel (1802–1829).

Example 1.2.1. The set $\mathbb{N}$ with operation $+$ is *not* a group, since there is no identity element for $+$ in $\mathbb{Z}^+$. The set $\mathbb{N}$ with operation $\cdot$ is *not* a group; there is an identity element 1, but no inverse of 3.

Example 1.2.2. The set of all non-negative integers, $\mathbb{Z}_{\geq 0}$, with operation $+$ is *not* a group; there is an identity element 0, but no inverse for 2.

Example 1.2.3. The sets $\mathbb{Q}^+$ and $\mathbb{R}^+$ of positive numbers and the sets $\mathbb{Q}^*$, $\mathbb{R}^*$ and $\mathbb{C}^*$ of nonzero numbers with operation $\cdot$ are Abelian groups.

Definition 1.2.2. If the binary operation of a group is denoted by $+$, then the identity of a group is denoted by 0 and the inverse a by $-a$; this group is said to be an *additive group*. If the binary operation of a group is denoted by $*$, then the identity of a group is denoted by 1 or e; this group is said to be a *multiplicative group*.

Definition 1.2.3. A group is called a *finite group* if it has a finite number of elements; otherwise it is called an *infinite group*. The number of elements in G is called the order of G and is denoted by $|G|$ or $\#(G)$.

Example 1.2.4. The order of $\mathbb{Z}$ is infinite, i.e., $|\mathbb{Z}| = \infty$. However, the order of $\mathbb{Z}_{11}$ is finite, since $|\mathbb{Z}_{11}| = 11$.

Definition 1.2.4. A nonempty set G' of a group G which is itself a group, under the same operation, is called a *subgroup* of G.

Definition 1.2.5. Let a be an element of a multiplicative group G. The elements a^r, where r is an integer, form a subgroup of G, called the subgroup generated by a. A group G is *cyclic* if there is an element $a \in G$ such that the subgroup generated by a is the whole of G. If G is a finite cyclic group with identity element e, the set of elements of G may be written $\{e, a, a^2, \ldots, a^{n-1}\}$, where $a^n = e$ and n is the smallest such positive integer. If G is an infinite cyclic group, the set of elements may be written $\{\ldots, a^{-2}, a^{-1}, e, a, a^2, \ldots\}$.

By making appropriate changes, a cyclic *additive group* can be defined. For example, the set $\{0, 1, 2, \ldots, n-1\}$ with addition modulo n is a cyclic group, and the set of all integers with addition is an infinite cyclic group.

Example 1.2.5. The congruences modulo n form a group. If we take $a + b \equiv c \ (\mathrm{mod}\ 6)$, then we get the following complete addition table for the additive group modulo 6 (see Table 1.7):

$\oplus$	0	1	2	3	4	5
0	0	1	2	3	4	5
1	1	2	3	4	5	0
2	2	3	4	5	0	1
3	3	4	5	0	1	2
4	4	5	0	1	2	3
5	5	0	1	2	3	4

Table 1.7. Additive group modulo 6

Definition 1.2.6. A *ring*, denoted $(R, \oplus, \odot)$, or simply R, is a set of at least two elements with *two* binary operations $\oplus$ and $\odot$, which we call addition and multiplication, defined on R such that the following axioms are satisfied:

(1) The set is *closed* under the operation $\oplus$:

$$a \oplus b \in R, \quad \forall a,\ b \in R, \tag{1.6}$$

(2) The associative law holds for $\oplus$:

$$a \oplus (b \oplus c) = (a \oplus b) \oplus c, \quad \forall a,\ b,\ c \in R, \tag{1.7}$$

(3) The commutative law holds for $\oplus$:

$$a \oplus b = b \oplus a, \quad \forall a,\ b \in R, \tag{1.8}$$

(4) There is a special (zero) element $0 \in R$, called the additive identity of R, such that

$$a \oplus 0 = 0 \oplus a = a, \quad \forall a \in R, \tag{1.9}$$

(5) For each $a \in R$, there is a corresponding element $-a \in R$, called the additive inverse of a, such that:

$$a \oplus (-a) = 0, \quad \forall a \in R, \tag{1.10}$$

(6) The set is closed under the operation $\odot$:

$$a \odot b \in R, \quad \forall a,\ b \in R, \tag{1.11}$$

(7) The associative law holds for $\odot$:

$$a \odot (b \odot c) = (a \odot b) \odot c, \quad \forall a, b, c \in R, \tag{1.12}$$

(8) The operation $\odot$ is distributive with respect to $\oplus$:

$$a \odot (b \oplus c) = a \odot b \oplus a \odot c, \quad \forall a, b, c \in R, \tag{1.13}$$

$$(a \oplus b) \odot c = a \odot c \oplus b \odot c, \quad \forall a, b, c \in R. \tag{1.14}$$

From a group theoretic point of view, a ring is an Abelian group, with the additional properties that the closure, associative and distributive laws hold for $\odot$.

Example 1.2.6. $(\mathbb{Z}, \oplus, \odot)$, $(\mathbb{Q}, \oplus, \odot)$, $(\mathbb{R}, \oplus, \odot)$, and $(\mathbb{C}, \oplus, \odot)$ are all rings.

Definition 1.2.7. A *commutative ring* is a ring that further satisfies:

$$a \odot b = b \odot a, \quad \forall a, b \in R. \tag{1.15}$$

Definition 1.2.8. A *ring with identity* is a ring that contains an element 1 satisfying:

$$a \odot 1 = a = 1 \odot a, \quad \forall a \in R. \tag{1.16}$$

Definition 1.2.9. An *integral domain* is a commutative ring with identity $1 \neq 0$ that satisfies:

$$a, b \in R \ \& \ ab = 0 \implies a = 0 \ \text{ or } \ b = 0. \tag{1.17}$$

Definition 1.2.10. A *division ring* is a ring R with identity $1 \neq 0$ that satisfies:

for each $a \neq 0 \in R$, the equation $ax = 1$ and $xa = 1$ have solutions in R.

Definition 1.2.11. A *field*, denoted by K, is a division ring with commutative multiplication.

Example 1.2.7. The integer set $\mathbb{Z}$, with the usual addition and multiplication, forms a commutative ring with identity, but is not a field.

It is clear that a field is a type of ring, which can be defined more generally as follows:

Definition 1.2.12. A *field*, denoted by $(K, \oplus, \odot)$, or simply K, is a set of at least two elements with *two* binary operations $\oplus$ and $\odot$, which we call addition and multiplication, defined on K such that the following axioms are satisfied:

(1) The set is *closed* under the operation $\oplus$:

$$a \oplus b \in K, \quad \forall a, \ b \in K, \tag{1.18}$$

(2) The associative law holds for $\oplus$:

$$a \oplus (b \oplus c) = (a \oplus b) \oplus c, \quad \forall a, \ b, \ c \in K, \tag{1.19}$$

(3) The commutative law holds for $\oplus$:

$$a \oplus b = b \oplus a, \quad \forall a, \ b \in K, \tag{1.20}$$

(4) There is a special (zero) element $0 \in K$, called the additive identity of K, such that

$$a \oplus 0 = 0 \oplus a = a, \quad \forall a \in K, \tag{1.21}$$

(5) For each $a \in K$, there is a corresponding element $-a \in K$, called the additive inverse of a, such that:

$$a \oplus (-a) = 0, \quad \forall a \in K, \tag{1.22}$$

(6) The set is closed under the operation $\odot$:

$$a \odot b \in K, \quad \forall a, \ b \in K, \tag{1.23}$$

(7) The associative law holds for $\odot$:

$$a \odot (b \odot c) = (a \odot b) \odot c, \quad \forall a, b, c \in K \tag{1.24}$$

(8) The operation $\odot$ is distributive with respect to $\oplus$:

$$a \odot (b \oplus c) = a \odot b \oplus a \odot c, \quad \forall a, b, c \in K, \tag{1.25}$$
$$(a \oplus b) \odot c = a \odot c \oplus b \odot c, \quad \forall a, \ b, \ c \in K. \tag{1.26}$$

(9) There is an element $1 \in K$, called the multiplicative identity of K, such that $1 \neq 0$ and

$$a \odot 1 = a, \quad \forall a \in K, \tag{1.27}$$

(10) For each nonzero element $a \in K$ there is a corresponding element $a^{-1} \in K$, called the multiplicative inverse of a, such that

$$a \odot a^{-1} = 1, \tag{1.28}$$

(11) The commutative law holds for $\odot$:

$$a \odot b = b \odot a, \quad \forall a, b \in K, \tag{1.29}$$

Again, from a group theoretic point of view, a field is an Abelian group with respect to addition and also the non-zero field elements form an Abelian group with respect to multiplication.

Remark 1.2.1. An alternative definition of a field is :

If all the elements of a ring, other than the zero, form a commutative group under $\odot$, then it is called a *field*.

Example 1.2.8. The integer set $\mathbb{Z}$, with the usual addition and multiplication, forms a commutative ring with identity.

Figure 1.4 gives a Venn diagram view of containment for algebraic structures having two binary operations.

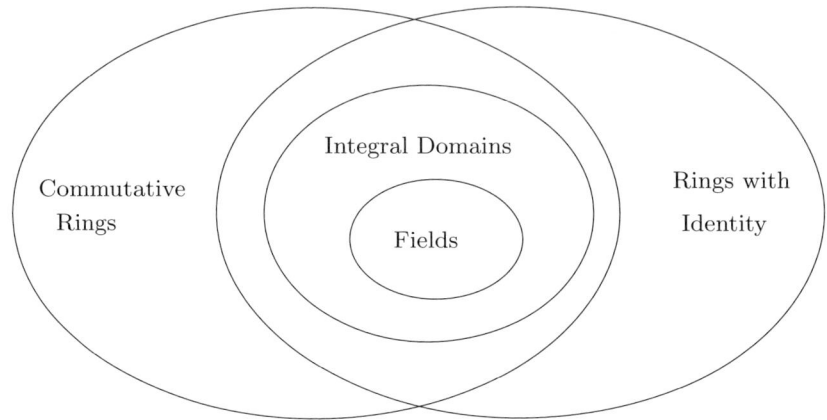

Figure 1.4. Containment of various rings

Example 1.2.9. Familiar examples of fields are the set of rational numbers, $\mathbb{Q}$, the set of real numbers, $\mathbb{R}$ and the set of complex numbers, $\mathbb{C}$; since $\mathbb{Q}$, $\mathbb{R}$ and $\mathbb{C}$ are all infinite sets, they are all infinite fields. The set of integers $\mathbb{Z}$ is a ring but *not* a field, since 2, for example, has no multiplicative inverse; 2 is not a unit in $\mathbb{Z}$. The only units (i.e., the invertible elements) in $\mathbb{Z}$ are 1 and -1. Another example of a ring which is not a field is the set $K[x]$ of polynomials in x with coefficients belonging to a field K.

Theorem 1.2.1. $\mathbb{Z}/n\mathbb{Z}$ is a field if and only if n is prime.

What this theorem says is that whenever n is prime, the set of congruence classes modulo n forms a field. This *prime field* $\mathbb{Z}/p\mathbb{Z}$ will be specifically denoted by $\mathbb{F}_p$.

Definition 1.2.13. A *finite field* is a field that has a finite number of elements in it; we call the number the *order* of the field.

The following fundamental result on finite fields was first proved by Évariste Galois (1811–1832):

Theorem 1.2.2. There exists a field of order q if and only if q is a prime power (i.e., $q = p^r$) with p prime and $r \in \mathbb{N}$. Moreover, if q is a prime power, then there is, up to relabelling, only one field of that order.

A finite field of order q with q a prime power is often called a *Galois field*, and is denoted by $\mathrm{GF}(q)$, or just $\mathbb{F}_q$.

Example 1.2.10. The finite field $\mathbb{F}_5$ has elements $\{0, 1, 2, 3, 4\}$ and is described by the following addition and multiplication tables (see Table 1.8):

$\oplus$	0	1	2	3	4
0	0	1	2	3	4
1	1	2	3	4	0
2	2	3	4	0	1
3	3	4	0	1	2
4	4	0	1	2	3

$\odot$	0	1	2	3	4
0	0	0	0	0	0
1	0	1	2	3	4
2	0	2	4	1	3
3	0	3	1	4	2
4	0	4	3	2	1

Table 1.8. Addition and multiplication for $\mathbb{F}_5$

Polynomial Arithmetic. Let F be a ring. A polynomial with coefficients in F is an expression

$$f(x) = a_n x^n + a_{n-1} x^{n-1} + \cdots + a_1 x + a_0$$

where $a_i \in F$ for $i = 0, 1, 2, \ldots, n$ and x is a variable. The set of all polynomials $f(x)$ with coefficients in F is denoted by $F[x]$. In particular, if F is taken to be $\mathbb{Z}_p$, $\mathbb{Z}$, $\mathbb{Q}$, $\mathbb{R}$, or $\mathbb{C}$, then the corresponding polynomial sets are denoted by $\mathbb{Z}_p[x]$, $\mathbb{Z}[x]$, $\mathbb{Q}[x]$, $\mathbb{R}[x]$, $\mathbb{C}[x]$, respectively. The degree of the polynomial $f(x) = a_n x^n + a_{n-1} x^{n-1} + \cdots + a_1 x + a_0$ is n if $a_n \neq 0$. a_n is called the leading coefficient, and if $a_n = 1$ then the polynomial is called monic. Two polynomials $f(x)$ and $g(x)$ in $F[x]$ are equal if they have the same degree and all their coefficients are identical. If $f(a) = 0$, then a is called a root of $f(x)$ or zero of $f(x)$. Two polynomials $f(x) = a_m x^m + a_{m-1} x^{m-1} + \cdots + a_1 x + a_0$ and $g(x) = b_n x^n + b_{n-1} x^{n-1} + \cdots + b_1 x + b_0$, with $n > m$, can be added, subtracted and multiplied as follows:

$$\begin{aligned} f(x) \pm g(x) &= (a_0 \pm b_0) + (a_1 \pm b_1)x + \cdots + (a_m \pm b_m)x^m + \\ & \qquad b_{m+1} x^{m+1} + \cdots + b_n x^n \\ &= \sum_{i=1}^{m} (a_i \pm b_i) x^i + \sum_{j=m+1}^{n} b_j x^j. \end{aligned}$$

$$\begin{aligned} f(x)g(x) &= a_0 b_0 + (a_0 b_1 + a_1 b_0)x + \cdots + a_m b_n x^{m+n} \\ &= \sum_{i=0}^{m} \sum_{j=0}^{n} a_i b_j x^{i+j}. \end{aligned}$$

Example 1.2.11. Let $f(x) = 2x^5 + x - 1$ and $g(x) = 3x^2 + 2$. Then

$$f(x) + g(x) = 2x^5 + 3x^2 + x + 1,$$

$$f(x) - g(x) = 2x^5 - 3x^2 + x - 3.$$

Let $f(x) = 1 + x - x^2$ and $g(x) = 2 + x^2 + x^3$. Then

$$f(x)g(x) = 2 + 2x - x^2 + 2x^3 - x^5.$$

The division algorithm and Euclid's algorithm for integers can be extended naturally to polynomials.

Theorem 1.2.3 (Division Algorithm for Polynomials). Let F be a field, $f(x)$ and $p(x)$ $(p(x) \neq 0)$ polynomials in $F[x]$. There are unique polynomials $q(x)$ and $r(x)$ such that

$$f(x) = p(x)q(x) + r(x)$$

where either $r(x) = 0$ or $\deg(r(x)) < \deg(p(x))$.

Example 1.2.12. Let $f(x) = 2x^5 + x - 1$ and $p(x) = 3x^2 + 2$. Then

$$2x^5 + x - 1 = (3x^2 + 2)\left(\frac{2}{3}x^3 - \frac{4}{9}x\right) + \left(-1 + \frac{17}{9}x\right) \quad \text{in } \mathbb{Q}[x],$$

$$2x^5 + x - 1 = (3x^2 + 2)(3x^3 + 5x) + 5x + 6 \quad \text{in } \mathbb{Z}_7[x].$$

Theorem 1.2.4 (Euclid's Algorithm for Polynomials). Let f and g be nonzero polynomials in $F[x]$. The Euclid's algorithm for polynomials runs in exactly the same way as that for integers

$$
\begin{aligned}
f &= gq_0 + r_1 \\
g &= r_1 q_1 + r_2 \\
r_1 &= r_2 q_2 + r_3 \\
r_2 &= r_3 q_3 + r_4 \\
&\vdots \\
r_{n-2} &= r_{n-1} q_{n-1} + r_n \\
r_{n-1} &= r_n q_n + 0
\end{aligned}
$$

Then, $\gcd(f, g) = r_n$. Moreover, if $d(x)$ is the greatest common divisor of $f(x)$ and $g(x)$, then there are polynomials $s(x)$ and $t(x)$ such that

$$d(x) = s(x)f(x) + t(x)g(x).$$

Example 1.2.13. Let

$$
\begin{aligned}
f(x) &= x^5 + x^3 - x^2 - 1, \\
g(x) &= x^3 + 3x^2 + x + 3.
\end{aligned}
$$

Then

$$
\begin{aligned}
d(x) &= x^2 + 1, \\
s(x) &= -\frac{1}{28}, \\
t(x) &= \frac{1}{28}x^2 - \frac{3}{28}x + \frac{9}{28}.
\end{aligned}
$$

For polynomials, the analog to *prime number* is that of *irreducible polynomials*. A polynomials $f(x)$ of degree at least one in $F[x]$ is called *irreducible over F* if it cannot be written as a product of two nonconstant polynomials in $F[x]$ of lower degree. For example, in $Q[x]$, $f(x) = x^2 + 1$ is irreducible, since there is no factorization of $f(x)$ into polynomials both of degree less than 2 (of course, $x^2 + 1 = \frac{1}{2}(2x^2 + 2)$, but $\frac{1}{2}$ is unit in $\mathbb{Q}$. $x^2 - 2$ is irreducible in $Q[x]$ since it has no rational root. However, $x^2 - 2$ is reducible in $\mathbb{R}[x]$ as $x^2 - 2 = (x - \sqrt{2})(x + \sqrt{2})$. Factoring polynomials over rings with zero divisors can lead to some strange behaviours. For example, in $\mathbb{Z}_6$, 3 is a zero divisor, not a unit, since $1/3 \mod 6$ does not exist. So if we consider the polynomial $3x + 3$ in $\mathbb{Z}_6[x]$, then we can factor it in several ways

$$3x + 3 = 3(x + 1) = (2x + 1)(3x + 3) = (2x^2 + 1)(3x + 3).$$

However, if F is a field, say, e.g., $\mathbb{Z}_5$, then $3x + 3$ can be uniquely factored into reducible polynomials in $\mathbb{Z}_5[x]$.

Theorem 1.2.5. (Unique Factorization in $F[x]$) Every nonconstant polynomial $f(x)$ in $F[x]$ with F a field is the product of irreducible polynomials

$$f(x) = c \prod_{i=1}^{k} p_i(x)$$

where c is the constant, $p_i(x)$ for $i = 1, 2, \ldots, k$ are irreducible polynomials in $F[x]$.

Arithmetic of Algebraic Numbers.

Definition 1.2.14. Let θ be a complex number and

$$f(x) = a_n x^n + a_{n-1} x^{n-1} + \cdots + a_1 x + a_0 \in \mathbb{Q}[x]. \tag{1.30}$$

If θ is the root of the polynomial $f(x)$, then θ is called an *algebraic number*. If $f(x)$ is irreducible over $\mathbb{Q}$ and $a_n \neq 0$, then θ is of degree n.

Example 1.2.14. $i = \sqrt{-1}$, $\sqrt{2}$ are the algebraic numbers of degree 2, since they are roots of the polynomials $x^2 + 1$ and $x^2 - 2$, whereas $\sqrt[5]{3}$ is an algebraic number with degree 5, since it is the root of the polynomial $x^5 - 3$.

Every rational number is an algebraic number since $\frac{a}{b}$ is the root of the linear polynomial $x - \frac{a}{b} \in \mathbb{Q}[x]$. The set of all algebraic numbers is a field with respect to the operations of complex addition and multiplication. In particular, if α and β are algebraic numbers, then $\alpha + \beta$, $\alpha - \beta$, and $\frac{\alpha}{\beta}$ with $\beta > 0$ are all algebraic numbers.

Requiring a number to be a root of a polynomial with *rational* coefficients is the same as asking for it to be a root of a polynomial with *integer* coefficients. The rational number $\frac{a}{b}$ is the root of $bx - a \in \mathbb{Z}[x]$ as well as of $x - \frac{a}{b} \in \mathbb{Q}[x]$. So every algebraic number α is a root of same polynomial

$$f(x) = a_n x^n + a_{n-1} x^{n-1} + \cdots + a_1 x + a_0 \in \mathbb{Z}[x]. \qquad (1.31)$$

If the leading coefficient of $f(x) \in \mathbb{Z}[x]$ is 1 (i.e., $a_n = 1$), then α is an *algebraic integer*.

Example 1.2.15. $\sqrt{2}$, $\dfrac{-1 + \sqrt{-3}}{2}$ and $\sqrt{7} + \sqrt{11}$ are algebraic integers. Every ordinary integer a is an algebraic integer since it is a root of $x - a \in \mathbb{Z}[x]$.

Let $a, b \in \mathbb{Z}$, then $a + bi$ is an algebra integer of degree 2 as it is the root of $x^2 - 2ax + (a^2 + b^2)$. The set of all $a + bi$ is denoted by $\mathbb{Z}[i]$ and is called *Gaussian integers*. Similarly, the elements in set $\mathbb{Z}$ are called *rational integers*, In $\mathbb{Z}$, the numbers $2, 3, 5, 7, 11, 13, 17$ are primes. However, in $\mathbb{Z}[i]$, the numbers $2, 5, 13, 17$ are not primes, since

$$
\begin{aligned}
2 &= (1+i)(1-i) \\
5 &= (2+i)(2-i) = (1+2i)(1-2i) = -i(2+i)(1-2i) \\
13 &= (3+2i)(3-2i) \\
17 &= (4+i)(4-i)
\end{aligned}
$$

In fact, any prime in $\mathbb{Z}$ of the form $p \equiv 1 \pmod 4$ can always be factored into the form $-i(a + bi)(b + ai)$. To distinguish, we call the primes in $\mathbb{Z}$ *rational primes*, and primes in $\mathbb{Z}[i]$ *Gaussian primes*. Also we define the *norm* of $a + b\sqrt{m}$ to be $N(a + b\sqrt{m}) = a^2 + mb^2$, so $N(-22 + 19i) = 845$.

Every algebraic integer is an algebraic number, but not vice versa.

Definition 1.2.15. Let α be algebraic over a field F. The unique, monic, irreducible polynomial f in $F[x]$ with α as a zero is called *minimal polynomial* of α over F. The degree of α over F is defined to be the degree of f. For example, the minimal polynomial $\sqrt[3]{2} \in \mathbb{Q}(\sqrt[3]{2})$ over $\mathbb{Q}$ is $x^3 - 2$.

Theorem 1.2.6. An algebraic number is an algebraic integer if and only if its minimal polynomial has integer coefficients.

Example 1.2.16. The number $\sqrt[3]{\dfrac{5}{7}}$ is an algebraic number but not an algebraic integer since its minimal polynomial is $x^3 - \dfrac{5}{7}$.

Remark 1.2.2. The elements of $\mathbb{Z}$ are the only rational numbers that are algebraic integers, since $\dfrac{a}{b}$ has minimal polynomial $x - \dfrac{a}{b}$ and this only has integer coefficients if $\dfrac{a}{b} \in \mathbb{Z}$.

Theorem 1.2.7. The set of algebraic numbers forms a field, and the set of algebraic integers forms a ring.

Problems for Section 1.2

Problem 1.2.1. Let $G = \{a, b, c, d, e, f\}$ and let $\oplus$ be defined as follows:

$\oplus$	e	a	b	c	d	f
e	e	a	b	c	d	f
a	a	e	d	f	b	c
b	b	f	e	d	c	a
c	c	d	f	e	a	b
d	d	c	a	b	f	c
f	f	b	c	a	e	d

Show that G is a noncommutative group.

Problem 1.2.2. Show that the set of all rational numbers forms a field.

Problem 1.2.3. Show that $\mathbb{Z} = \{0, \pm 1, \pm 2, \pm 3, \ldots\}$ is a ring.

Problem 1.2.4. Show that $\mathbb{Z}_n = \{0, 1, 2, 3, \ldots, n-1\}$ is a ring.

Problem 1.2.5. Show that $\mathbb{Z}_n^* = \{a : a \in \mathbb{Z}_n, \gcd(a, n) = 1\}$ is a multiplicative group.

Problem 1.2.6. Show if p is a prime, then $\mathbb{Z}_p$ is a field.

Problem 1.2.7. Show that the multiplicative group is isomorphic group modulo 9 to the additive group modulo 6.

Problem 1.2.8. Show that any two cyclic groups of order n are isomorphic.

Problem 1.2.9. Show that the set of all rational numbers forms a field.

Problem 1.2.10. Prove that for any prime $p > 2$, the sum

$$\frac{a}{b} = \frac{1}{1^3} + \frac{1}{2^3} + \frac{1}{3^3} + \cdots + \frac{1}{(p-1)^3}$$

has the property that

$$p \mid a.$$

1.3 Divisibility Properties

Definition 1.3.1. Let a and b be integers with $a \neq 0$. We say a divides b, denoted by $a \mid b$, if there exists an integer c such that $b = ac$. When a divides b, we say that a is a *divisor* (or *factor*) of b, and b is a *multiple* of a. If a does not divide b, we write $a \nmid b$. If $a \mid b$ and $0 < a < b$, then a is called a *proper divisor* of b.

Note that it is usually sufficient to consider only positive divisors of an integer.

Example 1.3.1. The integer 200 has the following divisors:

$$1, 2, 4, 5, 8, 10, 20, 25, 40, 50, 100, 200.$$

Thus, for example, we can write

$$8 \mid 200, \quad 50 \mid 200, \quad 7 \nmid 200, \quad 35 \nmid 200.$$

Definition 1.3.2. A divisor of n is called a *trivial divisor* of n if it is either 1 or n itself. A divisor of n is called a *nontrivial divisor* if it is a divisor of n, but is neither 1, nor n.

Theorem 1.3.1 (Division algorithm). For any integer a and any positive integer b, there exist unique integers q and r such that

$$a = bq + r, \quad 0 \le r < b, \tag{1.32}$$

where a is called the *dividend,* q the *quotient,* and r the *remainder.* If $b \nmid a$, then r satisfies the stronger inequalities $0 < r < b$.

Proof. Consider the arithmetic progression

$$\ldots, -3b, -2b, -b, 0, b, 2b, 3b, \ldots$$

then there must be an integer q such that

$$qb \le a < (q+1)b.$$

Let $a - qb = r$, then $a = bq + r$ with $0 \le r < b$. To prove the uniqueness of q and r, suppose there is another pair q_1 and r_1 satisfying the same condition in (1.32), then

$$a = bq_1 + r_1, \quad 0 \le r_1 < b.$$

We first show that $r_1 = r$. For if not, we may presume that $r < r_1$, so that $0 < r_1 - r < b$, and then we see that $b(q - q_1) = r_1 - r$, and so $b \mid (r_1 - r)$, which is impossible. Hence, $r = r_1$, and also $q = q_1$.   □

Definition 1.3.3. Consider the following equation

$$a = 2q + r, \quad a, q, r \in \mathbb{Z}, \ 0 \le r < 2. \tag{1.33}$$

Then if $r = 0$, then a is *even*, whereas if $r = 1$, then a is *odd*.

Definition 1.3.4. A positive integer n greater than 1 is called *prime* if its only divisors are n and 1. Otherwise, it is called *composite*.

Example 1.3.2. The integer 23 is prime since its only divisors are 1 and 23, whereas 22 is composite since it is divisible by 2 and 11.

Prime numbers have many special and nice properties, and play a central role in the development of number theory. Mathematicians throughout history have been fascinated by primes. The first result on prime numbers is due to Euclid:

Theorem 1.3.2 (Euclid). There are infinitely many primes.

Proof. Suppose that $p_1, p_2, \ldots, p_k$ are all the primes. Consider the number $N = p_1 p_2 \cdots p_k + 1$. If it is a prime, then it is a new prime. Otherwise, it has a prime factor q. If q were one of the primes p_i, $i = 1, 2, \ldots, k$, then $q \mid (p_1 p_2 \cdots p_k)$, and since $q \mid (p_1 p_2 \cdots p_k + 1)$, q would divide the difference of these numbers, namely 1, which is impossible. So q cannot be one of the p_i for $i = 1, 2, \ldots, k$, and must therefore be a new prime. This completes the proof. $\square$

Theorem 1.3.3. If n is a composite, then n has a prime divisor p such that $p \leq \sqrt{n}$.

Proof. Let p be the smallest prime divisor of n. If $n = rs$, then $p \leq r$ and $p \leq s$. Hence, $p^2 \leq rs = n$. That is, $p \leq \sqrt{n}$. $\square$

Theorem 1.3.3 can be used to find all the prime numbers up to a given positive integer x; this procedure is called the Sieve of Eratosthenes, attributed to the ancient Greek astronomer and mathematician Eratosthenes of Cyrene. To apply the sieve, list all the integers from 2 up to x in order:

$$2, 3, 4, 5, 6, 7, 8, 9, 10, 11, 12, 13, 14, 15, \ldots, x.$$

Starting from 2, delete all the multiples $2m$ of 2 such that $2 < 2m \leq x$:

$$2, 3, 5, 7, 9, 11, 13, 15, \ldots, x.$$

Starting from 3, delete all the multiples $3m$ of 3 such that $3 < 3m \leq x$:

$$2, 3, 5, 7, 11, 13, \ldots, x.$$

In general, if the resulting sequence at the kth stage is

$$2, 3, 5, 7, 11, 13, \ldots, p, \ldots, x.$$

then delete all the multiples pm of p such that $p < pm \leq x$. Continue this exhaustive computation, until $p \leq \lfloor \sqrt{x} \rfloor$, where $\lfloor \sqrt{x} \rfloor$ denotes the greatest integer $\leq \sqrt{x}$, e.g., $\lfloor 0.5 \rfloor = 0$ and $\lfloor 2.9 \rfloor = 2$. The remaining integers are all the primes between $\lfloor \sqrt{x} \rfloor$ and x and if we take care not to delete $2, 3, 5, \ldots, p \leq \lfloor \sqrt{x} \rfloor$, the sieve then gives all the primes less than or equal to x.

Algorithm 1.3.1 (The Sieve of Eratosthenes). Given a positive integer $n > 1$, this algorithm will find all prime numbers up to n.

[1] Create a list of integers from 2 to n;

[2] For prime numbers p_i ($i = 1, 2, \ldots$) from $2, 3, 5$ up to $\lfloor \sqrt{n} \rfloor$, delete all the multiples mp_i from the list, with $p_i < mp_i \leq n$, $m = 1, 2, \ldots$.

[3] Print the integers remaining in the list.

Example 1.3.3. Suppose we want to find all primes up to 100. First note that up to $\sqrt{100} = 10$, there are only 4 primes $2, 3, 5, 7$. Thus in a table containing all positive integers from 2 to 100. Retain 2,3,5,7, but cross all the multiples of 2,3,5,7. After the sieving steps, the remaining numbers are the primes up to 100, as shown in Table 1.9.

	2	3	4	5	6	7	8	9	10
11	12	13	14	15	16	17	18	19	20
21	22	23	24	25	26	27	28	29	30
31	32	33	34	35	36	37	38	39	40
41	42	43	44	45	46	47	48	49	50
51	52	53	54	55	56	57	58	59	60
61	62	63	64	65	66	67	68	69	70
71	72	73	74	75	76	77	78	79	80
81	82	83	84	85	86	87	88	89	90
91	92	93	94	95	96	97	98	99	100

Table 1.9. Sieve of Eratosthenes for numbers up to 100

Theorem 1.3.4. Every composite number has a prime factor.

Proof. Let n be a composite number. Then

$$n = n_1 n_2$$

where n_1 and n_2 are positive integers with $n_1, n_2 < n$. If either n_1 or n_2 is a prime, then the theorem is proved. If n_1 and n_2 are not prime, then

$$n_1 = n_3 n_4$$

where n_3 and n_4 are positive integers with $n_3, n_4 < n_1$. Again if n_3 or n_4 is a prime, then the theorem is proved. If n_3 and n_4 are not prime, then we can write

$$n_3 = n_5 n_6$$

where n_5 and n_6 are positive integers with $n_5, n_6 < n_3$. In general, after k steps we write

$$n_{2k-1} = n_{2k+1}n_{2k+2}$$

where n_{2k+1} and n_{2k+2} are positive integers with $n_{2k+1}, n_{2k+1} < n_{2k-1}$. Since

$$n > n_1 > n_3 > n_5 > \cdots n_{2k-1} > 0$$

for any value k, the process must terminate. So there must exist an n_{2k-1} for some value of k, that is prime. Hence, every composite has a prime factor. $\square$

Prime numbers are the building blocks of positive integers, as the following theorem shows:

Theorem 1.3.5 (Fundamental Theorem of Arithmetic). Every positive integer n greater than 1 can be written uniquely as the product of primes:

$$n = p_1^{\alpha_1} p_2^{\alpha_2} \cdots p_k^{\alpha_k} = \prod_{i=1}^{k} p_i^{\alpha_i} \tag{1.34}$$

where $p_1, p_2, \ldots, p_k$ are distinct primes, and $\alpha_1, \alpha_2, \ldots, \alpha_k$ are natural numbers.

Proof. We shall first show that a factorization exists. Starting from $n > 1$, if n is a prime, then it stands as a *product* with a single factor. Otherwise, n can be factored into, say, ab, where $a > 1$ and $b > 1$. Apply the same argument to a and b: each is either a prime or a product of two numbers both > 1. The numbers other than primes involved in the expression for n are greater than 1 and decrease at every step; hence eventually all the numbers must be prime.

Now we come to uniqueness. Suppose that the theorem is false and let $n > 1$ be the smallest number having more than one expression as the product of primes, say

$$n = p_1 p_2 \ldots p_r = q_1 q_2 \ldots q_s$$

where each p_i $(i = 1, 2, \ldots, r)$ and each q_j $(j = 1, 2, \ldots, s)$ is prime. Clearly both r and s must be greater than 1 (otherwise n is prime, or a prime is equal to a composite). If for example p_1 were one of the q_j $(j = 1, 2, \ldots, s)$, then n/p_1 would have two expressions as a product of primes, but $n/p_1 < n$ so this would contradict the definition of n. Hence p_1 is not equal to any of the q_j $(j = 1, 2, \ldots, s)$, and similarly none of the p_i $(i = 1, 2, \ldots, r)$ equals any of the q_j $(j = 1, 2, \ldots, s)$. Next, there is no loss of generality in presuming that $p_1 < q_1$, and we define the positive integer N as

$$N = (q_1 - p_1)q_2 q_3 \cdots q_s = p_1(p_2 p_3 \cdots p_r - q_2 q_3 \cdots q_s).$$

Certainly $1 < N < n$, so N is uniquely factorable into primes. However, $p_1 \nmid (q_1 - p_1)$, since $p_1 < q_1$ and q_1 is prime. Hence one of the above expressions for N contains p_1 and the other does not. This contradiction proves the result: there cannot be any exceptions to the theorem. $\square$

Definition 1.3.5. Let a and b be integers, not both zero. The largest divisor d such that $d \mid a$ and $d \mid b$ is called the *greatest common divisor* (gcd) of a and b. The greatest common divisor of a and b is denoted by $\gcd(a, b)$.

Example 1.3.4. The sets of positive divisors of 111 and 333 are as follows:

$$1, 3, 37, 111,$$
$$1, 3, 9, 37, 111, 333,$$

so $\gcd(111, 333) = 111$. But $\gcd(91, 111) = 1$, since 91 and 111 have no common divisors other than 1.

The next theorem indicates that $\gcd(a, b)$ can be represented as a linear combination of a and b.

Theorem 1.3.6. Let a and b be integers, not both zero. Then there exists integers x and y such that

$$d = \gcd(a, b) = ax + by. \tag{1.35}$$

Proof. Consider the set of all linear combinations $au + bv$, where u and v range over all integers. Clearly this set of integers $\{au + bv\}$ includes positive, negative as well as 0. It contains a smallest positive element, say, m, such that $m = ax + by$. Use the Division algorithm, to write $a = mq + r$, with $0 \leq r < m$. Then

$$r = a - mq = a - q(ax + by) = (1 - qx)a + (-qy)b$$

and hence r is also a linear combination of a and b. But $r < m$, so it follows from the definition of m that $r = 0$. Thus $a = mq$, that is, $m \mid a$; similarly, $m \mid b$. Therefore, m is a common divisor of a and b. Since $d \mid a$ and $d \mid b$, d divides any linear combination of a and b. Since $d = \gcd(a, b)$, we must have $d = m$.   □

Corollary 1.3.1. If a and b are integers, not both zero, then the set

$$S = \{ax + by : \ x, y \in \mathbb{Z}\}$$

is precisely the set of all multiples of $d = \gcd(a, b)$.

Proof. It follows from Theorem 1.3.6, because d is the smallest positive values of $ax + by$ where x and y range over all integers.   □

Definition 1.3.6. Two integers a and b are called *relatively prime* if $\gcd(a, b) = 1$. We say that integers $n_1, n_2, ..., n_k$ are *pairwise relatively prime* if, whenever $i \neq j$, we have $\gcd(n_i, n_j) = 1$.

Example 1.3.5. 91 and 111 are relatively prime, since $\gcd(91, 111) = 1$.

The following theorem characterizes relatively primes in terms of linear combinations.

Theorem 1.3.7. Let a and b be integers, not both zero, then a and b are relatively prime if and only if there exist integers x and y such that $ax + by = 1$.

Proof. If a and b are relatively prime, so that $\gcd(a, b) = 1$, then Theorem 1.3.6 guarantees the existence of integers x and y satisfying $ax + by = 1$. As for the converse, suppose that $ax + by = 1$ and that $d = \gcd(a, b)$. Since $d \mid a$ and $d \mid b$, $d \mid (ax + by)$, that is, $d \mid 1$. Thus $d = 1$. The result follows. □

Theorem 1.3.8. If $a \mid bc$ and $\gcd(a, b) = 1$, then $a \mid c$.

Proof. By Theorem 1.3.6, we can write $ax + by = 1$ for some choice of integers x and y. Multiplying this equation by c we get

$$acx + bcy = c.$$

Since $a \mid ac$ and $a \mid bc$, it follows that $a \mid (acx + bcy)$. The result thus follows. □

For the greatest common divisor of more than two integers, we have the following result.

Theorem 1.3.9. Let $a_1, a_2, \ldots, a_n$ be n integers. Let also

$$\left. \begin{array}{rcl} \gcd(a_1, a_2) & = & d_2 \\ \gcd(d_2, a_3) & = & d_3 \\ & \vdots & \\ \gcd(d_{n-1}, a_n) & = & d_n \end{array} \right\} \tag{1.36}$$

Then

$$\gcd(a_1, a_2, \ldots, a_n) = d_n. \tag{1.37}$$

Proof. By (1.36), we have $d_n \mid a_n$ and $d_n \mid d_{n-1}$. Since $d_{n-1} \mid a_{n-1}$ and $d_{n-1} \mid d_{n-2}$, $d_n \mid a_{n-1}$ and $d_n \mid d_{n-2}$. Continuing in this way, we finally have $d_n \mid a_n$, $d_n \mid a_{n-1}$, $\cdots$, $d_n \mid a_1$, so d_n is a common divisor of $a_1, a_2, \ldots, a_n$. Now suppose that d is any common divisor of $a_1, a_2, \ldots, a_n$, then $d \mid a_1$ and $d \mid d_2$. Observe the fact that the common divisor of a and b and the divisor of $\gcd(a, b)$ are the same, so $d \mid d_2$. Similarly, we have $d \mid d_3, \ldots, d \mid d_n$. Therefore, $d \leq |d| \leq d_n$. So, d_n is the greatest common divisor of $a_1, a_2, \ldots, a_n$. □

Definition 1.3.7. If d is a multiple of a and also a multiple of b, then d is a common multiple of a and b. The *least common multiple* (lcm) of two integers a and b, is the smallest of the common multiples of a and b. The least common multiple of a and b is denoted by $\mathrm{lcm}(a, b)$.

Theorem 1.3.10. Suppose a and b are not both zero (i.e., one of the a and b can be zero, but not both zero), and that $m = \text{lcm}(a, b)$. If x is a common multiple of a and b, then $m \mid x$. That is, every common multiple of a and b is a multiple of the least common multiple.

Proof. If any one of a and b is zero, then all common multiples of a and b are zero, so the statement is trivial. Now we assume that both a and b are not zero. Dividing x by m, we get

$$x = mq + r, \quad \text{where } 0 \le r < m.$$

Now $a \mid x$ and $b \mid x$ and also $a \mid m$ and $b \mid m$; so by Theorem 1.3.1, $a \mid r$ and $b \mid r$. That is, r is a common multiple of a and b. But m is the least common multiple of a and b, so $r = 0$. Therefore, $x = mq$ and the result follows. $\square$

For the lest common multiple of more than two integers, we have the following result.

Theorem 1.3.11. Let $a_1, a_2, \ldots, a_n$ be n integers. Let also

$$\left.\begin{aligned}
\text{lcm}(a_1, a_2) &= m_2, \\
\text{lcm}(m_2, a_3) &= m_3, \\
&\vdots \\
\text{lcm}(m_{n-1}, a_n) &= m_n.
\end{aligned}\right\} \tag{1.38}$$

Then

$$\text{lcm}(a_1, a_2, \ldots, a_n) = m_n. \tag{1.39}$$

Proof. By (1.38), we have $m_i \mid m_{i+1}$, $i = 2, 3, \ldots, n - 1$, and $a_1 \mid m_2$, $a_i \mid m_i$, $i = 2, 3, \ldots, n$. So, m_n is a common multiple of $a_1, a_2, \ldots, a_n$. Now let m be any common multiple of $a_1, a_2, \ldots, a_n$, then $a_1 \mid m$, $a_2 \mid m$. Observe the result that all the common multiples of a and b are the multiples of $\text{lcm}(a, b)$. So $m_1 \mid m$ and $a_3 \mid m$. Continuing the process in this way, we finally have $m_n \mid m$. Thus, $m_n \le |m|$. Therefore, $m_n = \text{lcm}(a_1, a_2, \ldots, a_n)$. $\square$

One way to calculate the $\gcd(a, b)$ or the $\text{lcm}(a, b)$ is to use the standard prime factorizations of a and b. That is:

Theorem 1.3.12. If

$$a = \prod_{i=1}^{k} p_i^{\alpha_i}, \quad \alpha_i \ge 0,$$

and

$$b = \prod_{i=1}^{k} p_i^{\beta_i}, \quad \beta_i \ge 0,$$

then

$$\gcd(a,b) = \prod_{i=1}^{k} p_i^{\gamma_i} \tag{1.40}$$

$$\operatorname{lcm}(a,b) = \prod_{i=1}^{k} p_i^{\delta_i} \tag{1.41}$$

where $\gamma_i = \min(\alpha_i, \beta_i)$ and $\delta_i = \max(\alpha_i, \beta_i)$ for $i = 1, 2, \ldots, k$.

Proof. It is easy to see that

$$\gcd(a,b) = \prod_{i=1}^{k} p_i^{\gamma_i}, \text{ where } \gamma_i \text{ is the lesser of } \alpha_i \text{ and } \beta_i,$$

$$\operatorname{lcm}(a,b) = \prod_{i=1}^{k} p_i^{\delta_i}, \text{ where } \delta_i \text{ is the greater of } \alpha_i \text{ and } \beta_i.$$

The result thus follows. $\square$

Corollary 1.3.2. Suppose a and b are positive integers, then

$$\operatorname{lcm}(a,b) = \frac{ab}{\gcd(a,b)}. \tag{1.42}$$

Proof. Since $\gamma_i + \delta_i = \alpha_i + \beta_i$, it is now obvious that

$$\gcd(a,b) \cdot \operatorname{lcm}(a,b) = ab.$$

The result thus follows. $\square$

Example 1.3.6. Find $\gcd(240, 560)$ and $\operatorname{lcm}(240, 560)$. Since the prime factorizations of 240 and 560 are

$$240 = 2^4 \cdot 3 \cdot 5 = 2^4 \cdot 3^1 \cdot 5^1 \cdot 7^0,$$
$$560 = 2^4 \cdot 5 \cdot 7 = 2^4 \cdot 3^0 \cdot 5^1 \cdot 7^1,$$

we get

$$\gcd(240, 560) = 2^{\min(4,4)} \cdot 3^{\min(1,0)} \cdot 5^{\min(1,1)} \cdot 7^{\min(0,1)}$$
$$= 2^4 \cdot 3^0 \cdot 5^1 \cdot 7^0 = 80,$$
$$\operatorname{lcm}(240, 560) = 2^{\max(4,4)} \cdot 3^{\max(1,0)} \cdot 5^{\max(1,1)} \cdot 7^{\max(0,1)}$$
$$= 2^4 \cdot 3^1 \cdot 5^1 \cdot 7^1 = 1680.$$

Of course, if we know $\gcd(240, 560) = 80$, then we can find $\operatorname{lcm}(240, 560)$ by

$$\operatorname{lcm}(240, 560) = 240 \cdot 560/80 = 1680.$$

Similarly, if we know $\operatorname{lcm}(240, 560)$, we can find $\gcd(240, 560)$ by

$$\gcd(240, 560) = 240 \cdot 560/1680 = 80.$$

Problems for Section 1.3

Problem 1.3.1. Note that in bases $2 \leq b \leq 12$, the number 1010101 are always composite:

$$
\begin{aligned}
1010101_2 &= 1 \cdot 2^6 + 1 \cdot 2^4 + 1 \cdot 2^2 + 1 \cdot 2^0 = 85 = 5 \cdot 17 \\
1010101_3 &= 1 \cdot 3^6 + 1 \cdot 3^4 + 1 \cdot 3^2 + 1 \cdot 3^0 = 820 = 2^2 \cdot 5 \cdot 41 \\
1010101_4 &= 1 \cdot 4^6 + 1 \cdot 4^4 + 1 \cdot 4^2 + 1 \cdot 4^0 = 4369 = 17 \cdot 257 \\
1010101_5 &= 1 \cdot 5^6 + 1 \cdot 5^4 + 1 \cdot 5^2 + 1 \cdot 5^0 = 16276 = 2^2 \cdot 13 \cdot 313
\end{aligned}
$$

$$\vdots$$

$$
\begin{aligned}
1010101_{10} &= 1 \cdot 10^6 + 1 \cdot 10^4 + 1 \cdot 10^2 + 1 \cdot 10^0 = 1010101 = 73 \cdot 101 \cdot 137 \\
1010101_{11} &= 1 \cdot 11^6 + 1 \cdot 11^4 + 1 \cdot 11^2 + 1 \cdot 11^0 = 1786324 = 2^2 \cdot 61 \cdot 7321 \\
1010101_{12} &= 1 \cdot 12^6 + 1 \cdot 12^4 + 1 \cdot 12^2 + 1 \cdot 12^0 = 3006865 = 5 \cdot 29 \cdot 89 \cdot 233
\end{aligned}
$$

(1) Show that in any basis the number 1010101 cannot be prime.
(2) How about the number 11010101? Can this number be always composite in any basis $b \geq 2$? For $2 \leq b \leq 100$, list the numbers 11010101_b which are not composite.

Problem 1.3.2. A number is a *perfect square*, sometimes also called a *square number*, if it is of the form n^2 for some integer n, e.g., 0, 1, 4, 9, 16, 25, 36, 49, 64, 81,100, 122,144,169,196 are the first 15 perfect squares.

(1) Show that the product of four consecutive positive integers $a, a+1, a+2, a+3$ cannot be a perfect square.

(2) Are there infinitely many primes p such that $p-1$ is a perfect square? This is one of the four problems proposed by he German number theorist Edmund Landau (1877–1938) in 1921; it is unsolved to this day.

(3) Show that there is a prime number between two consecutive perfect squares n^2 and $(n+1)^2$ for every positive integer n. This is the famous Legendre conjecture, unsolved to this day.

(4) Show that there is a prime number between consecutive perfect squares n^2 and $(n+1)^2$ for every positive integer n. (This is the famous Legendre Conjecture; it is unproven as of 2008. However, partial results are obtained. For example, a result due to Ingham shows that there is a prime between n^3 and $(n+1)^3$ for every positive integer n, and the Chinese Mathematician J R Chen showed in 1975 that there always exists a number P which is either a prime or product of two primes between the consecutive perfect squares n^2 and $(n+1)^2$.)

(5) Are there infinitely many primes p such that $p-1$ is a perfect square? In other words: Are there infinitely many primes (called generalized Fermat primes) of the form n^2+1? (This is one of the four problems about prime numbers proposed by the German mathematician Edmund Landau in the 1912 International Congress of Mathematicians. Although the problem has still not been settled, some progress are made, for example, the famous The Bombieri–Friedlander–Iwaniec theorem shows that infinitely many primes are of the form $x^2 + y^4$.)

(6) Show that a perfect square cannot be a perfect number.

Problem 1.3.3. A positive integer that has no perfect square divisors except 1 is called square-free, e.g.,10 is square-free but 18 is not, as it is divisible by $9 = 3^2$. The first 25 square-free numbers are as follows:

$$1, 2, 3, 5, 6, 7, 10, 11, 13, 14, 15, 17, 19, 21, 22, 23, 26, 29, 30, 31, 33, 34, 35, 37, 38.$$

(1) Show that n is square-free if and only if in every factorization $n = ab$, $\gcd(a,b) = 1$.

(2) The radical of an integer is always square-free. (The radical of a positive integer n is defined to be the product of the prime numbers dividing n:

$$\text{Rad}(n) = \prod_{p|n} p$$

e.g., $n = 600 = 2^3 \cdot 3 \cdot 5^2$, $\text{Rad}(n) = 2 \cdot 3 \cdot 5 = 30$.)

(3) Show that each odd prime p can be written as the difference of two perfect squares.

Problem 1.3.4. Show that $7 \mid \left(1^{47} + 2^{47} + 3^{47} + 4^{47} + 5^{47} + 6^{47}\right)$.

Problem 1.3.5. Let p_k be the kth prime. Prove that

$$p_k = 1 + \sum_{m=1}^{2^k} \left[\left| \frac{k}{1 + \sum_{j=2}^{m} \left[\left| \frac{(j-1)! + 1}{j} \right| - \left\lfloor \frac{(j-1)!}{j} \right\rfloor \right]} \right|^{1/k} \right].$$

Problem 1.3.6. Use mathematical induction to prove that when $n \geq 1$,

$$F_0 F_1 F_2 \cdots F_{n-1} = F_n - 2 \tag{1.43}$$

where $F_i = 2^{2^i} + 1$, $i = 0, 1, 2, 3, \ldots$ are the Fermat numbers. Use (1.43) to prove that if m and n are distinctive positive integers, then

$$\gcd(F_m, F_n) = 1. \tag{1.44}$$

Furthermore, use (1.44) to prove that there are infinitely many primes.

Problem 1.3.7. Let n be a positive integer. Find

$$\gcd\left(\binom{2n}{1}, \binom{2n}{3}, \binom{2n}{5}, \ldots, \binom{2n}{2n-1}\right)$$

where

$$\binom{n}{k} = \frac{n!}{k!(n-k)!}$$

is the binomial coefficient.

Problem 1.3.8. Find the inverse of the matrix

$$\begin{pmatrix} 1 & 1 \\ 6 & 1 \end{pmatrix} \pmod{26}.$$

Find also all the values of $b \bmod 26$ such that

$$\begin{pmatrix} 1 & 1 \\ b & 1 \end{pmatrix} \pmod{26}$$

is invertible.

Problem 1.3.9. Let p be prime and n a positive integer. An integer $n \geq 2$ is called a powerful number if $p \mid n$ implies $p^2 \mid n$. That is, n is of the form $n = a^2 b^3$, where a and b are positive integers. Find all the powerful numbers up to 1000, and prove that every sufficiently large integer is a sum of at most three *powerful numbers*. (this result was proved by Heath-Brown of Oxford University in 1987).

Problem 1.3.10. Prove that none of the following numbers is prime:

$$12321, \; 1234321, \; 123454321, \; 12345654321, \; 1234567654321,$$

$$123456787654321, 12345678987654321$$

1.4 Euclid's Algorithm and Continued Fractions

Theorem 1.4.1 (Division theorem). Let a, b, q, r be integers with $b > 0$ and $0 \leq r < b$ such that $a = bq + r$. Then $\gcd(a, b) = \gcd(b, r)$.

Proof. Let $X = \gcd(a, b)$ and $Y = \gcd(b, r)$, it suffices to show that $X = Y$. If integer c is a divisor of a and b, it follows from the equation $a = bq + r$ and the divisibility properties that c is a divisor of r also. By the same argument, every common divisor of b and r is a divisor of a. □

Theorem 1.4.1 can be used to reduce the problem of finding $\gcd(a, b)$ to the simpler problem of finding $\gcd(b, r)$. The problem is simpler because the numbers are smaller, but it has the same answer as the original one. The process of finding $\gcd(a, b)$ by repeated application of Theorem 1.4.1 is called Euclid's algorithm which proceeds as follows:

$$a = bq_0 + r_1, \qquad\qquad 0 \le r_1 < b \qquad \text{(dividing } b \text{ into } a\text{)},$$

$$b = r_1 q_1 + r_2, \qquad\qquad 0 \le r_2 < r_1 \qquad \text{(dividing } r_1 \text{ into } b\text{)},$$

$$r_1 = r_2 q_2 + r_3, \qquad\qquad 0 \le r_3 < r_2 \qquad \text{(dividing } r_2 \text{ into } r_1\text{)},$$

$$r_2 = r_3 q_3 + r_4, \qquad\qquad 0 \le r_4 < r_3 \qquad \text{(dividing } r_3 \text{ into } r_2\text{)},$$

$$\vdots \qquad\qquad\qquad\qquad \vdots \qquad\qquad\qquad\qquad \vdots$$

$$r_{n-2} = r_{n-1} q_{n-1} + r_n, \quad 0 \le r_n < r_{n-1} \quad \text{(dividing } r_{n-1} \text{ into } r_{n-2}\text{)},$$

$$r_{n-1} = r_n q_n + 0, \qquad\qquad r_{n+1} = 0 \qquad \text{(arriving at a zero-remainder)}$$

or, diagrammatically,

a			b
$-\ bq_0$	q_0		
r_1	q_1		$-\ r_1 q_1$
$-\ r_2 q_2$	q_2		r_2
r_3	q_3		$-\ r_3 q_3$
$\vdots$	$\vdots$		$\vdots$
r_{n-1}	q_{n-1}		$-\ r_{n-1} q_{n-1}$
$-\ r_n q_n$	q_n		$\boxed{r_n}$
$r_{n+1} = 0$			

Then the greatest common divisor gcd of a and b is r_n. That is,

$$d = \gcd(a, b) = r_n. \qquad\qquad (1.45)$$

We now restate it in a theorem form.

Theorem 1.4.2 (Euclid's algorithm). Let a and b be positive integers with $a \ge b$. If $b \mid a$, then $\gcd(a, b) = b$. If $b \nmid a$, then apply the division algorithm repeatedly as follows:

$$\left.\begin{array}{ll} a = bq_0 + r_1, & 0 < r_1 < b, \\[4pt] b = r_1 q_1 + r_2, & 0 < r_2 < r_1, \\[4pt] r_1 = r_2 q_2 + r_3, & 0 < r_3 < r_2, \\[4pt] r_2 = r_3 q_3 + r_4, & 0 < r_4 < r_3, \\[4pt] \quad\vdots & \quad\vdots \\[4pt] r_{n-2} = r_{n-1} q_{n-1} + r_n, & 0 < r_n < r_{n-1}, \\[4pt] r_{n-1} = r_n q_n + 0. & \end{array}\right\} \qquad (1.46)$$

Then r_n, the last nonzero remainder, is the *greatest common divisor* of a and b. That is,

$$\gcd(a, b) = r_n. \qquad (1.47)$$

Values of x and y in

$$\gcd(a, b) = ax + by \qquad (1.48)$$

can be obtained by writing each r_i as a linear combination of a and b.

Proof. The system of equations is obtained by the series divisions:

$$\frac{a}{b}, \; \frac{b}{r_1}, \; \frac{r_1}{r_2}, \; \cdots$$

The process stops whenever $r_i = 0$ for $i = 1, 2, \ldots, n$.

We now prove that r_n is the greatest common divisor of a and b. By Theorem 1.4.1, we have

$$\begin{aligned} \gcd(a, b) &= \gcd(a - bq_0, b) \\ &= \gcd(r_1, b) \\ &= \gcd(r_1, b - r_1 q_1) \\ &= \gcd(r_1, r_2) \\ &= \gcd(r_1 - r_2 q_2, r_2) \\ &= \gcd(r_3, r_2) \end{aligned}$$

Continuing by mathematical induction, we have

$$\gcd(a, b) = \gcd(r_{n-1}, r_n) = \gcd(r_n, 0) = r_n.$$

To see that r_n is a linear combination of a and b, we argue by induction that each r_i is a a linear combination of a and b. Clearly, r_1 is a linear combination of a and b, since $r_1 = a - bq_0$, so does r_2. In general, r_i is a linear combination of r_{i-1} and r_{i-2}. By the inductive hypothesis we may suppose that these latter two numbers are linear combinations of a and b, and it follows that r_i is also a linear combination of a and b. $\qquad\square$

Algorithm 1.4.1 (Euclid's algorithm). Given integers a and b with $a > b > 0$, this algorithm will compute $\gcd(a, b)$.

[1] (Initialization) Set

$$r_{-1} \leftarrow a$$
$$r_0 \leftarrow b,$$
$$i = 0.$$

[2] (Decision) If $r_i = 0$, Output $r_{i-1} = \gcd(a, b)$ and Exit.

[3] (Computation)

$$q_i \leftarrow \lfloor r_{i-1}/r_i \rfloor,$$
$$r_{i+1} \leftarrow r_{i-1} - q_i \cdot r_i,$$
$$i \leftarrow i + 1,$$
go to Step [2].

Remark 1.4.1. Euclid's algorithm is found in Book VII, Proposition 1 and 2 of his *Elements*, but it probably wasn't his own invention. Scholars believe that the method was known up to 200 years earlier. However, it first appeared in Euclid's *Elements*, and more importantly, it is the first nontrivial algorithm to have survived to this day.

Remark 1.4.2. It is evident that the algorithm cannot recur indefinitely, since the second argument strictly decreases in each recursive call. Therefore, the algorithm always terminates with the correct answer. More importantly, it can be performed in polynomial time. That is, if Euclid's algorithm is applied to two positive integers a and b with $a \geq b$, then the number of divisions required to find $\gcd(a, b)$ is $\mathcal{O}(\log b)$, a polynomial-time complexity.

Example 1.4.1. Use Euclid's algorithm to find the gcd of 1281 and 243. Since

1281			
$-$ 1215	5		243
66	3		$-$ 198
$-$ 45	1		45
21	2		$-$ 42
$-$ 21	7		$\boxed{3}$
0			

we have $\gcd(1281, 243) = 3$.

Theorem 1.4.3. If a and b are any two integers, then

$$Q_k a - P_k b = (-1)^{k-1} r_k, \quad k = 1, 2, \ldots, n \qquad (1.49)$$

where

$$\left.\begin{array}{l} P_0 = 1, \ P_1 = q_0, \ P_k = q_{k-1}P_{k-1} + P_{k-2} \\[2mm] Q_0 = 0, \ Q_1 = 1, \ Q_k = q_{k-1}Q_{k-1} + Q_{k-2} \end{array}\right\} \qquad (1.50)$$

for $k = 2, 3, \ldots, n$.

Proof. When $k = 1$, (1.49) is clearly true, since $Q_1 a - P_1 b = (-1)^{1-1} r_1$ implies $a - q_0 b = r_1$. When $k = 2$, $r_2 = -(aq_1 - b(1 + q_0 q_1))$. But $1 + q_0 q_1 = q_2 P_1 + P_0$, $q_1 = q_1 \cdot 1 + 0 = q_1 Q_1 + Q_0$, therefore, $Q_2 a - P_2 b = (-1)^{2-1} r_2$, $P_2 = q_1 P_1 + P_0$, $Q_2 = q_1 Q_1 + Q_0$. Assume (1.49) and (1.50) hold for all positive integers $\leq k$, then

$$\begin{aligned} (-1)^k r_{k+1} &= (-1)^k (r_{k-1} - q_k r_k) \\ &= (Q_{k-1}a - P_k b) + q_k(Q_k a - P_k b) \\ &= (q_k Q_k + Q_{k-1})a - (q_{k+1}P_k + P_{k+1})b. \end{aligned}$$

Thus, $Q_{k+1}a - P_{k+1}b = (-1)^k r_{k+1}$, where $P_{k+1} = q_k P_k + P_{k-1}$, $Q_{k+1} = q_{k+1}Q_k + Q_{k-1}$. By induction, the result is true for all positive integers. $\square$

Euclid's algorithm for computing the greatest common divisor of two integers is intimately connected with continued fractions.

Definition 1.4.1. Let a and b be integers and let Euclid's algorithm run as

$$\begin{aligned} a &= bq_0 + r_1, \\[1mm] b &= r_1 q_1 + r_2, \\[1mm] r_1 &= r_2 q_2 + r_3, \\[1mm] r_2 &= r_3 q_3 + r_4, \\ &\ \ \vdots \\ r_{n-2} &= r_{n-1} q_{n-1} + r_n, \\[1mm] r_{n-1} &= r_n q_n + 0. \end{aligned}$$

That is,

a			
$-\,bq_0$	q_0		b
r_1	q_1		$-\,r_1q_1$
$-\,r_2q_2$	q_2		r_2
r_3	q_3		$-\,r_3q_3$
$\vdots$	$\vdots$		$\vdots$
r_{n-1}	q_{n-1}		$-\,r_{n-1}q_{n-1}$
$-\,r_nq_n$	q_n		r_n
$r_{n+1}=0$			

Then the fraction $\dfrac{a}{b}$ can be expressed as a simple continued fraction:

$$\frac{a}{b} = q_0 + \cfrac{1}{q_1 + \cfrac{1}{q_2 + \cfrac{1}{\ddots\, q_{n-1} + \cfrac{1}{q_n}}}} \tag{1.51}$$

where $q_0, q_1, \ldots, q_{n-1}, q_n$ are taken directly from Euclid's algorithm expressed in (1.46), and are called the *partial quotients* of the continued fraction. For simplicity, the continued fraction expansion (1.51) of $\dfrac{a}{b}$ is usually written as

$$\frac{a}{b} = q_0 + \frac{1}{q_1+}\,\frac{1}{q_2+} \cdots \frac{1}{q_{n-1}+}\,\frac{1}{q_n} \tag{1.52}$$

or even more briefly as

$$\frac{a}{b} = [q_0, q_1, q_2, \ldots q_{n-1}, q_n]. \tag{1.53}$$

If each q_i is an integer, the continued fraction is called *simple*; a simple continued fraction can either be *finite* or *infinite*. A continued fraction formed from $[q_0, q_1, q_2, \ldots q_{n-1}, q_n]$ by neglecting all of the terms after a given term is called a *convergent* of the original continued fraction. If we denote the kth convergent by $C_k = \dfrac{P_k}{Q_k}$, then

$$(1) \begin{cases} C_0 = \dfrac{P_0}{Q_0} = \dfrac{q_0}{1}; \\[2mm] C_1 = \dfrac{P_1}{Q_1} = \dfrac{q_0 q_1 + 1}{q_1}; \\[2mm] \vdots \\[2mm] C_k = \dfrac{P_k}{Q_k} = \dfrac{q_k P_{k-1} + P_{k-2}}{q_k Q_{k-1} + Q_{k-2}}, \text{ for } k \geq 2. \end{cases}$$

(2) If $P_k = q_k Q_{k-1} + Q_{k-2}$ and $Q_k = q_k P_{k-1} + P_{k-2}$, then $\gcd(P_k, Q_k) = 1$.

(3) $P_k Q_{k-1} - P_{k-1} Q_k = (-1)^{k-1}$, for $k \geq 1$.

The following example shows how to use Euclid's algorithm to express a rational number as a finite simple continued fraction.

Example 1.4.2. Expand the rational number $\dfrac{1281}{243}$ as a simple continued fraction. First let $a = 1281$ and $b = 243$, and then let Euclid's algorithm run as follows:

1281			
$-\ 1215$	5		243
66	3	$-\ 198$	
$-\ 45$	1	45	
21	2	$-\ 42$	
$-\ 21$	7	3	
0			

So $\dfrac{1281}{243} = [5, 3, 1, 2, 7]$. Thus

$$\frac{1281}{243} = 5 + \cfrac{1}{3 + \cfrac{1}{1 + \cfrac{1}{2 + \cfrac{1}{7}}}}.$$

Of course, as a by-product, we also find that $\gcd(1281, 243) = 3$.

Theorem 1.4.4. Any finite simple continued fraction represents a rational number. Conversely, any rational number can be expressed as a finite simple continued fraction, in exactly two ways, one with an odd number of terms and one with an even number of terms.

Proof. The first assertion is proved by induction. When $n = 1$, we have

$$[q_0, q_1] = q_0 + \frac{1}{q_1} = \frac{q_0 q_1 + 1}{q_1}$$

which is rational. Now we assume for $n = k$ the simple continued fraction $[q_0, q_1, \ldots, q_k]$ is rational whenever $q_0, q_1, \ldots, q_k$ are integers with $q_1, \ldots, q_k$ positive. Let $q_0, q_1, \ldots, q_{k+1}$ be integers with $q_1, \ldots, q_{k+1}$ positive. Note that

$$[q_0, q_1, \ldots, q_k, q_{k+1}] = a_0 + \frac{1}{[q_1, \ldots, q_k, q_{k+1}]}.$$

By the induction hypothesis, $[q_1, q_2, \ldots, q_k, q_{k+1}]$ is rational. That is, there exist two integers r and s with $s \neq 0$ such that

$$[q_1, q_2, \ldots, q_k, q_{k+1}] = \frac{r}{s}.$$

Thus,

$$[q_0, q_1, \ldots, q_k, q_{k+1}] = a_0 + \frac{1}{r/s} = \frac{q_0 r + s}{r}$$

which is rational.

Now we use Euclid's algorithm to show that every rational number can be written as a finite simple continued fraction. Let a/b be a rational number with $b > 0$. Euclid's algorithm tells us that

$$
\begin{aligned}
a &= bq_0 + r_1, & 0 < r_1 < b, \\
b &= r_1 q_1 + r_2, & 0 < r_2 < r_1, \\
r_1 &= r_2 q_2 + r_3, & 0 < r_3 < r_2, \\
r_2 &= r_3 q_3 + r_4, & 0 < r_4 < r_3, \\
&\ \ \vdots & \vdots \\
r_{n-2} &= r_{n-1} q_{n-1} + r_n, & 0 < r_n < r_{n-1}, \\
r_{n-1} &= r_n q_n + 0.
\end{aligned}
$$

In these equations, $q_1, q_2, \ldots, q_n$ are positive integers. Rewriting these equations, we obtain

$$
\begin{aligned}
\frac{a}{b} &= q_0 + \frac{r_1}{b} \\
\frac{b}{r_1} &= q_1 + \frac{r_2}{r_1} \\
\frac{r_1}{r_2} &= q_2 + \frac{r_3}{r_2} \\
&\ \ \vdots \\
\frac{r_{n-1}}{r_n} &= q_n
\end{aligned}
$$

By successive substitution

$$\frac{a}{b} = q_0 + \frac{1}{\frac{b}{r_1}}$$

$$= q_0 + \frac{1}{q_1 + \frac{1}{\frac{r_1}{r_2}}}$$

$$\vdots$$

$$= q_0 + \cfrac{1}{q_1 + \cfrac{1}{q_2 + \cfrac{1}{\ddots\, q_{n-1} + \cfrac{1}{q_n}}}}$$

This shows that every rational number can be written as a finite simple continued fraction.

Further, it can be shown that any rational number can be expressed as a finite simple continued fraction in exactly two ways, one with an odd number of terms and one with an even number of terms; we leave this as an exercise.

□

Definition 1.4.2. Let $q_0, q_1, q_2, \ldots$ be a sequence of integers, all positive except possibly q_0. Then the expression $[q_0, q_1, q_2, \ldots]$ is called an *infinite* simple continued fraction and is defined to be equal to the number $\lim_{n \to \infty} [q_0, q_1, q_2, \ldots, q_{n-1}, q_n]$.

Theorem 1.4.5. Any *irrational* number can be written uniquely as an *infinite* simple continued fraction. Conversely, if α is an infinite simple continued fraction, then α is irrational.

Proof. Let α be an irrational number. We write

$$\alpha = [\alpha] + \{\alpha\} = [\alpha] + \frac{1}{\frac{1}{\{\alpha\}}}$$

where $[\alpha]$ is the integral part and $\{\alpha\}$ the fractional part of α, respectively. Because α is irrational, $1/\{\alpha\}$ is irrational and greater than 1. Let

$$q_0 = [\alpha], \quad \text{and} \quad \alpha_1 = \frac{1}{\{\alpha\}}.$$

We now write

$$\alpha_1 = [\alpha_1] + \{\alpha_1\} = [\alpha_1] + \frac{1}{\frac{1}{\{\alpha_1\}}}$$

where $1/\{\alpha_1\}$ is irrational and greater than 1. Let

$$q_1 = [\alpha_1], \quad \text{and} \quad \alpha_2 = \frac{1}{\{\alpha_1\}}.$$

We continue inductively

$$q_2 = [\alpha_2], \quad \text{and} \quad \alpha_3 = \frac{1}{\{\alpha_2\}} > 1 \quad (\alpha_3 \text{ irrational})$$

$$q_3 = [\alpha_3], \quad \text{and} \quad \alpha_4 = \frac{1}{\{\alpha_3\}} > 1 \quad (\alpha_3 \text{ irrational})$$

$$\vdots$$

$$q_n = [\alpha_n], \quad \text{and} \quad \alpha_n = \frac{1}{\{\alpha_{n-1}\}} > 1 \quad (\alpha_3 \text{ irrational})$$

$$\vdots$$

Since each α_n, $n = 2, 3 \cdots$ is greater than 1, then $q_{n-1} \geq 1$, $n = 2, 3, \ldots$. If we substitute successively, we obtain

$$\begin{aligned}
\alpha &= [q_0, \alpha_1] \\
&= [q_0, q_1, \alpha_2] \\
&= [q_0, q_1, q_2, \alpha_3] \\
&\vdots \\
&= [q_0, q_1, q_2, \ldots, q_n, \alpha_{n+1}]
\end{aligned}$$

$$\vdots$$

Next we shall show that $\alpha = [q_0, q_1, q_2, \ldots]$. Note that C_n, the nth convergent to $[q_0, q_1, q_2, \ldots]$ is also the nth convergent to $[q_0, q_1, q_2, \ldots, q_n, \alpha_{n+1}]$. If we denote the $(n+1)$st convergent to this finite continued fraction by $P'_{n+1}/Q'_{n+1} = \alpha$, then

$$\alpha - C_n = \frac{P'_{n+1}}{Q'_{n+1}} - \frac{P_n}{Q_n} = \frac{(-1)^{n+1}}{Q'_{n+1}Q_n}.$$

Since Q_n and Q'_{n+1} become infinite as $n \to \infty$, then

$$\lim_{n \to \infty} (\alpha - C_n) = \lim_{n \to \infty} \frac{(-1)^{n+1}}{Q'_{n+1}Q_n} = 0$$

and

$$\alpha = \lim_{n \to \infty} C_n = [q_0, q_1, \ldots].$$

The uniqueness of the representation, as well as the second assertion are left as an exercise.   □

Definition 1.4.3. A real irrational number which is the root of a quadratic equation $ax^2+bx+c = 0$ with integer coefficients is called *quadratic irrational*.

For example, $\sqrt{3}$, $\sqrt{5}$, $\sqrt{7}$ are quadratic irrationals. For convenience, we shall denote $\sqrt{N}$, with N not a perfect square, as a quadratic irrational. Quadratic irrationals are the simplest possible irrationals.

Definition 1.4.4. An infinite simple continued fraction is said to be *periodic* if there exists integers k and m such that $q_{i+m} = q_i$ for all $i \geq k$. The periodic simple continued fraction is usually denoted by $[q_0, q_1, \ldots, q_k, \overline{q_{k+1}, q_{k+2}, \ldots, q_{k+m}}]$. If it is of the form $[\overline{q_0, q_1, \ldots, q_{m-1}}]$, then it is called *purely periodic*. The smallest positive integer m satisfying the above relationship is called the *period* of the expansion.

Theorem 1.4.6. Any *periodic* simple continued fraction is a quadratic irrational. Conversely, any *quadratic irrational* has a periodic expansion as a simple continued fraction.

Proof. The proof is rather lengthy and left as an exercise; a complete proof can be found on pp 224–226 in Redmond [192].   □

We are now in a position to present an algorithm for finding the simple continued fraction expansion of a *real number*.

Theorem 1.4.7 (Continued fraction algorithm). Suppose x is irrational, and let $x_0 = x$. Then x can be expressed as a simple continued fraction

$$[q_0, q_1, q_2, \ldots, q_n, q_{n+1}, \ldots]$$

by the following process:

$$
\left.
\begin{aligned}
& x_0 = x \\[1em]
& q_0 = \lfloor x_0 \rfloor, \qquad x_1 = \frac{1}{x_0 - q_0} \\[1em]
& q_1 = \lfloor x_1 \rfloor, \qquad x_2 = \frac{1}{x_1 - q_1} \\[0.5em]
& \quad\vdots \qquad\qquad\qquad \vdots \\[0.5em]
& q_n = \lfloor x_n \rfloor, \qquad x_{n+1} = \frac{1}{x_n - q_n} \\[1em]
& q_{n+1} = \lfloor x_{n+1} \rfloor, \quad x_{n+2} = \frac{1}{x_{n+1} - q_{n+1}} \\[0.5em]
& \quad\vdots \qquad\qquad\qquad \vdots
\end{aligned}
\right\}
\qquad (1.54)
$$

Proof. Follows from Theorem 1.4.5.   □

Algorithm 1.4.2 (Continued fraction algorithm). Given a real number x, this algorithm will compute and output the partial quotients $q_0, q_1, q_2, \ldots, q_n$ of the continued fraction x.

[1] (Initialization) Set

$i \leftarrow 0$,
$x_i \leftarrow x$,
$q_i \leftarrow \lfloor x_i \rfloor$,
print(q_i).

[2] (Decision) If $x_i = q_i$, Exit.

[3] (Computation)

$x_{i+1} \leftarrow \dfrac{1}{x_i - q_i}$,
$i \leftarrow i + 1$,
$q_i \leftarrow \lfloor x_i \rfloor$,
print(q_i),
go to Step [2].

Example 1.4.3. Let $x = 160523347/60728973$. Then by applying Algorithm 1.4.2, we get $160523347/60728973 = [2, 1, 1, 1, 4, 12, 102, 1, 1, 2, 3, 2, 2, 36]$. That is,

$$\frac{160523347}{60728973} = 2 + \cfrac{1}{1 + \cfrac{1}{1 + \cfrac{1}{1 + \cfrac{1}{4 + \cfrac{1}{12 + \cfrac{1}{102 + \cfrac{1}{1 + \cfrac{1}{1 + \cfrac{1}{2 + \cfrac{1}{3 + \cfrac{1}{2 + \cfrac{1}{2 + \cfrac{1}{36}}}}}}}}}}}}$$

Theorem 1.4.8. Each quadratic irrational number $\sqrt{N}$ has a periodic expansion as an infinite simple continued fraction of the form

$$[q_0, q_1, q_2, \ldots, q_k, \overline{q_{k+1}, \ldots, q_{k+m}}].$$

Example 1.4.4. Expand $\sqrt{3}$ as a periodic simple continued fraction. Let $x_0 = \sqrt{3}$. Then we have

$$q_0 = \lfloor x_0 \rfloor = \lfloor \sqrt{3} \rfloor = 1$$

$$x_1 = \frac{1}{x_0 - q_0} = \frac{1}{\sqrt{3} - 1} = \frac{\sqrt{3} + 1}{2}$$

$$q_1 = \lfloor x_1 \rfloor = \lfloor \frac{\sqrt{3} + 1}{2} \rfloor = \lfloor 1 + \frac{\sqrt{3} - 1}{2} \rfloor = 1$$

$$x_2 = \frac{1}{x_1 - q_1} = \frac{1}{\frac{\sqrt{3} + 1}{2} - 1} = \frac{1}{\frac{\sqrt{3} - 1}{2}} = \frac{2(\sqrt{3} + 1)}{(\sqrt{3} - 1)(\sqrt{3} + 1)} = \sqrt{3} + 1$$

$$q_2 = \lfloor x_2 \rfloor = \lfloor \sqrt{3} + 1 \rfloor = 2$$

$$x_3 = \frac{1}{x_2 - q_2} = \frac{1}{\sqrt{3} + 1 - 2} = \frac{1}{\sqrt{3} - 1} = \frac{\sqrt{3} + 1}{2} = x_1$$

$$q_3 = \lfloor x_3 \rfloor = \lfloor \frac{\sqrt{3} + 1}{2} \rfloor = \lfloor 1 + \frac{\sqrt{3} - 1}{2} \rfloor = 1 = q_1$$

$$x_4 = \frac{1}{x_3 - q_3} = \frac{1}{\frac{\sqrt{3} + 1}{2} - 1} = \frac{1}{\frac{\sqrt{3} - 1}{2}} = \frac{2(\sqrt{3} + 1)}{(\sqrt{3} - 1)(\sqrt{3} + 1)} = \sqrt{3} + 1 = x_2$$

$$q_4 = \lfloor x_3 \rfloor = \lfloor \sqrt{3} + 1 \rfloor = 2 = q_2$$

$$x_5 = \frac{1}{x_4 - q_4} = \frac{1}{\sqrt{3} + 1 - 2} = \frac{1}{\sqrt{3} - 1} = \frac{\sqrt{3} + 1}{2} = x_3 = x_1$$

$$q_5 = \lfloor x_5 \rfloor = \lfloor x_3 \rfloor = 1 = q_3 = q_1$$

$$\vdots$$

So, for $n = 1, 2, 3, \ldots$, we have $q_{2n-1} = 1$ and $q_{2n} = 2$. Thus, the *period* of the continued fraction expansion of $\sqrt{3}$ is 2. Therefore, we finally get

$$\sqrt{3} = 1 + \cfrac{1}{1 + \cfrac{1}{2 + \cfrac{1}{1 + \cfrac{1}{2 + \cfrac{1}{\ddots}}}}} = [1, \overline{1, 2}].$$

Definition 1.4.5. The algebraic equation with two variables

$$ax + by = c \tag{1.55}$$

is called a *linear Diophantine equation*, for which we wish to find integer solutions in x and y.

Theorem 1.4.9. Let a, b, c be integers with not both a and b equal to 0. If $d \nmid c$, then the linear Diophantine equation

$$ax + by = c$$

has no integer solution. The equation has an integer solution in x and y if and only if $d \mid c$. Moreover, if (x_0, y_0) is a solution of the equation, then the general solution of the equation is

$$(x, y) = \left(x_0 + \frac{b}{d} \cdot t, \ y_0 - \frac{a}{d} \cdot t \right), \quad t \in \mathbb{Z}. \tag{1.56}$$

Proof.   Assume that x and y are integers such that $ax + by = c$. Since $d \mid a$ and $d \mid b$, $d \mid c$. Hence, if $d \nmid c$, there is no integer solutions of the equation.

Now suppose $d \mid c$. There is an integer k such that $c = kd$. Since d is a sum of multiples of a and b, we may write

$$am + bn = d.$$

Multiplying this equation by k, we get

$$a(mk) + b(nk) = dk = c$$

so that $x = mk$ and $y = nk$ is a solution.

For the "only if" part, suppose x_0 and y_0 is a solution of the equation. Then

$$ax_0 + by_0 = c.$$

Since $d \mid a$ and $d \mid b$, then $d \mid c$.   □

Theorem 1.4.10. Let the convergents of the finite continued fraction of a/b be as follows:

$$\left[\frac{P_0}{Q_0}, \frac{P_1}{Q_1}, \dots, \frac{P_{n-1}}{Q_{n-1}}, \frac{P_n}{Q_n} \right] = \frac{a}{b}. \tag{1.57}$$

Then the integer solution in x and y of the equation $ax - by = d$ is

$$\left. \begin{array}{l} x = (-1)^{n-1} Q_{n-1}, \\ y = (-1)^{n-1} P_{n-1}. \end{array} \right\} \tag{1.58}$$

Remark 1.4.3. We have already seen a method to solve the linear Diophantine equations by applying Euclid's algorithm to a and b and working backwards through the resulting equations (the so-called extended Euclid's algorithm). Our new method here turns out to be equivalent to this since the continued fraction for a/b is derived from Euclid's algorithm. However, it is quicker to generate the convergents P_i/Q_i using the recurrence relations than to work backwards through the equations in Euclid's algorithm.

Example 1.4.5. Use the continued fraction method to solve the following linear Diophantine equation:

$$364x - 227y = 1.$$

Since $364/227$ can be expanded as a finite continued fraction with convergents

$$\left[1, \ 2, \ \frac{3}{2}, \ \frac{5}{3}, \ \frac{8}{5}, \ \frac{85}{53}, \ \frac{93}{58}, \ \frac{364}{227} \right]$$

we have

$$x = (-1)^{n-1} q_{n-1} = (-1)^{7-1} 58 = 58,$$
$$y = (-1)^{n-1} p_{n-1} = (-1)^{7-1} 93 = 93.$$

That is,

$$364 \cdot 58 - 227 \cdot 93 = 1.$$

Example 1.4.6. Use the continued fraction method to solve the following linear Diophantine equation:

$$20719x + 13871y = 1.$$

Note first that

$$20719x + 13871y = 1 \iff 20719x - (-13871y) = 1.$$

Now since $20719/13871$ can be expanded as a finite simple continued fraction with convergents

$$\left[1, \ \frac{3}{2}, \ \frac{118}{79}, \ \frac{829}{555}, \ \frac{947}{634}, \ \frac{1776}{1189}, \ \frac{2723}{1823}, \ \frac{4499}{3012}, \ \frac{20719}{13871} \right],$$

we have

$$x = (-1)^{n-1} q_{n-1} = (-1)^{8-1} 3012 = -3012,$$
$$y = (-1)^{n-1} p_{n-1} = (-1)^{8-1} 4499 = -4499.$$

That is,

$$20719 \cdot (-3012) - 13871 \cdot (-4499) = 1.$$

Remark 1.4.4. To find the integral solution to equation $ax + by = d$, the equation

$$(-1)^{n-1} a q_{n-1} - (-1)^{n-1} b p_{n-1} = d$$

for $ax - by = d$ must be changed to

$$(-1)^{n-1} a q_{n-1} + (-1)(-1)^{n-1} b p_{n-1} = d.$$

That is,

$$(-1)^{n-1} a q_{n-1} + (-1)^n b p_{n-1} = d \qquad (1.59)$$

Thus a solution to equation $ax + by = d$ is given by

$$\begin{cases} x = (-1)^{n-1}q_{n-1}, \\ y = (-1)^{n}p_{n-1}. \end{cases} \tag{1.60}$$

Generally, we have the following four cases:

$$\begin{cases} x = (-1)^{n-1}q_{n-1}, \\ y = (-1)^{n-1}p_{n-1} \end{cases} \quad \text{for} \quad ax - by = d. \tag{1.61}$$

$$\begin{cases} x = (-1)^{n-1}q_{n-1}, \\ y = (-1)^{n}p_{n-1} \end{cases} \quad \text{for} \quad ax + by = d. \tag{1.62}$$

$$\begin{cases} x = (-1)^{n}q_{n-1}, \\ y = (-1)^{n-1}p_{n-1} \end{cases} \quad \text{for} \quad -ax - by = d. \tag{1.63}$$

$$\begin{cases} x = (-1)^{n}q_{n-1}, \\ y = (-1)^{n}p_{n-1} \end{cases} \quad \text{for} \quad -ax + by = d. \tag{1.64}$$

All the above four cases are, in fact, of the same type of linear Diophantine equations.

Problems for Section 1.4

Problem 1.4.1. For any positive integers a and b, prove that

$$ab = \gcd(a, b)\operatorname{lcm}(a, b).$$

Problem 1.4.2. Prove that if Euclid's algorithm runs with $a = f_{k+2}$ and $b = f_{k+1}$, then exactly k divisions are needed for computing $\gcd(a, b)$, where $f_0 = 0$, $f_1 = 1$, and $f_n = f_{n-1} + f_{n-2}$ for $n \geq 2$ are defined to be the Fibonacci numbers beginning with numbers $0, 1, 1, 2, 3, 5, 8, 13, \ldots$.

Problem 1.4.3. Use the continued fraction method to solve $377x - 120y = -3$ and $314x \equiv 271 \pmod{11111}$.

Problem 1.4.4. Prove that if α is an irrational number, then there exist infinitely many rational numbers $\dfrac{p}{q}$ such that

$$\left| \alpha - \frac{p}{q} \right| < \frac{1}{q^2}.$$

Problem 1.4.5. Prove that if α is an irrational number and $\dfrac{P_i}{Q_i}$ the ith convergent of the continued fraction of α, then

$$\left| \alpha - \frac{P_i}{Q_i} \right| < \frac{1}{Q_i Q_{i+1}}.$$

Problem 1.4.6. Prove that if α is an irrational number and $\dfrac{c}{d}$ is a rational number with $d > 1$ such that

$$\left| \alpha - \frac{c}{d} \right| < \frac{1}{2d^2},$$

then $\dfrac{c}{d}$ is one of the convergents of the infinite continued fraction of α.

Problem 1.4.7. Let $\pi = 3.14159926\cdots$. Prove that the first three convergents to π are $\dfrac{22}{7}, \dfrac{333}{106}$ and $\dfrac{355}{113}$. Verify that

$$\left| \pi - \frac{355}{113} \right| < 10^{-6}.$$

Problem 1.4.8. Prove that the denominators Q_n in the convergents to any real number θ satisfy that

$$Q_n \leq \left(\frac{1 + \sqrt{5}}{2} \right)^{n-1}.$$

Problem 1.4.9. Prove that if $m < n$, then

$$(2^{2^m} + 1) \nmid (2^{2^n} + 1), \quad \gcd((2^{2^m} + 1), \, (2^{2^n} + 1)) = 1.$$

Problem 1.4.10. Find the integer solution (x, y, z) to the Diophantine equation $35x + 55y + 77z = 1$.

1.5 Arithmetic Functions $\sigma(n), \tau(n), \phi(n), \lambda(n), \mu(n)$

Definition 1.5.1. A function f is called an *arithmetic function* or a *number-theoretic function* if it assigns to each positive integer n a unique real or complex number $f(n)$. Typically, an arithmetic function is a real-valued function whose domain is the set of positive integers.

Example 1.5.1. The equation

$$f(n) = \sqrt{n}, \quad n \in \mathbb{Z}^+ \tag{1.65}$$

defines an arithmetic function f which assigns the real number $\sqrt{n}$ to each positive integer n.

Definition 1.5.2. A real function f defined on the positive integers is said to be *multiplicative* if

$$f(m)f(n) = f(mn), \quad \forall m, n \in \mathbb{Z}^+, \tag{1.66}$$

where $\gcd(m, n) = 1$. If

$$f(m)f(n) = f(mn), \quad \forall m, n \in \mathbb{Z}^+, \tag{1.67}$$

then f is *completely multiplicative*. Every completely multiplicative function is multiplicative.

Theorem 1.5.1. Let

$$n = \prod_{i=1}^{k} p_i^{\alpha_i}$$

be the prime factorization of n and let f be a multiplicative function, then

$$f(n) = \prod_{i=1}^{k} f(p_i^{\alpha_i}).$$

Proof. Clearly, if $k = 1$, we have the identity, $f(p_i^{\alpha_i}) = f(p_i^{\alpha_i})$. Assume that the representation is valid whenever n has r or fewer distinct prime factors, and consider $n = \prod_{i=1}^{r+1} f(p_i^{\alpha_i})$. Since $\gcd\left(\prod_{i=1}^{r} p_i^{\alpha_i}, p_{r+1}^{\alpha_{r+1}}\right) = 1$ and f is multiplicative, we have

$$
\begin{aligned}
f(n) &= f\left(\prod_{i=1}^{r+1} p_i^{\alpha_i}\right) \\
&= f\left(\prod_{i=1}^{r} p_i^{\alpha_i} \cdot p_{r+1}^{\alpha_{r+1}}\right) \\
&= f\left(\prod_{i=1}^{r} p_i^{\alpha_i}\right) \cdot f\left(p_{r+1}^{\alpha_{r+1}}\right) \\
&= \prod_{i=1}^{r} f(p_i^{\alpha_i}) \cdot f(p_{r+1}^{\alpha_{r+1}}) \\
&= \prod_{i=1}^{r+1} f(p_i^{\alpha_i}).
\end{aligned}
$$

$\square$

Theorem 1.5.2. If f is multiplicative and if g is given by

$$g(n) = \sum_{d|n} f(d) \tag{1.68}$$

where the sum is over all divisors d of n, then g is also multiplicative.

Proof. Since f is multiplicative, if $\gcd(m,n) = 1$, then

$$
\begin{aligned}
g(mn) &= \sum_{d|mn} f(d) \\
&= \sum_{d_1|m \ d_2|n} f(d_1 d_2) \\
&= \sum_{d_1|m \ d_2|n} f(d_1)f(d_2) \\
&= \sum_{d_1|m} f(d_1) \sum_{d_2|n} f(d_2) \\
&= g(m)g(n).
\end{aligned}
$$

$\square$

Theorem 1.5.3. If f and g are multiplicative, then so is

$$F(n) = \sum_{d|m} f(d)g\left(\frac{n}{d}\right).$$

Proof. If $\gcd(m,n) = 1$, then $d \mid mn$ if and only if $d = d_1 d_2$, where $d_1 \mid m$ and $d_2 \mid n$, $\gcd(d_1, d_2) = 1$ and $\gcd(m/d_1, n/d_2) = 1$. Thus,

$$
\begin{aligned}
F(mn) &= \sum_{d|mn} f(d)g\left(\frac{mn}{d}\right) \\
&= \sum_{d_1|m}\sum_{d_2|n} f(d_1 d_2)g\left(\frac{mn}{d_1 d_2}\right) \\
&= \sum_{d_1|m}\sum_{d_2|n} f(d_1)f(d_2)g\left(\frac{m}{d_1}\right)g\left(\frac{n}{d_2}\right) \\
&= \left[\sum_{d_1|m} f(d_1)g\left(\frac{m}{d_1}\right)\right]\left[\sum_{d_2|m} f(d_2)g\left(\frac{n}{d_2}\right)\right] \\
&= F(m)F(n).
\end{aligned}
$$

$\square$

Definition 1.5.3. Let n be a positive integer. Then the arithmetic functions $\tau(n)$ and $\sigma(n)$ are defined as follows:

$$\tau(n) = \sum_{d|n} 1, \qquad \sigma(n) = \sum_{d|n} d. \tag{1.69}$$

That is, $\tau(n)$ designates the number of all positive divisors of n, and $\sigma(n)$ designates the sum of all positive divisors of n.

Example 1.5.2. By Definition 1.5.3, we have

n	1	2	3	4	5	6	7	8	9	10	100	101	220	284
$\tau(n)$	1	2	2	3	2	4	2	4	3	4	9	2	12	6
$\sigma(n)$	1	3	4	7	6	12	8	15	13	18	217	102	504	504

Lemma 1.5.1. If n is a positive integer greater than 1 and has the following standard prime factorization form

$$n = \prod_{i}^{k} p_i^{\alpha_i},$$

then the positive divisors of n are precisely those integers d of the form

$$d = \prod_{i}^{k} p_i^{\beta_i},$$

where $0 \le \beta_i \le \alpha_i$.

Proof. If $d \mid n$, then $n = dq$. By the Fundamental Theorem of Arithmetic, the prime factorization of n is unique, so the prime numbers in the prime factorization of d must occur in p_j, $(j = 1, 2, \dots, k)$. Furthermore, the power β_j of p_j occurring in the prime factorization of d cannot be greater than α_j, that is, $\beta_j \le \alpha_j$. Conversely, when $\beta_j \le \alpha_j$, d clearly divides n. $\square$

Theorem 1.5.4. Let n be a positive integer. Then

(1) $\tau(n)$ is multiplicative. That is,

$$\tau(mn) = \tau(m)\tau(n) \tag{1.70}$$

where $\gcd(m, n) = 1$.

(2) If n is a prime, say p, then $\tau(p) = 2$. More generally, if n is a prime power p^α, then

$$\tau(p^\alpha) = \alpha + 1. \tag{1.71}$$

(3) If n is a composite and has the standard prime factorization form, then

$$\tau(n) = (\alpha_1 + 1)(\alpha_2 + 1) \cdots (\alpha_k + 1) = \prod_{i=1}^{k}(\alpha_i + 1). \tag{1.72}$$

Proof.

(1) Since the constant function $f(n) = 1$ is multiplicative and $\tau(n) = \sum_{d|n} 1$, the result follows immediately from Theorem 1.5.2.

(2) Clearly, if n is a prime, there are only two divisors, namely, 1 and n itself. If $n = p^\alpha$, then by Lemma 1.5.1, the positive divisors of n are precisely those integers $d = p^\beta$, with $0 \le \beta \le \alpha$. Since there are $\alpha + 1$ choices for the exponent β, there are $\alpha + 1$ possible positive divisors of n.

(3) By Lemma 1.5.1 and Part (2) of this theorem, there are $\alpha_1 + 1$ choices for the exponent β_1, $\alpha_2 + 1$ choices for the exponent β_2, $\cdots$, $\alpha_k + 1$ choices for the exponent β_k. From the multiplication principle it follows that there are $(\alpha_1 + 1)(\alpha_2 + 1) \cdots (\alpha_k + 1)$ different choices for the $\beta_1, \beta_2, \ldots, \beta_k$, thus that many divisors of n. Therefore, $\tau(n) = (\alpha_1 + 1)(\alpha_2 + 1) \cdots (\alpha_k + 1)$.

$\square$

Theorem 1.5.5. The product of all divisors of a number n is

$$\prod_{d|n} d = n^{\tau(n)/2}. \tag{1.73}$$

Proof. Let d denote an arbitrary positive divisor of n, so that

$$n = dd'$$

for some d'. As d ranges over all $\tau(n)$ positive divisors of n, there are $\tau(n)$ such equations. Multiplying these together, we get

$$n^{\tau(n)} = \prod_{d|n} d \prod_{d'|n} d'.$$

But as d runs through the divisors of n, so does d', hence

$$\prod_{d|n} d = \prod_{d'|n} d'.$$

So,

$$n^{\tau(n)} = \left(\prod_{d|n} d\right)^2,$$

or equivalently

$$n^{\tau(n)/2} = \prod_{d\mid n} d.$$

□

Example 1.5.3. Let $n = 1371$, then

$$\tau(1371) = 4.$$

Therefore

$$\prod d = 1371^{4/2} = 1879641.$$

It is of course true, since

$$d(1371) = \{1, 3, 457, 1371\}$$

implies that

$$\prod d = 1 \cdot 3 \cdot 457 \cdot 1371 = 1879641.$$

Theorem 1.5.6. Let n be a positive integer. Then

(1) $\sigma(n)$ is multiplicative. That is,

$$\sigma(mn) = \sigma(m)\sigma(n) \tag{1.74}$$

where $\gcd(m, n) = 1$.

(2) If n is a prime, say p, then $\sigma(p) = p + 1$. More generally, if n is a prime power p^α, then

$$\sigma(p^\alpha) = \frac{p^{\alpha+1} - 1}{p - 1}. \tag{1.75}$$

(3) If n is a composite and has the standard prime factorization form, then

$$\sigma(n) = \frac{p_1^{\alpha_1+1} - 1}{p_1 - 1} \cdot \frac{p_2^{\alpha_2+1} - 1}{p_2 - 1} \cdots \frac{p_k^{\alpha_k+1} - 1}{p_k - 1}$$

$$= \prod_{i=1}^{k} \frac{p_i^{\alpha_i+1} - 1}{p_i - 1}. \tag{1.76}$$

Proof.

(1) The results follows immediately from Theorem 1.5.2 since the identity function $f(n) = n$ and $\sigma(n)$ can be represented in the form $\sigma(n) = \sum_{d\mid n} d$.

(2) Left as an exercise; we prove the most general case in Part (3).

(3) The sum of the divisors of the positive integer

$$n = p_1^{\alpha_1} p_2^{\alpha_2} \cdots p_k^{\alpha_k}$$

can be expressed by the product

$$\left(1 + p_1 + p_1^2 + \cdots + p_1^{\alpha_1}\right)\left(1 + p_2 + p_2^2 + \cdots + p_2^{\alpha_2}\right)$$
$$\cdots \left(1 + p_k + p_k^2 + \cdots + p_k^{\alpha_k}\right).$$

Using the finite geometric series

$$1 + x + x^2 + \cdots + x^n = \frac{x^{n+1} - 1}{x - 1},$$

we simplify each of the k sums in the above product to find that the sum of the divisors can be expressed as

$$
\begin{aligned}
\sigma(n) &= \frac{p_1^{\alpha_1+1} - 1}{p_1 - 1} \cdot \frac{p_2^{\alpha_2+1} - 1}{p_2 - 1} \cdots \frac{p_k^{\alpha_k+1} - 1}{p_k - 1} \\
&= \prod_{i=1}^{k} \frac{p_i^{\alpha_i+1} - 1}{p_i - 1}.
\end{aligned}
\tag{1.77}
$$

$\square$

Definition 1.5.4. Let n be a positive integer. *Euler's (totient) ϕ-function,* $\phi(n)$, is defined to be the number of positive integers k less than n which are relatively prime to n:

$$\phi(n) = \sum_{\substack{0 \le k < n \\ \gcd(k,n)=1}} 1. \tag{1.78}$$

Example 1.5.4. By Definition 1.5.4, we have

n	1	2	3	4	5	6	7	8	9	10	100	101	102	103
$\phi(n)$	1	1	2	2	4	2	6	4	6	4	40	100	32	102

Lemma 1.5.2. For any positive integer n,

$$\sum_{d|n} \phi(d) = n. \tag{1.79}$$

Proof. Let n_d denote the number of elements in the set $\{1, 2, \ldots, n\}$ having a greatest common divisor of d with n. Then

$$n = \sum_{d|n} n_d = \sum_{d|n} \phi\left(\frac{n}{d}\right) = \sum_{d|n} \phi(d).$$

$\square$

Theorem 1.5.7. Let n be a positive integer and $\gcd(m, n) = 1$. Then

(1) Euler's ϕ-function is multiplicative. That is,

$$\phi(mn) = \phi(m)\phi(n) \tag{1.80}$$

where $\gcd(m, n) = 1$.

(2) If n is a prime, say p, then

$$\phi(p) = p - 1. \tag{1.81}$$

(Conversely, if p is a positive integer with $\phi(p) = p - 1$, then p is prime.)

(3) If n is a prime power p^α with $\alpha > 1$, then

$$\phi(p^\alpha) = p^\alpha - p^{\alpha-1}. \tag{1.82}$$

(4) If n is a composite and has the standard prime factorization form, then

$$
\begin{aligned}
\phi(n) &= p_1^{\alpha_1}\left(1 - \frac{1}{p_1}\right) p_2^{\alpha_2}\left(1 - \frac{1}{p_2}\right) \cdots p_k^{\alpha_k}\left(1 - \frac{1}{p_k}\right) \\
&= n \prod_{i=1}^{k}\left(1 - \frac{1}{p_i}\right).
\end{aligned}
\tag{1.83}
$$

Proof.

(1) Use Theorem 1.5.3 and Lemma 1.5.2. (A nicer way to prove this result is to use the Chinese Remainder Theorem, which will be discussed in Section 1.6.)

(2) If n is prime, then $1, 2, \ldots, n - 1$ are relatively prime to n, so it follows from the definition of Euler's ϕ-function that $\phi(n) = n - 1$. Conversely, if n is not prime, n has a divisor d such that $\gcd(d, n) \neq 1$. Thus, there is at least one positive integer less than n that is not relatively prime to n, and hence $\phi(n) \leq n - 2$.

(3) Note that $\gcd(n, p^\alpha) = 1$ if and only if $p \nmid n$. There are exactly $p^{\alpha-1}$ integers between 1 and p^α divisible by p, namely,

$$p, \ 2p, \ 3p, \ldots, \ (p^{\alpha-1})p.$$

Thus, the set $\{1, 2, \ldots, p^\alpha\}$ contains exactly $p^\alpha - p^{\alpha-1}$ integers that are relatively prime to p^α, and so by the definition of the ϕ-function, $\phi(p^\alpha) = p^\alpha - p^{\alpha-1}$.

(4) By Part (1) of this theorem, ϕ-function is multiplicative, thus

$$\phi(n) = \phi\left(p_1^{\alpha_1}\right)\phi\left(p_2^{\alpha_2}\right)\cdots\phi\left(p_k^{\alpha_k}\right).$$

In addition, by Part (3) of this theorem and Theorem 1.5.1, we have

$$
\begin{aligned}
\phi(n) &= p_1^{\alpha_1}\left(1-\frac{1}{p_1}\right)p_2^{\alpha_2}\left(1-\frac{1}{p_2}\right)\cdots p_k^{\alpha_k}\left(1-\frac{1}{p_k}\right) \\
&= p_1^{\alpha_1}p_2^{\alpha_2}\cdots p_k^{\alpha_k}\left(1-\frac{1}{p_1}\right)\left(1-\frac{1}{p_2}\right)\cdots\left(1-\frac{1}{p_k}\right) \\
&= n\left(1-\frac{1}{p_1}\right)\left(1-\frac{1}{p_2}\right)\cdots\left(1-\frac{1}{p_k}\right) \\
&= n\prod_{i=1}^{k}\left(1-\frac{1}{p_i}\right).
\end{aligned}
$$

$\square$

Definition 1.5.5. Carmichael's λ-function, $\lambda(n)$, is defined as follows

$$
\left.
\begin{aligned}
\lambda(p) &= \phi(p) = p-1 && \text{for prime } p, \\
\lambda(p^{\alpha}) &= \phi(p^{\alpha}) && \text{for } p=2 \text{ and } \alpha \leq 2, \\
& && \text{and for } p \geq 3 \\
\lambda(2^{\alpha}) &= \frac{1}{2}\phi(2^{\alpha}) && \text{for } \alpha \geq 3 \\
\lambda(n) &= \mathrm{lcm}\left(\lambda(p_1^{\alpha_1}),\lambda(p_2^{\alpha_2}),\ldots,\lambda(p_k^{\alpha_k})\right) && \text{if } n=\prod_{i=1}^{k}p_i^{\alpha_i}.
\end{aligned}
\right\}
\quad (1.84)
$$

Example 1.5.5. By Definition 1.5.5, we have

n	1	2	3	4	5	6	7	8	9	10	100	101	102	103
$\lambda(n)$	1	1	2	2	4	2	6	2	6	4	20	100	16	102

Example 1.5.6. Let $n = 65520 = 2^4 \cdot 3^2 \cdot 5 \cdot 7 \cdot 13$, and $a = 11$. Then $\gcd(65520, 11) = 1$ and we have

$$
\begin{aligned}
\phi(65520) &= 8 \cdot 6 \cdot 4 \cdot 6 \cdot 12 = 13824, \\
\lambda(65520) &= \mathrm{lcm}(4, 6, 4, 6, 12) = 12.
\end{aligned}
$$

Definition 1.5.6. Let n be a positive integer. Then the *Möbius μ-function*, $\mu(n)$, is defined as follows:

$$
\mu(n) = \begin{cases}
1, & \text{if } n = 1, \\
0, & \text{if } n \text{ contains a squared factor}, \\
(-1)^k, & \text{if } n = p_1 p_2 \cdots p_k \text{ is the product of} \\
& \quad k \text{ distinct primes}.
\end{cases}
\quad (1.85)
$$

Example 1.5.7. By Definition 1.85, we have

n	1	2	3	4	5	6	7	8	9	10	100	101	102
$\mu(n)$	1	-1	-1	0	-1	1	-1	0	0	1	0	-1	-1

Theorem 1.5.8. Let $\mu(n)$ be the Möbius function. Then

(1) $\mu(n)$ is multiplicative, i.e., for $\gcd(m, n) = 1$,

$$\mu(mn) = \mu(m)\mu(n). \tag{1.86}$$

(2) Let

$$\nu(n) = \sum_{d|n} \mu(d). \tag{1.87}$$

Then

$$\nu(n) = \begin{cases} 1, & \text{if } n = 1, \\ 0, & \text{if } n > 1. \end{cases} \tag{1.88}$$

Proof.

(1) If either $p^2 \mid m$ or $p^2 \mid n$, p is a prime, then $p^2 \mid mn$. Hence, $\mu(mn) = 0 = \mu(m)\mu(n)$. If both m and n are square-free integers, say, $m = p_1 p_2 \cdots p_s$ and $n = q_1 q_2 \cdots q_t$. Then

$$\begin{aligned}
\mu(mn) &= \mu(p_1 p_2 \cdots p_s q_1 q_2 \cdots q_t) \\
&= (-1)^{s+t} \\
&= (-1)^s (-1)^t \\
&= \mu(m)\mu(n).
\end{aligned}$$

(2) If $n = 1$, then $\nu(1) = \sum_{d|n} \nu(d) = \mu(1) = 1$. If $n > 1$, since $\nu(n)$ is multiplicative, we need only to evaluate ν on prime to powers. In addition, if p is prime,

$$\begin{aligned}
\nu(p^\alpha) &= \sum_{d|p^\alpha} \mu(d) \\
&= \mu(1) + \mu(p) + \mu(p^2) + \cdots + \mu(p^\alpha) \\
&= 1 + (-1) + 0 + \cdots + 0 \\
&= 0.
\end{aligned}$$

Thus, $\nu(n) = 0$ for any positive integer n greater than 1. $\square$

The importance of the Möbius function lies in the fact that it plays an important role in the inversion formula given in the following theorem. The formula involves a general arithmetic function f which is not necessarily multiplicative.

Theorem 1.5.9 (The Möbius inversion formula). If f is any arithmetic function and if

$$g(n) = \sum_{d|n} f(d), \tag{1.89}$$

then

$$f(n) = \sum_{d|n} \mu\left(\frac{n}{d}\right) g(d) = \sum_{d|n} \mu(d)\, g\left(\frac{n}{d}\right). \tag{1.90}$$

Proof. If f is an arithmetic function and $g(n) = \sum_{d|n} f(d)$. Then

$$
\begin{aligned}
\sum_{d|n} \mu(d)\, g\left(\frac{n}{d}\right) &= \sum_{d|n} \mu(d) \sum_{a|(n/d)} f(a) \\
&= \sum_{d|n} \sum_{a|(n/d)} \mu(d) f(a) \\
&= \sum_{a|n} \sum_{d|(n/a)} f(a)\mu(d) \\
&= \sum_{a|n} f(a) \sum_{d|(n/a)} \mu(d) \\
&= f(n) \cdot 1 \\
&= f(n).
\end{aligned}
$$

$\square$

The converse of Theorem 1.5.9 is also true and can be stated as follows:

Theorem 1.5.10 (The converse of the Möbius inversion formula). If

$$f(n) = \sum_{d|n} \mu\left(\frac{n}{d}\right) g(d), \tag{1.91}$$

then

$$g(n) = \sum_{d|n} f(d). \tag{1.92}$$

Note that the functions τ and σ

$$\tau(n) = \sum_{d|n} 1 \quad \text{and} \quad \sigma(n) = \sum_{d|n} d$$

may be inverted to give

$$1 = \sum_{d|n} \mu\left(\frac{n}{d}\right) \tau(d) \quad \text{and} \quad n = \sum_{d|n} \mu\left(\frac{n}{d}\right) \sigma(d)$$

for all $n \geq 1$. The relationship between Euler's ϕ-function and Möbius' μ-function is given by the following theorem.

Theorem 1.5.11. For any positive integer n,

$$\phi(n) = n \sum_{d|n} \frac{\mu(d)}{d}. \tag{1.93}$$

Proof. Apply Möbius inversion formula to

$$g(n) = n = \sum_{d|n} \phi(d)$$

we get

$$\begin{aligned} \phi(n) &= \sum_{d|n} \mu(d)\, g\left(\frac{n}{d}\right) \\ &= \sum_{d|n} \frac{\mu(d)}{d} n. \end{aligned}$$

$\square$

Problems for Section 1.5

Problem 1.5.1. Let

$$\Lambda(n) = \begin{cases} \log p, & \text{if } n \text{ is a power of a prime } p \\ 0, & \text{otherwise} \end{cases}$$

Evaluate

$$\sum_{d|n} \Lambda(d).$$

Problem 1.5.2. Evaluate

$$\sum_{d|n} \mu(d)\sigma(d)$$

in terms of the distinctive prime factors of n.

Problem 1.5.3. Let $n > 1$ and a run over all integers with $1 \leq a \leq n$ and $\gcd(a, n) = 1$. Prove that

$$\frac{1}{n^3} \sum a^3 = \frac{1}{4} \phi(n) \left(1 + \frac{(-1)^k p_1 p_2 \cdots p_k}{n^2} \right),$$

where $p_1, p_2 \cdots p_k$ are the distinct prime factors of n.

Problem 1.5.4. (Ramanujan sum) Let m, n be positive integers and d run over all divisors of $\gcd(m, n)$. Prove that

$$\sum d \mu \left(\frac{n}{d} \right) = \frac{\mu \left(\dfrac{n}{\gcd(m, n)} \right) \phi(n)}{\phi \left(\dfrac{n}{\gcd(m, n)} \right)}$$

Problem 1.5.5. (Lambert series) Prove that

$$\sum_{n=1}^{\infty} \frac{\phi(n) x^n}{1 - x^n} = \frac{x}{(1 - x)^2}.$$

Problem 1.5.6. Prove that

$$\sum_{n \leq x} \frac{\phi(n)}{n} = \frac{6x}{\pi^2} + \mathcal{O}(\log x).$$

Problem 1.5.7. Let $p_1, p_2, \ldots, p_k$ be distinct primes. Show that

$$\frac{(p_1 + 1)(p_2 + 1) \cdots (p_k + 1)}{p_1 p_2 \cdots p_k} \leq 2 \leq \frac{p_1 p_2 \cdots p_k}{(p_1 - 1)(p_2 - 1) \cdots (p_k - 1)}$$

is the necessary condition for

$$n = p_1^{\alpha_1} p_2^{\alpha_2} \cdots p_k^{\alpha_k}$$

to be a perfect number.

Problem 1.5.8. Show that $\tau(n)$ is odd if and only if n is a perfect square, and that $\sigma(n)$ is odd if and only if n is a square or two times a square.

Problem 1.5.9. Show that for $n > 2$,

$$\sum_{\substack{k=1 \\ \gcd(k,n)=1}}^{\phi(n)} \frac{1}{k}$$

cannot be an integer.

Problem 1.5.10. Prove that for each positive integer n,

$$\sum_{\substack{k=1 \\ \gcd(k,n)=1}}^{n} k = \frac{n}{2}\phi(n) + \frac{n}{2}\sum_{d|n}\mu(d).$$

$$\sum_{\substack{k=1 \\ \gcd(k,n)=1}}^{n} k^2 = \frac{n^2}{3}\phi(n) + \frac{n^2}{2}\sum_{d|n}\mu(d) + \frac{n}{6}\prod_{p|n}(1-p).$$

$$\sum_{\substack{k=1 \\ \gcd(k,n)=1}}^{n} k^3 = \frac{n^3}{4}\phi(n) + \frac{n^3}{2}\sum_{d|n}\mu(d) + \frac{n^2}{4}\prod_{p|n}(1-p).$$

1.6 Linear Congruences

Definition 1.6.1. Let a be an integer and n a positive integer greater than 1. We define "$a \bmod n$" to be the remainder r when a is divided by n, that is

$$r = a \bmod n = a - \lfloor a/n \rfloor n. \tag{1.94}$$

We may also say that "r is equal to a reduced modulo n".

Remark 1.6.1. It follows from the above definition that $a \bmod n$ is the integer r such that $a = \lfloor a/n \rfloor n + r$ and $0 \le r < n$, which was known to the ancient Greeks 2000 years ago.

Example 1.6.1. The following are some examples of $a \bmod n$:

$$35 \bmod 12 = 11,$$

$$-129 \bmod 7 = 4,$$

$$3210 \bmod 101 = 79,$$

$$1412^{13115} \bmod 12349 = 1275.$$

Given the well-defined notion of the remainder of one integer when divided by another, it is convenient to provide a special notion to indicate equality of remainders.

Definition 1.6.2. Let a and b be integers and n a positive integer. We say that "a is *congruent* to b modulo n", denoted by

$$a \equiv b \pmod{n} \tag{1.95}$$

if n is a divisor of $a - b$, or equivalently, if $n \mid (a - b)$. Similarly, we write

$$a \not\equiv b \pmod{n} \tag{1.96}$$

if a is not congruent (or incongruent) to b modulo n, or equivalently, if $n \nmid (a - b)$. Clearly, for $a \equiv b \pmod{n}$ (resp. $a \not\equiv b \pmod{n}$), we can write $a = kn + b$ (resp. $a \neq kn + b$) for some integer k. The integer n is called the *modulus*.

Clearly,

$$a \equiv b \pmod{n} \iff n \mid (a - b) \iff a = kn + b, \quad k \in \mathbb{Z}$$

and

$$a \not\equiv b \pmod{n} \iff n \nmid (a - b) \iff a \neq kn + b, \quad k \in \mathbb{Z}$$

So, the above definition of congruences, introduced by Gauss in his *Disquisitiones Arithmeticae*, does not offer any new idea than the divisibility relation, since "$a \equiv b \pmod{n}$" and "$n \mid (a - b)$" (resp. "$a \not\equiv b \pmod{n}$" and "$n \nmid (a - b)$") have the same meaning, although each of them has its own advantages. However, Gauss did present a *new* way (i.e., congruences) of looking at the old things (i.e., divisibility); this is exactly what we are interested in. It is interesting to note that the ancient Chinese mathematician Ch'in Chiu-Shao (1202–1261) already had the idea of congruences in his famous book *Mathematical Treatise in Nine Chapters* in 1247.

Definition 1.6.3. If $a \equiv b \pmod{n}$, then b is called a *residue* of a modulo n. If $0 \leq b \leq n - 1$, b is called the *least non-negative residue* of a modulo n.

Remark 1.6.2. It is common, particularly in computer programs, to denote the least non-negative residue of a modulo n by $a \bmod n$. Thus, $a \equiv b \pmod{n}$ if and only if $a \bmod n = b \bmod n$, and, of course, $a \not\equiv b \pmod{n}$ if and only if $a \bmod n \neq b \bmod n$.

Example 1.6.2. The following are some examples of congruences or incongruences.

$$35 \equiv 11 \pmod{12} \qquad \text{since} \qquad 12 \mid (35 - 11)$$

$$\not\equiv 12 \pmod{11} \qquad \text{since} \qquad 11 \nmid (35 - 12)$$

$$\equiv 2 \pmod{11} \qquad \text{since} \qquad 11 \mid (35 - 2).$$

The congruence relation has many properties in common with the of equality relation. For example, we know from high-school mathematics that equality is

(1) reflexive: $a = a$, $\forall a \in \mathbb{Z}$;

(2) symmetric: if $a = b$, then $b = a$, $\forall a, b \in \mathbb{Z}$;

(3) transitive: if $a = b$ and $b = c$, then $a = c$, $\forall a, b, c \in \mathbb{Z}$.

We shall see that congruence modulo n has the same properties:

Theorem 1.6.1. Let n be a positive integer. Then the congruence modulo n is

(1) reflexive: $a \equiv a \pmod{n}$, $\forall a \in \mathbb{Z}$;

(2) symmetric: if $a \equiv b \pmod{n}$, then $b \equiv a \pmod{n}$, $\forall a, b \in \mathbb{Z}$;

(3) transitive: if $a \equiv b \pmod{n}$ and $b \equiv c \pmod{n}$, then $a \equiv c \pmod{n}$, $\forall a, b, c \in \mathbb{Z}$.

Proof.

(1) For any integer a, we have $a = 0 \cdot n + a$, hence $a \equiv a \pmod{n}$.

(2) For any integers a and b, if $a \equiv b \pmod{n}$, then $a = kn + b$ for some integer k. Hence $b = a - kn = (-k)n + a$, which implies $b \equiv a \pmod{n}$, since $-k$ is an integer.

(3) If $a \equiv b \pmod{n}$ and $b \equiv c \pmod{n}$, then $a = k_1 n + b$ and $b = k_2 n + c$. Thus, we can get

$$a = k_1 n + k_2 n + c = (k_1 + k_2)n + c = k'n + c$$

which implies $a \equiv c \pmod{n}$, since k' is an integer. $\qquad \square$

Theorem 1.6.1 shows that congruence modulo n is an equivalence relation on the set of integers $\mathbb{Z}$. But note that the divisibility relation $a \mid b$ is reflexive, and transitive but not symmetric; in fact if $a \mid b$ and $b \mid a$ then $a = b$, so it is not an equivalence relation. The congruence relation modulo n partitions $\mathbb{Z}$ into n *equivalence classes*. In number theory, we call these classes *congruence classes*, or *residue classes*.

Definition 1.6.4. If $x \equiv a \pmod{n}$, then a is called a *residue* of x modulo n. The *residue class* of a modulo n, denoted by $[a]_n$ (or just $[a]$ if no confusion will be caused), is the set of all those integers that are congruent to a modulo n. That is,

$$[a]_n = \{x : x \in \mathbb{Z} \text{ and } x \equiv a \pmod{n}\} = \{a + kn : k \in \mathbb{Z}\}. \qquad (1.97)$$

Note that writing $a \in [b]_n$ is the same as writing $a \equiv b \pmod{n}$.

Example 1.6.3. Let $n = 5$. Then there are five residue classes, modulo 5, namely the sets:

$$[0]_5 = \{\ldots, -15, -10, -5, 0, 5, 10, 15, 20, \ldots\},$$
$$[1]_5 = \{\ldots, -14, -9, -4, 1, 6, 11, 16, 21, \ldots\},$$
$$[2]_5 = \{\ldots, -13, -8, -3, 2, 7, 12, 17, 22, \ldots\},$$
$$[3]_5 = \{\ldots, -12, -7, -2, 3, 8, 13, 18, 23, \ldots\},$$
$$[4]_5 = \{\ldots, -11, -6, -1, 4, 9, 14, 19, 24, \ldots\}.$$

The first set contains all those integers congruent to 0 modulo 5, the second set contains all those congruent to 1 modulo 5, $\cdots$, and the fifth (i.e., the last) set contains all those congruent to 4 modulo 5. So, for example, the residue class $[2]_5$ can be represented by any one of the elements in the set

$$\{\ldots, -13, \ -8, -3, 2, 7, 12, 17, 22, \ldots\}.$$

Clearly, there are infinitely many elements in the set $[2]_5$.

Example 1.6.4. In residue classes modulo 2, $[0]_2$ is the set of all even integers, and $[1]_2$ is the set of all odd integers:

$$[0]_2 = \{\ldots, -6, -4, -2, 0, 2, 4, 6, 8, \ldots\},$$
$$[1]_2 = \{\ldots, -5, -3, -1, 1, 3, 5, 7, 9, \ldots\}.$$

Example 1.6.5. In congruence modulo 5, we have

$$
\begin{aligned}
[9]_5 &= \{9 + 5k : \ k \in \mathbb{Z}\} = \{9, 9 \pm 5, 9 \pm 10, 9 \pm 15, \ldots\} \\
&= \{\ldots, -11, -6, -1, 4, 9, 14, 19, 24, \ldots\}.
\end{aligned}
$$

We also have

$$
\begin{aligned}
[4]_5 &= \{4 + 5k : \ k \in \mathbb{Z}\} = \{4, 4 \pm 5, 4 \pm 10, 4 \pm 15, \ldots\} \\
&= \{\ldots, -11, -6, -1, 4, 9, 14, 19, 24, \ldots\}.
\end{aligned}
$$

So, clearly, $[4]_5 = [9]_5$.

Example 1.6.6. Let $n = 7$. There are seven residue classes, modulo 7. In each of these seven residue classes, there is exactly one least residue of x modulo 7. So the complete set of all least residues x modulo 7 is $\{0, 1, 2, 3, 4, 5, 6\}$.

Definition 1.6.5. The set of all residue classes modulo n, often denoted by $\mathbb{Z}/n\mathbb{Z}$ or $\mathbb{Z}/n\mathbb{Z}$, is

$$\mathbb{Z}/n\mathbb{Z} = \{[a]_n : \ 0 \le a \le n - 1\}. \tag{1.98}$$

Remark 1.6.3. One often sees the definition

$$\mathbb{Z}/n\mathbb{Z} = \{0, 1, 2, \ldots, n - 1\}, \tag{1.99}$$

which should be read as equivalent to (1.98) with the understanding that 0 represents $[0]_n$, 1 represents $[1]_n$, 2 represents $[2]_n$, and so on; each class is represented by its least non-negative residue, but the underlying residue classes must kept in mind. For example, a reference to $-a$ as a member of $\mathbb{Z}/n\mathbb{Z}$ is a reference to $[n - a]_n$, provided $n \ge a$, since $-a \equiv n - a \pmod{n}$.

The following theorem gives some elementary properties of residue classes:

Theorem 1.6.2. Let n be a positive integer. Then we have

(1) $[a]_n = [b]_n$ if and only if $a \equiv b \pmod{n}$;

(2) Two residue classes modulo n are either disjoint or identical;

(3) There are exactly n distinct residue classes modulo n, namely, $[0]_n, [1]_n, [2]_n, [3]_n, \ldots, [n-1]_n$, and they contain all of the integers.

Proof.

(1) If $a \equiv b \pmod{n}$, it follows from the transitive property of congruence that an integer is congruent to a modulo n if and only if it is congruent to b modulo n. Thus, $[a]_n = [b]_n$. To prove the converse, suppose $[a]_n = [b]_n$. Because $a \in [a]_n$ and $a \in [b]_n$, Thus, $a \equiv b \pmod{n}$.

(2) Suppose $[a]_n$ and $[b]_n$ have a common element c. Then $c \equiv a \pmod{n}$ and $c \equiv b \pmod{n}$. From the symmetric and transitive properties of congruence, it follows that $a \equiv b \pmod{n}$. From part (1) of this theorem, it follows that $[a]_n = [b]_n$. Thus, either $[a]_n$ and $[b]_n$ are disjoint or identical.

(3) If a is an integer, we can divide a by n to get

$$a = kn + r, \quad 0 \leq r < k.$$

Thus, $a \equiv r \pmod{n}$ and so $[a]_n = [r]_n$. This implies that a is in one of the residue classes $[0]_n, [1]_n, [2]_n, \ldots, [n-1]_n$, Because the integers $0, 1, 2, \ldots, n-1$ are incongruent modulo n, it follows that there are exactly n residue classes modulo n. □

Definition 1.6.6. Let n be a positive integer. A set of integers $a_1, a_2, \ldots, a_n$ is called a *complete system of residues* modulo n, if the set contains exactly one element from each residue class modulo n.

Example 1.6.7. Let $n = 4$. Then $\{-12, 9, -6, -1\}$ is a complete system of residues modulo 4, since $-12 \in [0]$, $9 \in [1]$, $-6 \in [2]$ and $-1 \in [3]$. Of course, it can be easily verified that $\{12, -7, 18, -9\}$ is another complete system of residues modulo 4. It is clear that the simplest complete system of residues modulo 4 is $\{0, 1, 2, 3\}$, the set of all non-negative least residues modulo 4.

Example 1.6.8. Let $n = 7$. Then

$$\{x,\ x+3,\ x+3^2,\ x+3^3,\ x+3^4,\ x+3^5,\ x+3^6\}$$

is a complete system of residues modulo 7, for any $x \in \mathbb{Z}$. To see this let us first evaluate the powers of 3 modulo 7:

3	$3^2 \equiv 2 \pmod{7}$	$3^3 \equiv 6 \pmod{7}$
$3^4 \equiv 4 \pmod{7}$	$3^5 \equiv 5 \pmod{7}$	$3^6 \equiv 1 \pmod{7}$

hence, the result follows from $x = 0$. Now the general result follows immediately, since $(x + 3^i) - (x + 3^j) = 3^i - 3^j$.

Theorem 1.6.3. Let n be a positive integer and S a set of integers. S is a complete system of residues modulo n if and only if S contains n elements and no two elements of S are congruent, modulo n.

Proof. If S is a complete system of residues, then the two conditions are satisfied. To prove the converse, we note that if no two elements of S are congruent, the elements of S are in different residue classes modulo n. Since S has n elements, all the residue classes must be represented among the elements of S. Thus, S is a complete system of residues modulo n □

We now introduce one more type of system of residues, the *reduced* system of residues modulo n.

Definition 1.6.7. Let $[a]_n$ be a residue class modulo n. We say that $[a]_n$ is relatively prime to n if each element in $[a]_n$ is relatively prime to n.

Example 1.6.9. Let $n = 10$. Then the ten residue classes, modulo 10, are as follows:

$$[0]_{10} = \{\ldots, -30, -20, -10, 0, 10, 20, 30, \ldots\}$$
$$[1]_{10} = \{\ldots, -29, -19, -9, 1, 11, 21, 31, \ldots\}$$
$$[2]_{10} = \{\ldots, -28, -18, -8, 2, 12, 22, 32, \ldots\}$$
$$[3]_{10} = \{\ldots, -27, -17, -7, 3, 13, 23, 33, \ldots\}$$
$$[4]_{10} = \{\ldots, -26, -16, -6, 4, 14, 24, 34, \ldots\}$$
$$[5]_{10} = \{\ldots, -25, -15, -5, 5, 15, 25, 35, \ldots\}$$
$$[6]_{10} = \{\ldots, -24, -14, -4, 6, 16, 26, 36, \ldots\}$$
$$[7]_{10} = \{\ldots, -23, -13, -3, 7, 17, 27, 37, \ldots\}$$
$$[8]_{10} = \{\ldots, -22, -12, -2, 8, 18, 28, 38, \ldots\}$$
$$[9]_{10} = \{\ldots, -21, -11, -1, 9, 19, 29, 39, \ldots\}.$$

Clearly, $[1]_{10}$, $[3]_{10}$, $[7]_{10}$, and $[9]_{10}$ are residue classes that are relatively prime to 10.

Proposition 1.6.1. If a residue class modulo n has *one* element which is relatively prime to n, then every element in that residue class is relatively prime to n.

Proposition 1.6.2. If n is prime, then every residue class modulo n (except $[0]_n$) is relatively prime to n.

Definition 1.6.8. Let n be a positive integer, then $\phi(n)$ is the number of residue classes modulo n, which is relatively prime to n. A set of integers $\{a_1, a_2, \ldots, a_{\phi(n)}\}$ is called a *reduced system of residues*, if the set contains exactly one element from each residue class modulo n which is relatively prime to n.

Example 1.6.10. In Example 1.6.9, we know that $[1]_{10}$, $[3]_{10}$, $[7]_{10}$ and $[9]_{10}$ are residue classes that are relatively prime to 10, so by choosing -29 from $[1]_{10}$, -17 from $[3]_{10}$, 17 from $[7]_{10}$ and 39 from $[9]_{10}$, we get a reduced system of residues modulo 10: $\{-29, -17, 17, 39\}$. Similarly, $\{31, 3, -23, -1\}$ is another reduced system of residues modulo 10.

One method to obtain a reduced system of residues is to start with a complete system of residues and delete those elements that are not relatively prime to the modulus n. Thus, the simplest reduced system of residues (mod n) is just the collections of all integers in the set $\{0, 1, 2, \ldots, n-1\}$ that are relatively prime to n.

Theorem 1.6.4. Let n be a positive integer, and S a set of integers. Then S is a reduced system of residues (mod n) if and only if

(1) S contains exactly $\phi(n)$ elements;

(2) no two elements of S are congruent (mod n);

(3) each element of S is relatively prime to n.

Proof. It is obvious that a reduced system of residues satisfies the three conditions. To prove the converse, we suppose that S is a set of integers having the three properties. Because no two elements of S are congruent, the elements are in different residues modulo n. Since the elements of S are relatively prime n, there are in residue classes that are relatively prime n. Thus, the $\phi(n)$ elements of S are distributed among the $\phi(n)$ residue classes that are relatively prime n, one in each residue class. Therefore, S is a reduced system of residues modulo n. ☐

Corollary 1.6.1. Let $\{a_1, a_2, \ldots, a_{\phi(n)}\}$ be a reduced system of residues modulo m, and suppose that $\gcd(k, n) = 1$. Then $\{ka_1, ka_2, \ldots, ka_{\phi(n)}\}$ is also a reduced system of residues modulo n.

Proof. Left as an exercise. ☐

The finite set $\mathbb{Z}/n\mathbb{Z}$ is closely related to the infinite set $\mathbb{Z}$. So it is natural to ask if it is possible to define addition and multiplication in $\mathbb{Z}/n\mathbb{Z}$ and do some reasonable kind of arithmetic there. Surprisingly, the addition, subtraction and multiplication in $\mathbb{Z}/n\mathbb{Z}$ will be much the same as that in $\mathbb{Z}$.

Theorem 1.6.5. For all $a, b, c, d \in \mathbb{Z}$ and $n \in \mathbb{Z}_{>1}$, if $a \equiv b$ (mod n) and $c \equiv d$ (mod n). then

(1) $a \pm b \equiv c \pm d$ (mod n);

(2) $a \cdot b \equiv c \cdot d$ (mod n);

(3) $a^m \equiv b^m$ (mod n), $\quad \forall m \in \mathbb{Z}^+$.

Proof.

(1) Write $a = kn + b$ and $c = ln + d$ for some $k, l \in \mathbb{Z}$. Then $a + c = (k+l)n + b + d$. Therefore, $a + c = b + d + tn$, $t = k + l \in \mathbb{Z}$. Consequently, $a + c \equiv b + d \pmod{n}$, which is what we wished to show. The case for subtraction is left as an exercise.

(2) Similarly,
$$
\begin{aligned}
ac &= bd + bln + knd + kln^2 \\
&= bd + n(bl + k(d + ln)) \\
&= bd + n(bl + kc) \\
&= bd + sn
\end{aligned}
$$
where $s = bl + kc \in \mathbb{Z}$. Thus, $a \cdot b \equiv c \cdot d \pmod{n}$.

(3) We prove Part (3) by induction. We have $a \equiv b \pmod{n}$ (base step) and $a^m \equiv b^m \pmod{n}$ (inductive hypothesis). Then by Part (2) we have $a^{m+1} \equiv aa^m \equiv bb^m \equiv b^{m+1} \pmod{n}$. $\qquad\square$

Theorem 1.6.5 is equivalent to the following theorem, since
$$ a \equiv b \pmod{n} \iff a \bmod n = b \bmod n, $$

$$ a \bmod n \iff [a]_n, $$

$$ b \bmod n \iff [b]_n. $$

Theorem 1.6.6. For all $a, b, c, d \in \mathbb{Z}$, if $[a]_n = [b]_n$, $[c]_n = [d]_n$, then

(1) $[a \pm b]_n = [c \pm d]_n$,

(2) $[a \cdot b]_n = [c \cdot d]_n$,

(3) $[a^m]_n = [b^m]_n$, $\quad \forall m \in \mathbb{Z}^+$.

The fact that the congruence relation modulo n is stable for addition (subtraction) and multiplication means that we can define binary operations, again called addition (subtraction) and multiplication on the set of $\mathbb{Z}/n\mathbb{Z}$ of equivalence classes modulo n as follows (in case only one n is being discussed, we can simply write $[x]$ for the class $[x]_n$):

$$ [a]_n + [b]_n = [a+b]_n \tag{1.100} $$

$$ [a]_n - [b]_n = [a-b]_n \tag{1.101} $$

$$ [a]_n \cdot [b]_n = [a \cdot b]_n \tag{1.102} $$

Example 1.6.11. Let $n = 12$, then
$$ [7]_{12} + [8]_{12} = [7+8]_{12} = [15]_{12} = [3]_{12}, $$
$$ [7]_{12} - [8]_{12} = [7-8]_{12} = [-1]_{12} = [11]_{12}, $$
$$ [7]_{12} \cdot [8]_{12} = [7 \cdot 8]_{12} = [56]_{12} = [8]_{12}. $$

In many cases, we may still prefer to write the above operations as follows:

$$7 + 8 = 15 \equiv 3 \pmod{12},$$
$$7 - 8 = -1 \equiv 11 \pmod{12},$$
$$7 \cdot 8 = 56 \equiv 8 \pmod{12}.$$

We summarize the properties of addition and multiplication modulo n in the following two theorems.

Theorem 1.6.7. The set $\mathbb{Z}/n\mathbb{Z}$ of integers modulo n has the following properties with respect to addition:

(1) Closure: $[x] + [y] \in \mathbb{Z}/n\mathbb{Z}$, for all $[x], [y] \in \mathbb{Z}/n\mathbb{Z}$;

(2) Associative: $([x] + [y]) + [z] = [x] + ([y] + [z])$, for all $[x], [y], [z] \in \mathbb{Z}/n\mathbb{Z}$;

(3) Commutative: $[x] + [y] = [y] + [x]$, for all $[x], [y] \in \mathbb{Z}/n\mathbb{Z}$;

(4) Identity, namely, $[0]$;

(5) Additive inverse: $-[x] = [-x]$, for all $[x] \in \mathbb{Z}/n\mathbb{Z}$.

Proof. These properties follow directly from the stability and the definition of the operation in $\mathbb{Z}/n\mathbb{Z}$. □

Theorem 1.6.8. The set $\mathbb{Z}/n\mathbb{Z}$ of integers modulo n has the following properties with respect to multiplication:

(1) Closure: $[x] \cdot [y] \in \mathbb{Z}/n\mathbb{Z}$, for all $[x], [y] \in \mathbb{Z}/n\mathbb{Z}$;

(2) Associative: $([x] \cdot [y]) \cdot [z] = [x] \cdot ([y] \cdot [z])$, for all $[x], [y], [z] \in \mathbb{Z}/n\mathbb{Z}$;

(3) Commutative: $[x] \cdot [y] = [y] \cdot [x]$, for all $[x], [y] \in \mathbb{Z}/n\mathbb{Z}$;

(4) Identity, namely, $[1]$;

(5) Distributivity of multiplication over addition: $[x] \cdot ([y]) + [z]) = ([x] \cdot [y]) + ([x] \cdot [z])$, for all $[x], [y], [z] \in \mathbb{Z}/n\mathbb{Z}$.

Proof. These properties follow directly from the stability of the operation in $\mathbb{Z}/n\mathbb{Z}$ and the corresponding properties of $\mathbb{Z}$. □

The division a/b (we assume a/b is in lowest terms and $b \not\equiv 0 \pmod{n}$) in $\mathbb{Z}/n\mathbb{Z}$, however, will be more of a problem; sometimes you can divide, sometimes you cannot. For example, let $n = 12$ again, then

$$3/7 \equiv 9 \pmod{12} \qquad \text{(no problem)},$$
$$3/4 \equiv \perp \pmod{12} \qquad \text{(impossible)}.$$

Why is division sometimes possible (e.g., $3/7 \equiv 9 \pmod{12}$) and sometimes impossible (e.g., $3/8 \equiv \perp \pmod{12}$)? The problem is with the modulus n; if n is a prime number, then the division $a/b \pmod{n}$ is always possible and unique, whilst if n is a composite then the division $a/b \pmod{n}$ may be not possible or the result may be not unique. Let us observe two more examples, one with $n = 13$ and the other with $n = 14$. First note that $a/b \equiv a \cdot 1/b \pmod{n}$ if and only if $1/b \pmod{n}$ is possible, since multiplication modulo n is always possible. We call $1/b \pmod{n}$ the *multiplicative inverse* (or the *modular inverse*) of b modulo n. Now let $n = 13$ be a prime, then the following table gives all the values of the multiplicative inverses $1/x \pmod{13}$ for $x = 1, 2, \ldots, 12$:

x	1	2	3	4	5	6	7	8	9	10	11	12
$1/x \pmod{13}$	1	7	9	10	8	11	2	5	3	4	6	12

This means that division in $\mathbb{Z}/13\mathbb{Z}$ is always possible and unique. On the other hand, if $n = 14$ (the n now is a composite), then

x	1	2	3	4	5	6	7	8	9	10	11	12	13
$1/x \pmod{14}$	1	$\perp$	5	$\perp$	3	$\perp$	$\perp$	$\perp$	11	$\perp$	9	$\perp$	13

This means that only the numbers $1, 3, 5, 9, 11$ and 13 have multiplicative inverses modulo 14, or equivalently only those divisions by $1, 3, 5, 9, 11$ and 13 modulo 14 are possible. This observation leads to the following important results:

Theorem 1.6.9. The multiplicative inverse $1/b$ modulo n exists if and only if $\gcd(b, n) = 1$.

But how many b's satisfy $\gcd(b, n) = 1$? The following result answers this question.

Corollary 1.6.2. There are $\phi(n)$ numbers b for which $1/b \pmod{n}$ exists.

Example 1.6.12. Let $n = 21$. Since $\phi(21) = 12$, there are twelve values of b for which $1/b \pmod{21}$ exists. In fact, the multiplicative inverse modulo 21 only exists for each of the following b:

b	1	2	4	5	8	10	11	13	16	17	19	20
$1/b \pmod{21}$	1	11	16	17	8	19	2	13	4	5	10	20

Corollary 1.6.3. The division a/b modulo n (assume that a/b is in lowest terms) is possible if and only if $1/b \pmod{n}$ exists, i.e., if and only if $\gcd(b, n) = 1$.

Example 1.6.13. Compute $6/b \pmod{21}$ whenever it is possible. By the multiplicative inverses of $1/b \pmod{21}$ in the previous table, we just need to calculate $6 \cdot 1/b \pmod{21}$:

b	1	2	4	5	8	10	11	13	16	17	19	20
$6/b \pmod{21}$	6	3	12	18	6	9	12	15	3	9	18	15

As can be seen, addition (subtraction) and multiplication are always possible in $\mathbb{Z}/n\mathbb{Z}$, with $n > 1$, since $\mathbb{Z}/n\mathbb{Z}$ is a ring. Note also that $\mathbb{Z}/n\mathbb{Z}$ with n prime is an Abelian group with respect to addition, and all the non-zero elements in $\mathbb{Z}/n\mathbb{Z}$ form an Abelian group with respect to multiplication (i.e., a division is always possible for any two non-zero elements in $\mathbb{Z}/n\mathbb{Z}$ if n is prime); hence $\mathbb{Z}/n\mathbb{Z}$ with n prime is a field. That is,

Theorem 1.6.10. $\mathbb{Z}/n\mathbb{Z}$ is a field if and only if n is prime.

The above results only tell us when the multiplicative inverse $1/a$ modulo n is possible, without mentioning how to find the inverse. To actually find the multiplicative inverse, we let

$$1/a \pmod{n} = x, \tag{1.103}$$

which is equivalent to

$$ax \equiv 1 \pmod{n}. \tag{1.104}$$

Since

$$ax \equiv 1 \pmod{n} \iff ax - ny = 1. \tag{1.105}$$

Thus, finding the multiplicative inverse $1/a \pmod{n}$ is the same as finding the solution of the linear Diophantine equation $ax - ny = 1$, which, as we kown, can be solved by using the continued fraction expansion of a/n or by using Euclid's algorithm.

Example 1.6.14. Find

(1) $1/154 \pmod{801}$,

(2) $4/154 \pmod{801}$.

Solution

(1) Since

$$1/a \pmod{n} = x \iff ax \equiv 1 \pmod{n} \iff ax - ny = 1,$$

we only need to find x and y in

$$154x - 801y = 1.$$

To do so, we first use the Euclid's algorithm to find $\gcd(154, 801)$ as follows:

$$\begin{aligned}
801 &= 154 \cdot 5 + 31 \\
154 &= 31 \cdot 4 + 30 \\
31 &= 30 \cdot 1 + 1 \\
3 &= 1 \cdot 3.
\end{aligned}$$

Since $\gcd(154, 801) = 1$, by Theorem 1.6.9, the equation $154x - 801y = 1$ is soluble. We now rewrite the above resulting equations

$$\begin{aligned}
31 &= 801 - 154 \cdot 5 \\
30 &= 154 - 31 \cdot 4 \\
1 &= 31 - 30 \cdot 1
\end{aligned}$$

and work backwards on the above new equations

$$\begin{aligned}
1 &= 31 - 30 \cdot 1 \\
&= 31 - (154 - 31 \cdot 4) \cdot 1 \\
&= 31 - 154 + 4 \cdot 31 \\
&= 5 \cdot 31 - 154 \\
&= 5 \cdot (801 - 154 \cdot 5) - 154 \\
&= 5 \cdot 801 - 26 \cdot 154 \\
&= 801 \cdot 5 - 154 \cdot 26.
\end{aligned}$$

So, $x \equiv -26 \equiv 775 \pmod{801}$. That is,

$$1/154 \bmod 801 = 775.$$

(2) By Part (1) above, we have

$$\begin{aligned}
4/154 &\equiv 4 \cdot 1/154 \\
&\equiv 4 \cdot 775 \\
&\equiv 697 \pmod{801}.
\end{aligned}$$

The above procedure used to find the x and y in $ax + by = 1$ can be generalized to find the x and y in $ax + by = c$; this procedure is usually called the *extended Euclid's algorithm*.

Congruences have much in common with equations. In fact, the linear congruence $ax \equiv b \pmod{n}$ is equivalent to the linear Diophantine equation $ax - ny = b$. That is,

$$ax \equiv b \pmod{n} \iff ax - ny = b. \qquad (1.106)$$

Thus, linear congruences can be solved by using the continued fraction method just as for linear Diophantine equations.

Theorem 1.6.11. Let $\gcd(a, n) = d$. If $d \nmid b$, then the linear congruence

$$ax \equiv b \ (\mathrm{mod}\ n) \tag{1.107}$$

has no solutions.

Proof. We will prove the contrapositive of the assertion: if $ax \equiv b \ (\mathrm{mod}\ n)$ has a solution, then $\gcd(a, n) \mid b$. Suppose that s is a solution. Then $as \equiv b \ (\mathrm{mod}\ n)$, and from the definition of the congruence, $n \mid (as - b)$, or from the definition of divisibility, $as - b = kn$ for some integer k. Since $\gcd(a, m) \mid a$ and $\gcd(a, n) \mid kn$, it follows that $\gcd(a, n) \mid b$. □

Theorem 1.6.12. Let $\gcd(a, n) = d$. Then the linear congruence $ax \equiv b \ (\mathrm{mod}\ n)$ has solutions if and only if $d \mid b$.

Proof. Follows from Theorem 1.6.11. □

Theorem 1.6.13. Let $\gcd(a, n) = 1$. Then the linear congruence $ax \equiv b \ (\mathrm{mod}\ n)$ has exactly one solution.

Proof. If $\gcd(a, n) = 1$, then there exist x and y such that $ax + ny = 1$. Multiplying by b gives

$$a(xb) + n(yb) = b.$$

As $a(xb) - b$ is a multiple of n, or $a(xb) \equiv b \ (\mathrm{mod}\ n)$, the least residue of xb modulo n is then a solution of the linear congruence. The uniqueness of the solution is left as an exercise. □

Theorem 1.6.14. Let $\gcd(a, n) = d$ and suppose that $d \mid b$. Then the linear congruence

$$ax \equiv b \ (\mathrm{mod}\ n). \tag{1.108}$$

has exactly d solutions modulo n. These are given by

$$t, \ t + \frac{n}{d}, \ t + \frac{2n}{d}, \ \ldots, \ t + \frac{(d-1)n}{d} \tag{1.109}$$

where t is the solution, unique modulo n/d, of the linear congruence

$$\frac{a}{d}x \equiv \frac{b}{d} \ \left(\mathrm{mod}\ \frac{n}{d}\right). \tag{1.110}$$

Proof. By Theorem 1.6.12, the linear congruence has solutions since $d \mid b$. Now let t be be such a solution, then $t + k(n/d)$ for $k = 1, 2, \ldots, d - 1$ are also solutions, since $a(t + k(n/d)) \equiv at + kn(a/d) \equiv at \equiv b \ (\mathrm{mod}\ n)$. □

Example 1.6.15. Solve the linear congruence $154x \equiv 22 \pmod{803}$. Notice first that

$$154x \equiv 22 \pmod{803} \iff 154x - 803y = 22.$$

Now we use the Euclid's algorithm to find $\gcd(154, 803)$ as follows:

$$\begin{aligned}
803 &= 154 \cdot 5 + 33 \\
154 &= 33 \cdot 4 + 22 \\
33 &= 22 \cdot 1 + 11 \\
22 &= 11 \cdot 2 + 0.
\end{aligned}$$

Since $\gcd(154, 803) = 11$ and $11 \mid 22$, by Theorem 1.6.12, the equation $154x - 801y = 22$ is soluble. Now we rewrite the above resulting equations

$$\begin{aligned}
33 &= 803 - 154 \cdot 5 \\
22 &= 154 - 33 \cdot 4 \\
11 &= 33 - 22 \cdot 1
\end{aligned}$$

and work backwards on the above new equations

$$\begin{aligned}
11 &= 33 - 22 \cdot 1 \\
&= 33 - (154 - 33 \cdot 4) \cdot 1 \\
&= 33 - 154 + 4 \cdot 33 \\
&= 5 \cdot 33 - 154 \\
&= 5 \cdot (803 - 154 \cdot 5) - 154 \\
&= 5 \cdot 803 - 26 \cdot 154 \\
&= 803 \cdot 5 - 154 \cdot 26.
\end{aligned}$$

So, $x \equiv -26 \equiv 777 \pmod{803}$. By Theorems 1.6.13 and 1.6.14, $x \equiv -26 \equiv 47 \pmod{73}$ is the only solution to the simplified congruence:

$$154/11 \equiv 22/11 \pmod{803/11} \implies 14x \equiv 2 \pmod{73},$$

since $\gcd(14, 73) = 1$. By Theorem 1.6.14, there are, in total, eleven solutions to the congruence $154x \equiv 11 \pmod{803}$, as follows:

$$x = \begin{pmatrix} 777 \\ 47 \\ 120 \\ 193 \\ 266 \\ 339 \\ 412 \\ 485 \\ 558 \\ 631 \\ 704 \end{pmatrix}.$$

Thus,

$$x = \begin{pmatrix} 751 \\ 94 \\ 240 \\ 386 \\ 532 \\ 678 \\ 21 \\ 167 \\ 313 \\ 459 \\ 605 \end{pmatrix}$$

are the eleven solutions to the original congruence $154x \equiv 22 \pmod{803}$.

Remark 1.6.4. To find the solution for the linear Diophantine equation

$$ax \equiv b \pmod{n} \tag{1.111}$$

is equivalent to finding the quotient of the modular division

$$x \equiv \frac{b}{a} \pmod{n} \tag{1.112}$$

which is, again, equivalent to finding the multiplicative inverse

$$x \equiv \frac{1}{a} \pmod{n} \tag{1.113}$$

because if $\frac{1}{a}$ modulo n exists, the multiplication $b \cdot \frac{1}{a}$ is always possible.

Theorem 1.6.15 (Fermat's little theorem). Let a be a positive integer and $\gcd(a, p) = 1$. If p is prime, then

$$a^{p-1} \equiv 1 \pmod{p}. \tag{1.114}$$

Proof. First notice that the residues modulo p of $a, 2a, \ldots, (p-1)a$ are $1, 2, \ldots, (p-1)$ in some order, because no two of them can be equal. So, if we multiply them together, we get

$$\begin{aligned} a \cdot 2a \cdots (p-1)a &\equiv [(a \bmod p) \cdot (2a \bmod p) \cdots (p-1)a \bmod p] \pmod{p} \\ &\equiv (p-1)! \pmod{p}. \end{aligned}$$

This means that

$$(p-1)! a^{p-1} \equiv (p-1)! \pmod{p}.$$

Now we can cancel the $(p-1)!$ since $p \nmid (p-1)!$, and the result thus follows.

$\square$

There is a more convenient and more general form of Fermat's little theorem:

$$a^p \equiv a \ (\mathrm{mod} \ p), \tag{1.115}$$

for $a \in \mathbb{N}$. The proof is easy: if $\gcd(a,p) = 1$, we simply multiply (1.114) by a. If not, then $p \mid a$. So $a^p \equiv 0 \equiv a \ (\mathrm{mod} \ p)$.

Fermat's theorem has several important consequences which are very useful in compositeness; one of the these consequences is as follows:

Corollary 1.6.4 (Converse of the Fermat little theorem, 1640). Let n be an odd positive integer. If $\gcd(a,n) = 1$ and

$$a^{n-1} \not\equiv 1 \ (\mathrm{mod} \ n), \tag{1.116}$$

then n is composite.

Remark 1.6.5. Fermat in 1640 made a false conjecture that all the numbers of the form $F_n = 2^{2^n} + 1$ were prime. Fermat really should not have made such a "stupid" conjecture, since F_5 can be relatively easily verified to be composite, by just using his own recently discovered theorem – Fermat's little theorem:

$$
\begin{aligned}
3^{2^2} &\equiv 81 & (\mathrm{mod} \ 4294967297) \\
3^{2^3} &\equiv 6561 & (\mathrm{mod} \ 4294967297) \\
3^{2^4} &\equiv 43046721 & (\mathrm{mod} \ 4294967297) \\
3^{2^5} &\equiv 3793201458 & (\mathrm{mod} \ 4294967297) \\
&\vdots & \\
3^{2^{32}} &\equiv 3029026160 & (\mathrm{mod} \ 4294967297) \\
&\not\equiv 1 & (\mathrm{mod} \ 4294967297).
\end{aligned}
$$

Thus, by Fermat's little theorem, $2^{32} + 1$ is not prime!

Based on Fermat's little theorem, Euler established a more general result in 1760:

Theorem 1.6.16 (Euler's theorem). Let a and n be positive integers with $\gcd(a,n) = 1$. Then

$$a^{\phi(n)} \equiv 1 \ (\mathrm{mod} \ n). \tag{1.117}$$

Proof. Let $r_1, r_2, \ldots, r_{\phi(n)}$ be a reduced residue system modulo n. Then $ar_1, ar_2, \ldots, ar_{\phi(n)}$ is also a residue system modulo n. Thus we have

$$(ar_1)(ar_2)\cdots(ar_{\phi(n)}) \equiv r_1 r_2 \cdots r_{\phi(n)} \ (\mathrm{mod} \ n),$$

since $ar_1, ar_2, \ldots, ar_{\phi(n)}$, being a reduced residue system, must be congruent in some order to $r_1, r_2, \ldots, r_{\phi(n)}$. Hence,

$$a^{\phi(n)} r_1 r_2 \cdots r_{\phi(n)} \equiv r_1 r_2 \cdots r_{\phi(n)} \pmod{n},$$

which implies that $a^{\phi(n)} \equiv 1 \pmod{n}$.   □

It can be difficult to find the order[1] of an element a modulo n but sometimes it is possible to improve (1.117) by proving that every integer a modulo n must have an order smaller than the number $\phi(n)$ – this order is actually a number that is a factor of $\lambda(n)$.

Theorem 1.6.17 (Carmichael's theorem). Let a and n be positive integers with $\gcd(a, n) = 1$. Then

$$a^{\lambda(n)} \equiv 1 \pmod{n},   \tag{1.118}$$

where $\lambda(n)$ is Carmichael's function, given in Definition 1.5.5.

Proof. Let $n = p_1^{\alpha_1} p_2^{\alpha_2} \cdots p_k^{\alpha_k}$. We shall show that

$$a^{\lambda(n)} \equiv 1 \pmod{p_i^{\alpha_i}}$$

for $1 \le i \le k$, since this implies that $a^{\lambda(n)} \equiv 1 \pmod{n}$. If $p_k^{\alpha_k} = 2, 4$ or a power of an odd prime, then by Definition 1.5.5, $\lambda(\alpha_k) = \phi(\alpha_k)$, so $a^{\lambda(p_i^{\alpha_i})} \equiv 1 \pmod{p_i^{\alpha_i}}$. Since $\lambda(p_i^{\alpha_i}) \mid \lambda(n)$, $a^{\lambda(n)} \equiv 1 \pmod{p_i^{\alpha_i}}$. The case that $p_i^{\alpha_i}$ is a power of 2 greater than 4 is left as an exercise.   □

Note that $\lambda(n)$ will never exceed $\phi(n)$ and is often much smaller than $\phi(n)$; it is the value of the largest order it is possible to have.

Example 1.6.16. Let $a = 11$ and $n = 24$. Then $\phi(24) = 8$, $\lambda(24) = 2$. So,

$$11^{\phi(24)} = 11^8 \equiv 1 \pmod{24},$$

$$11^{\lambda(24)} = 11^2 \equiv 1 \pmod{24}.$$

That is, $\mathrm{ord}_{24}(11) = 2$.

In 1770 Edward Waring (1734–1793) published the following result, which is attributed to John Wilson (1741–1793).

Theorem 1.6.18 (Wilson's theorem). If p is a prime, then

$$(p-1)! \equiv -1 \pmod{p}.   \tag{1.119}$$

---

[1] The order of an element a modulo n is the smallest integer r such that $a^r \equiv 1 \pmod{n}$; we shall discuss this later in Section 1.8.

Proof. It suffices to assume that p is odd. Now to every integer a with $0 < a < p$ there is a unique integer a' with $0 < a' < p$ such that $aa' \equiv 1 \pmod{p}$. Further if $a = a'$ then $a^2 \equiv 1 \pmod{p}$ whence $a = 1$ or $a = p-1$. Thus the set $2, 3, \ldots, p-2$ can be divided into $(p-3)/2$ pairs a, a' with $aa' \equiv 1 \pmod{p}$. Hence we have $2 \cdot 3 \cdots (p-2) \equiv 1 \pmod{p}$, and so $(p-1)! \equiv -1 \pmod{p}$, as required.   □

Theorem 1.6.19 (Converse of Wilson's theorem). If n is an odd positive integer greater than 1 and

$$(n-1)! \equiv -1 \pmod{n}, \tag{1.120}$$

then n is a prime.

Remark 1.6.6. Prime p is called a *Wilson prime* if

$$W(p) \equiv 0 \pmod{p}, \tag{1.121}$$

where

$$W(p) = \frac{(p-1)! + 1}{p}$$

is an integer, or equivalently if

$$(p-1)! \equiv -1 \pmod{p^2}. \tag{1.122}$$

For example, $p = 5, 13, 563$ are Wilson primes, but 599 is not since

$$\frac{(599-1)! + 1}{599} \bmod 599 = 382 \neq 0.$$

It is not known whether there are infinitely many Wilson primes; to date, the only known Wilson primes for $p < 5 \cdot 10^8$ are $p = 5, 13, 563$. A prime p is called a *Wieferich prime*, named after A. Wieferich, if

$$2^{p-1} \equiv 1 \pmod{p^2}. \tag{1.123}$$

To date, the only known Wieferich primes for $p < 4 \cdot 10^{12}$ are $p = 1093$ and 3511.

In what follows, we shall show how to use Euler's theorem to calculate the multiplicative inverse modulo n, and hence the solutions of a linear congruence.

Theorem 1.6.20. Let x be the multiplicative inverse $1/a$ modulo n. If $\gcd(a, n) = 1$, then

$$x \equiv \frac{1}{a} \pmod{n} \tag{1.124}$$

is given by

$$x \equiv a^{\phi(n)-1} \pmod{n}. \tag{1.125}$$

Proof. By Euler's theorem, we have $a^{\phi(n)} \equiv 1 \pmod{n}$. Hence

$$aa^{\phi(n)-1} \equiv 1 \pmod{n},$$

and $a^{\phi(n)-1}$ is the multiplicative inverse of a modulo n, as desired. □

Corollary 1.6.5. Let x be the division b/a modulo n (b/a is assumed to be in lowest terms). If $\gcd(a, n) = 1$, then

$$x \equiv \frac{b}{a} \pmod{n} \tag{1.126}$$

is given by

$$x \equiv b \cdot a^{\phi(n)-1} \pmod{n}. \tag{1.127}$$

Corollary 1.6.6. If $\gcd(a, n) = 1$, then the solution of the linear congruence

$$ax \equiv b \pmod{n} \tag{1.128}$$

is given by

$$x \equiv ba^{\phi(n)-1} \pmod{n}. \tag{1.129}$$

Example 1.6.17. Solve the congruence $5x \equiv 14 \pmod{24}$. First note that because $\gcd(5, 24) = 1$, the congruence has exactly one solution. Using (1.129) we get

$$x \equiv 14 \cdot 5^{\phi(24)-1} \pmod{24} = 22.$$

Example 1.6.18. Solve the congruence $20x \equiv 15 \pmod{135}$. First note that as $d = \gcd(20, 135) = 5$ and $d \mid 15$, the congruence has exactly five solutions modulo 135. To find these five solutions, we divide by 5 and get a new congruence

$$4x' \equiv 3 \pmod{27}.$$

To solve this new congruence, we get

$$x' \equiv 3 \cdot 4^{\phi(27)-1} \equiv 21 \pmod{27}.$$

Therefore, the five solutions are as follows:

$$
\begin{aligned}
(x_0, x_1, x_2, x_3, x_4) &\equiv \left(x', \ x' + \frac{n}{d}, \ x' + \frac{2n}{d}, \ x' + \frac{3n}{d}, \ x' + \frac{4n}{d} \right) \\
&\equiv (21, \ 21 + 27, \ 21 + 2 \cdot 27, \ 21 + 3 \cdot 27, \ 21 + 4 \cdot 27) \\
&\equiv (21, 48, 75, 102, 129) \pmod{135}.
\end{aligned}
$$

Next we shall introduce a method for solving systems of linear congruences. The method, widely known as the Chinese Remainder Theorem (or just CRT, for short), was discovered by the ancient Chinese mathematician Sun Tsu (lived sometime between 200 B.C. and 200 A.D.).

Theorem 1.6.21 (The Chinese Remainder Theorem CRT). If m_1, m_2, $\cdots$, m_n are pairwise relatively prime and greater than 1, and a_1, a_2, $\cdots$, a_n are any integers, then there is a solution x to the following simultaneous congruences:

$$\left.\begin{array}{l} x \equiv a_1 \ (\mathrm{mod}\ m_1), \\[4pt] x \equiv a_2 \ (\mathrm{mod}\ m_2), \\[2pt] \vdots \\[4pt] x \equiv a_n \ (\mathrm{mod}\ m_n). \end{array}\right\} \tag{1.130}$$

If x and x' are two solutions, then $x \equiv x' \ (\mathrm{mod}\ M)$, where $M = m_1 m_2 \cdots m_n$.

Proof. Existence: Let us first solve a special case of the simultaneous congruences (1.130), where i is some fixed subscript,

$$a_i = 1, \ a_1 = a_2 = \cdots = a_{i-1} = a_{i+1} = \cdots = a_n = 0.$$

Let $k_i = m_1 m_2 \cdots m_{i-1} m_{i+1} \cdots m_n$. Then k_i and m_i are relatively prime, so we can find integers r and s such that $rk_i + sm_i = 1$. This gives the congruences:

$$rk_i \equiv 0 \ (\mathrm{mod}\ k_i),$$
$$rk_i \equiv 1 \ (\mathrm{mod}\ m_i).$$

Since $m_1, m_2, \ldots, m_{i-1}, m_{i+1}, \ldots m_n$ all divide k_i, it follows that $x_i = rk_i$ satisfies the simultaneous congruences:

$$x_i \equiv 0 \ (\mathrm{mod}\ m_1),$$
$$x_i \equiv 0 \ (\mathrm{mod}\ m_2),$$
$$\vdots$$
$$x_i \equiv 0 \ (\mathrm{mod}\ m_{i-1}).$$
$$x_i \equiv 1 \ (\mathrm{mod}\ m_i).$$
$$x_i \equiv 0 \ (\mathrm{mod}\ m_{i+1}).$$
$$\vdots$$
$$x_i \equiv 0 \ (\mathrm{mod}\ m_n).$$

For each subscript i, $1 \le i \le n$, we find such an x_i. Now to solve the system of the simultaneous congruences (1.130), set $x = a_1 x_1 + a_2 x_2 + \cdots + a_n x_n$. Then $x \equiv a_i x_i \equiv a_i \ (\mathrm{mod}\ m_i)$ for each i, $1 \le i \le n$, such that x is a solution of the simultaneous congruences.

Uniqueness: Let x' be another solution to the simultaneous congruences (1.130), but different from the solution x, so that $x' \equiv x \ (\mathrm{mod}\ m_i)$ for each x_i. Then $x - x' \equiv 0 \ (\mathrm{mod}\ m_i)$ for each i. So m_i divides $x - x'$ for each i; hence the least common multiple of all the m_j's divides $x - x'$. But since the

m_i are pairwise relatively prime, this least common multiple is the product M. So $x \equiv x' \pmod{M}$. $\qquad\square$

Remark 1.6.7. If the system of the linear congruences (1.130) is soluble, then its solution can be conveniently described as follows:

$$x \equiv \sum_{i=1}^{n} a_i M_i M_i' \pmod{m} \tag{1.131}$$

where

$$m = m_1 m_2 \cdots m_n,$$
$$M_i = m/m_i,$$
$$M_i' = M_i^{-1} \pmod{m_i},$$

for $i = 1, 2, \ldots, n$.

Example 1.6.19. Consider the Sun Zi problem:

$$x \equiv 2 \pmod{3},$$
$$x \equiv 3 \pmod{5},$$
$$x \equiv 2 \pmod{7}.$$

By (1.131), we have

$$m = m_1 m_2 m_3 = 3 \cdot 5 \cdot 7 = 105,$$
$$M_1 = m/m_1 = 105/3 = 35,$$
$$M_1' = M_1^{-1} \pmod{m_1} = 35^{-1} \pmod{3} = 2,$$
$$M_2 = m/m_2 = 105/5 = 21,$$
$$M_2' = M_2^{-1} \pmod{m_2} = 21^{-1} \pmod{5} = 1,$$
$$M_3 = m/m_3 = 105/7 = 15,$$
$$M_3' = M_3^{-1} \pmod{m_3} = 15^{-1} \pmod{7} = 1.$$

Hence,

$$
\begin{aligned}
x &= a_1 M_1 M_1' + a_2 M_2 M_2' + a_3 M_3 M_3' \pmod{m} \\
&= 2 \cdot 35 \cdot 2 + 3 \cdot 21 \cdot 1 + 2 \cdot 15 \cdot 1 \pmod{105} \\
&= 23.
\end{aligned}
$$

Problems for Section 1.6

Problem 1.6.1. Solve the following system of linear congruences:
$$\begin{cases} 2x \equiv 1 \ (\text{mod } 3) \\ 3x \equiv 1 \ (\text{mod } 5) \\ 5x \equiv 1 \ (\text{mod } 7). \end{cases}$$

Problem 1.6.2. Prove that n is prime if $\gcd(a, n) = 1$ and
$$a^{n-1} \equiv 1 \ (\text{mod } n)$$
but
$$a^m \not\equiv 1 \ (\text{mod } n)$$
for each divisor m of $n - 1$.

Problem 1.6.3. Show that the congruence
$$x^{p-1} \equiv 1 \ (\text{mod } p^k)$$
has just $p - 1$ solutions modulo p^k for every prime power p^k.

Problem 1.6.4. Show that for any positive integer n, either there is no primitive root modulo n or there are $\phi(\phi(n))$ primitive roots modulo n. (Note: primitive roots are defined in Definition 1.8.2.)

Problem 1.6.5. Let D be the sum of all the distinct primitive roots modulo a prime p. Show that
$$D \equiv \mu(p - 1) \ (\text{mod } n).$$

Problem 1.6.6. Let n be a positive integer such that $n \equiv 3 \ (\text{mod } 4)$. Show that there are no integer solutions in x for
$$x^2 \equiv -1 \ (\text{mod } n).$$

Problem 1.6.7. Show that for any prime p,
$$\sum_{j=1}^{p-1} \equiv -1 \ (\text{mod } 4).$$

Problem 1.6.8. Suppose $p \equiv 3 \ (\text{mod } 4)$ is prime. Show that
$$\left(\frac{p - 1}{2} \right) \equiv \pm 1 \ (\text{mod } n).$$

Problem 1.6.9. Let p be a prime. Show that for all positive integer $j \leq p-1$, we have
$$\binom{p}{j} \equiv 0 \ (\text{mod } p).$$

Problem 1.6.10. Prove if $\gcd(n_i, n_j) = 1$, $i, j = 1, 2, 3, \ldots, k$, $i \neq j$, then
$$\mathbb{Z}/n\mathbb{Z} \cong \mathbb{Z}/n_1\mathbb{Z} \oplus \mathbb{Z}/n_2\mathbb{Z} \oplus \cdots \oplus \mathbb{Z}/n_k\mathbb{Z}.$$

1.7 Quadratic Congruences

The congruences $ax \equiv b \pmod{m}$ we have studied so far are a special type of congruence; they are all linear congruences. In this section, we shall study the higher degree congruences, particularly the quadratic congruences.

Definition 1.7.1. Let m be a positive integer, and let

$$f(x) = a_0 + a_1 x + a_2 x^2 + \cdots + a_n x^n$$

be any polynomial with integer coefficients. Then a *high-order congruence* or a *polynomial congruence* is a congruence of the form

$$f(x) \equiv 0 \pmod{n}. \qquad (1.132)$$

A polynomial congruence is also called a *polynomial congruential equation*.

Let us consider the polynomial congruence

$$f(x) = x^3 + 5x - 4 \equiv 0 \pmod{7}.$$

This congruence holds when $x = 2$, since

$$f(2) = 2^3 + 5 \cdot 2 - 4 \equiv 0 \pmod{7}.$$

Just as for algebraic equations, we say that $x = 2$ is a root or a solution of the congruence. In fact, any value of x which satisfies the following condition

$$x \equiv 2 \pmod{7}$$

is also a solution of the congruence. In general, as in linear congruence, when a solution x_0 has been found, all values x for which

$$x \equiv x_0 \pmod{n}$$

are also solutions. But by convention, we still consider them as a *single* solution. Thus, our problem is to find all incongruent (different) solutions of $f(x) \equiv 0 \pmod{n}$. In general, this problem is very difficult, and many techniques of solution depend partially on trial-and-error methods. For example, to find all solutions of the congruence $f(x) \equiv 0 \pmod{n}$, we could certainly try all values $0, 1, 2, \ldots, n - 1$ (or the numbers in the complete residue system modulo n), and determine which of them satisfy the congruence; this would give us the total number of *incongruent* solutions modulo n.

Theorem 1.7.1. Let $M = m_1 m_2 \cdots m_n$, where $m_1, m_2, \ldots, m_n$ are pairwise relatively prime. Then the integer x_0 is a solution of

$$f(x) \equiv 0 \pmod{M} \qquad (1.133)$$

if and only if x_0 is a solution of the system of polynomial congruences:

$$\left.\begin{array}{rcl} f(x) & \equiv & 0 \ (\mathrm{mod}\ m_1) \\ f(x) & \equiv & 0 \ (\mathrm{mod}\ m_2) \\ & \vdots & \\ f(x) & \equiv & 0 \ (\mathrm{mod}\ m_n). \end{array}\right\} \qquad (1.134)$$

If x and x' are two solutions, then $x \equiv x' \ (\mathrm{mod}\ M)$, where $M = m_1 m_2 \cdots m_n$.

Proof. If $f(a) \equiv 0 \ (\mathrm{mod}\ M)$, then obviously $f(a) \equiv 0 \ (\mathrm{mod}\ m_i)$, for $i = 1, 2, \ldots, n$. Conversely, suppose a is a solution of the system

$$f(x) \equiv 0 \ (\mathrm{mod}\ m_i), \quad \text{for } i = 1, 2, \ldots, n.$$

Then $f(a)$ is a solution of the system

$$\left.\begin{array}{rcl} y & \equiv & 0 \ (\mathrm{mod}\ m_1) \\ y & \equiv & 0 \ (\mathrm{mod}\ m_2) \\ & \vdots & \\ y & \equiv & 0 \ (\mathrm{mod}\ m_n) \end{array}\right\}$$

and it follows from the Chinese Remainder Theorem that $f(a) \equiv 0 \ (\mathrm{mod}\ m_1 m_2 \cdots m_n)$. Thus, a is a solution of $f(x) \equiv 0 \ (\mathrm{mod}\ M)$. $\qquad\square$

We now restrict ourselves to quadratic congruences, the simplest possible nonlinear polynomial congruences.

Definition 1.7.2. A quadratic congruence is a congruence of the form:

$$x^2 \equiv a \ (\mathrm{mod}\ n) \qquad (1.135)$$

where $\gcd(a, n) = 1$. To solve the congruence is to find an integral solution for x which satisfies the congruence.

In most cases, it is sufficient to study the above congruence rather than the following more general quadratic congruence

$$ax^2 + bx + c \equiv 0 \ (\mathrm{mod}\ n) \qquad (1.136)$$

since if $\gcd(a, n) = 1$ and b is even or n is odd, then the congruence (1.136) can be reduced to a congruence of type (1.135). The problem can even be further reduced to solving a congruence of the type (if $n = p_1^{\alpha_1} p_2^{\alpha_2} \cdots p_k^{\alpha_k}$, where $p_1, p_2, \ldots p_k$ are distinct primes, and $\alpha_1, \alpha_2, \ldots, \alpha_k$ are positive integers):

$$x^2 \equiv a \ (\mathrm{mod}\ p_1^{\alpha_1} p_2^{\alpha_2} \cdots p_k^{\alpha_k}) \qquad (1.137)$$

because solving the congruence (1.137) is equivalent to solving the following system of congruences:

$$\left. \begin{array}{rcl} x^2 & \equiv & a \ (\text{mod } p_1^{\alpha_1}) \\ x^2 & \equiv & a \ (\text{mod } p_2^{\alpha_2}) \\ & \vdots & \\ x^2 & \equiv & a \ (\text{mod } p_k^{\alpha_k}). \end{array} \right\} \tag{1.138}$$

In what follows, we shall be only interested in quadratic congruences of the form

$$x^2 \equiv a \ (\text{mod } p) \tag{1.139}$$

where p is an odd prime and $a \not\equiv 0 \ (\text{mod } p)$.

Definition 1.7.3. Let a be any integer and n a natural number, and suppose that $\gcd(a, n) = 1$. Then a is called a *quadratic residue* modulo n if the congruence

$$x^2 \equiv a \ (\text{mod } n)$$

is soluble. Otherwise, it is called a *quadratic non-residue* modulo n.

Remark 1.7.1. Similarly, we can define the cubic residues, and fourth-power residues, etc. For example, a is a *kth power residue* modulo n if the congruence

$$x^k \equiv a \ (\text{mod } n) \tag{1.140}$$

is soluble. Otherwise, it is a *kth power non-residue* modulo n.

Theorem 1.7.2. Let p be an odd prime and a an integer not divisible by p. Then the congruence

$$x^2 \equiv a \ (\text{mod } p) \tag{1.141}$$

has either no solution or exactly two congruence solutions modulo p.

Proof. If x and y are solutions to $x^2 \equiv a \ (\text{mod } p)$, then $x^2 \equiv y^2 \ (\text{mod } p)$, that is, $p \mid (x^2 - y^2)$. Since $x^2 - y^2 = (x+y)(x-y)$, we must have $p \mid (x - y)$ or $p \mid (x + y)$, that is, $x \equiv \pm y \ (\text{mod } p)$. Hence, any two distinct solutions modulo p differ only be a factor of -1. $\qquad \square$

Example 1.7.1. Find the quadratic residues and quadratic non-residues for moduli $5, 7, 11, 15, 23$, respectively.

(1) Modulo 5, the integers $1, 4$ are quadratic residues, whilst $2, 3$ are quadratic non-residues, since

$$1^2 \equiv 4^2 \equiv 1, \qquad\qquad 2^2 \equiv 3^2 \equiv 4.$$

(2) Modulo 7, the integers $1, 2, 4$ are quadratic residues, whilst $3, 5, 6$ are quadratic non-residues, since

$$1^2 \equiv 6^2 \equiv 1, \qquad 2^2 \equiv 5^2 \equiv 4, \qquad 3^2 \equiv 4^2 \equiv 2.$$

(3) Modulo 11, the integers $1, 3, 4, 5, 9$ are quadratic residues, whilst $2, 6, 7, 8, 10$ are quadratic non-residues, since

$$1^2 \equiv 10^2 \equiv 1, \qquad 2^2 \equiv 9^2 \equiv 4, \qquad 3^2 \equiv 8^2 \equiv 9,$$
$$4^2 \equiv 7^2 \equiv 5, \qquad 5^2 \equiv 6^2 \equiv 3.$$

(4) Modulo 15, only the integers 1 and 4 are quadratic residues, whilst $2, 3, 5, 6, 7, 8, 9, 10, 11, 12, 13, 14$ are all quadratic non-residues, since

$$1^2 \equiv 4^2 \equiv 11^2 \equiv 14^2 \equiv 1, \qquad 2^2 \equiv 7^2 \equiv 8^2 \equiv 13^2 \equiv 4.$$

(5) Modulo 23, the integers $1, 2, 3, 4, 6, 8, 9, 12, 13, 16, 18$ are quadratic residues, whilst $5, 7, 10, 11, 14, 15, 17, 19, 20, 21, 22$ are quadratic non-residues, since

$$1^2 \equiv 22^2 \equiv 1, \qquad 5^2 \equiv 18^2 \equiv 2, \qquad 7^2 \equiv 16^2 \equiv 3,$$
$$2^2 \equiv 21^2 \equiv 4, \qquad 11^2 \equiv 12^2 \equiv 6, \qquad 10^2 \equiv 13^2 \equiv 8,$$
$$3^2 \equiv 20^2 \equiv 9, \qquad 9^2 \equiv 14^2 \equiv 12, \qquad 6^2 \equiv 17^2 \equiv 13,$$
$$4^2 \equiv 19^2 \equiv 16, \qquad 8^2 \equiv 15^2 \equiv 18.$$

The above example illustrates the following two theorems:

Theorem 1.7.3. Let p be an odd prime and $N(p)$ the number of consecutive pairs of quadratic residues modulo p in the interval $[1, p - 1]$. Then

$$N(p) = \frac{1}{4}\left(p - 4 - (-1)^{(p-1)/2}\right). \qquad (1.142)$$

Proof. (Sketch) The complete proof of this theorem can be found in [8] and [64]; here we only give the sketch of the proof. Let (**RR**), (**RN**), (**NR**) and (**NN**) denote the number of pairs of two quadratic residues, of a quadratic residue followed by a quadratic non-residue, of a quadratic non-residue followed by a quadratic residue, of two quadratic non-residues, among pairs of consecutive positive integers less than p, respectively. Then

$$(\mathbf{RR}) + (\mathbf{RN}) = \frac{1}{2}\left(p - 2 - (-1)^{(p-1)/2}\right)$$

$$(\mathbf{NR}) + (\mathbf{NN}) = \frac{1}{2}\left(p - 2 + (-1)^{(p-1)/2}\right)$$

$$(\mathbf{RR}) + (\mathbf{NR}) = \frac{1}{2}(p - 1) - 1$$

$$(\mathbf{RN}) + (\mathbf{NN}) = \frac{1}{2}(p - 1)$$

$$(\mathbf{RR}) + (\mathbf{NN}) - (\mathbf{RN}) - (\mathbf{NR}) = -1$$

$$(\mathbf{RR}) + (\mathbf{NN}) = \frac{1}{2}(p - 3)$$

$$(\mathbf{RR}) - (\mathbf{NN}) = -\frac{1}{2}\left(1 + (-1)^{(p-1)/2}\right)$$

Hence $(\mathbf{RR}) = \frac{1}{4}\left(p - 4 - (-1)^{(p-1)/2}\right)$. $\qquad\qquad\qquad\qquad\square$

Remark 1.7.2. Similarly, let $\nu(p)$ denote the number of consecutive triples of quadratic residues in the interval $[1, p - 1]$, where p is odd prime. Then

$$\nu(p) = \frac{1}{8}p + E_p, \qquad\qquad (1.143)$$

where $|E_p| < \frac{1}{8}\sqrt{p} + 2$.

Example 1.7.2. For $p = 23$, there are five consecutive pairs of quadratic residues, namely, $(1, 2)$, $(2, 3)$, $(3, 4)$, $(8, 9)$ and $(12, 13)$, modulo 23; there is also one consecutive triple of quadratic residues, namely, $(1, 2, 3)$, modulo 23.

Theorem 1.7.4. Let p be an odd prime. Then there are exactly $(p - 1)/2$ quadratic residues and exactly $(p - 1)/2$ quadratic non-residues modulo p.

Proof. Consider the $p - 1$ congruences:

$$x^2 \equiv 1 \pmod{p}$$
$$x^2 \equiv 2 \pmod{p}$$

$$\vdots$$

$$x^2 \equiv p - 1 \pmod{p}.$$

Since each of the above congruences has either no solution or exactly two congruence solutions modulo p, there must be exactly $(p - 1)/2$ quadratic residues modulo p among the integers $1, 2, \ldots, p - 1$. The remaining

$$p - 1 - (p - 1)/2 = (p - 1)/2$$

positive integers less than $p - 1$ are quadratic non-residues modulo p. $\qquad\square$

Example 1.7.3. Again for $p = 23$, there are eleven quadratic residues, and eleven quadratic non-residues modulo 23.

Euler devised a simple criterion for deciding whether an integer a is a quadratic residue modulo a prime number p.

Theorem 1.7.5 (Euler's criterion). Let p be an odd prime and $\gcd(a, p) = 1$. Then a is a quadratic residue modulo p if and only if

$$a^{(p-1)/2} \equiv 1 \pmod{p}.$$

Proof. Using Fermat's little theorem, we find that

$$(a^{(p-1)/2} - 1)(a^{(p-1)/2} + 1) \equiv a^{p-1} - 1 \equiv 0 \pmod{p}$$

and thus $a^{(p-1)/2} \equiv 1 \pmod{p}$. If a is a quadratic residue modulo p, then there exists an integer x_0 such that $x_0^2 \equiv a \pmod{p}$. By Fermat's little theorem, we have

$$a^{(p-1)/2} \equiv (x_0^2)^{(p-1)/2} \equiv x_0^{p-1} \equiv 1 \pmod{p}.$$

To prove the converse, we assume that $a^{(p-1)/2} \equiv 1 \pmod{p}$. If g is a primitive root modulo p (g is a primitive root modulo p if $\mathrm{order}(g, p) = \phi(p)$; we shall formally define primitive roots in Section 1.8), then there exists a positive integer t such that $g^t \equiv a \pmod{p}$. Then

$$g^{t(p-1)/2} \equiv a^{(p-1)/2} \equiv 1 \pmod{p}$$

which implies that

$$t(p-1)/2 \equiv 0 \pmod{p-1}.$$

Thus, t is even, and so

$$(g^{t/2})^2 \equiv g^t \equiv a \pmod{p}$$

which implies that a is a quadratic residue modulo p.   □

Euler's criterion is not very useful as a practical test for deciding whether or not an integer is a quadratic residue, unless the modulus is small. Euler's studies on quadratic residues were further developed by Legendre, who introduced the Legendre symbol.

Definition 1.7.4. Let p be an odd prime and a an integer. Suppose that $\gcd(a, p) = 1$. Then the *Legendre symbol*, $\left(\dfrac{a}{p}\right)$, is defined by

$$\left(\frac{a}{p}\right) = \begin{cases} 1, & \text{if } a \text{ is a quadratic residue modulo } p, \\ -1, & \text{if } a \text{ is a quadratic non-residue modulo } p. \end{cases} \tag{1.144}$$

We shall use the notation $a \in Q_p$ to denote that a is a quadratic residue modulo p; similarly, $a \in \overline{Q}_p$ will be used to denote that a is a quadratic non-residue modulo p.

Example 1.7.4. Let $p = 7$ and

$$1^2 \equiv 1 \pmod{7}, \qquad 2^2 \equiv 4 \pmod{7}, \qquad 3^2 \equiv 2 \pmod{7},$$

$$4^2 \equiv 2 \pmod{7}, \qquad 5^2 \equiv 4 \pmod{7}, \qquad 6^2 \equiv 1 \pmod{7}.$$

Then

$$\left(\frac{1}{7}\right) = \left(\frac{2}{7}\right) = \left(\frac{4}{7}\right) = 1, \qquad \left(\frac{3}{7}\right) = \left(\frac{5}{7}\right) = \left(\frac{6}{7}\right) = -1.$$

Some elementary properties of the Legendre symbol, which can be used to evaluate it, are given in the following theorem.

Theorem 1.7.6. Let p be an odd prime, and a and b integers that are relatively prime to p. Then

(1) If $a \equiv b \pmod{p}$, then $\left(\dfrac{a}{p}\right) = \left(\dfrac{b}{p}\right)$;

(2) $\left(\dfrac{a^2}{p}\right) = 1$, and so $\left(\dfrac{1}{p}\right) = 1$;

(3) $\left(\dfrac{a}{p}\right) \equiv a^{(p-1)/2} \pmod{p}$;

(4) $\left(\dfrac{ab}{p}\right) = \left(\dfrac{a}{p}\right)\left(\dfrac{b}{p}\right)$;

(5) $\left(\dfrac{-1}{p}\right) = (-1)^{(p-1)/2}$.

Proof. Assume p is an odd prime and $\gcd(p, a) = \gcd(p, b) = 1$.

(1) If $a \equiv b \pmod{p}$, then $x^2 \equiv a \pmod{p}$ has solution if and only if $x^2 \equiv b \pmod{p}$ has a solution. Hence $\left(\dfrac{a}{p}\right) = \left(\dfrac{b}{p}\right)$.

(2) The quadratic congruence $x^2 \equiv a^2 \pmod{p}$ clearly has a solution, namely a, so $\left(\dfrac{a^2}{p}\right) = 1$.

(3) This is Euler's criterion in terms of Legendre's symbol.

(4) We have

$$\left(\frac{ab}{p}\right) \equiv (ab)^{(p-1)/2} \pmod{p} \quad \text{(by Euler's criterion)} \quad (1.145)$$

$$\equiv a^{(p-1)/2} b^{(p-1)/2} \pmod{p} \quad\quad\quad\quad\quad (1.146)$$

$$\equiv \left(\frac{a}{p}\right)\left(\frac{b}{p}\right) \quad\quad\quad\quad\quad\quad\quad\quad\quad (1.147)$$

(5) By Euler's criterion, we have

$$\left(\frac{-1}{p}\right) = (-1)^{(p-1)/2}.$$

This completes the proof.   □

Corollary 1.7.1. Let p be an odd prime. Then

$$\left(\frac{-1}{p}\right) = \begin{cases} 1 & \text{if } p \equiv 1 \pmod 4 \\ -1 & \text{if } p \equiv 3 \pmod 4. \end{cases} \tag{1.148}$$

Proof. If $p \equiv 1 \pmod 4$, then $p = 4k + 1$ for some integer k. Thus,

$$(-1)^{(p-1)/2} = (-1)^{((4k+1)-1)/2} = (-1)^{2k} = 1,$$

so that $\left(\frac{-1}{p}\right) = 1$. The proof for $p \equiv 3 \pmod 4$ is similar.   □

Example 1.7.5. Does $x^2 \equiv 63 \pmod{11}$ have a solution? We first evaluate the Legendre symbol $\left(\frac{63}{11}\right)$ corresponding to the quadratic congruence as follows:

$$\begin{aligned}
\left(\frac{63}{11}\right) &= \left(\frac{8}{11}\right) && \text{by (1) of Theorem 1.7.6} \\
&= \left(\frac{2}{11}\right)\left(\frac{2^2}{11}\right) && \text{by (2) of Theorem 1.7.6} \\
&= \left(\frac{2}{11}\right) \cdot 1 && \text{by (2) of Theorem 1.7.6} \\
&= -1 && \text{by "trial and error".}
\end{aligned}$$

Therefore, the quadratic congruence $x^2 \equiv 63 \pmod{11}$ has no solution.

To avoid the "trial-and-error" in the above and similar examples, we introduce in the following the so-called Gauss lemma for evaluating the Legendre symbol.

Definition 1.7.5. Let $a \in \mathbb{Z}$ and $n \in \mathbb{N}$. Then the *least residue* of a modulo n is the integer a' in the interval $(-n/2, n/2]$ such that $a \equiv a' \pmod n$. We denote the least residue of a modulo n by $\mathrm{LR}_n(a)$.

Example 1.7.6. The set $\{-5, -4, -3, -2, -1, 0, 1, 2, 3, 4, 5\}$ is a complete set of of the least residues modulo 11. Thus, $\mathrm{LR}_{11}(21) = -1$ since $21 \equiv 10 \equiv -1 \pmod{11}$; similarly, $\mathrm{LR}_{11}(99) = 0$ and $\mathrm{LR}_{11}(70) = 4$.

Lemma 1.7.1 (Gauss's lemma). Let p be an odd prime number and suppose that $\gcd(a, p) = 1$. Further let ω be the number of integers in the set

$$\left\{ 1a, \ 2a, \ 3a, \ \ldots, \ \left(\frac{p-1}{2}\right)a \right\}$$

whose least residues modulo p are negative, then

$$\left(\frac{a}{p}\right) = (-1)^\omega. \tag{1.149}$$

Proof. When we reduce the following numbers (modulo p)

$$\left\{ a, \ 2a, \ 3a, \ \ldots, \ \left(\frac{p-1}{2}\right)a \right\}$$

to lie in set

$$\left\{ \pm 1, \pm 2, \ldots, \ \pm \left(\frac{p-1}{2}\right) \right\},$$

then no two different numbers ma and na can go to the same numbers. Further, it cannot happen that ma goes to k and na goes to $-k$, because then $ma + na \equiv k + (-k) \equiv 0 \pmod{p}$, and hence (multiplying by the inverse of a), $m + n \equiv 0 \pmod{p}$, which is impossible. Hence, when reducing the numbers

$$\left\{ a, \ 2a, \ 3a, \ \ldots, \ \left(\frac{p-1}{2}\right)a \right\}$$

we get exactly one of -1 and 1, exactly one of -2 and 2, $\cdots$, exactly one of $-(p-1)/2$ and $(p-1)/2$. Hence, modulo p, we get

$$a \cdot 2a \cdots \left(\frac{p-1}{2}\right) a \equiv 1 \cdot 2 \cdots \left(\frac{p-1}{2}\right) (-1)^\omega \pmod{p}.$$

Cancelling the numbers $1, 2, \ldots, (p-1)/2$, we have

$$a^{(p-1)/2} \equiv (-1)^\omega \pmod{p}.$$

By Euler's criterion, we have $\left(\frac{a}{p}\right) \equiv (-1)^\omega \pmod{p}$. Since $\left(\frac{a}{p}\right) \equiv \pm 1$, we must have $\left(\frac{a}{p}\right) = (-1)^\omega$. $\qquad \square$

Example 1.7.7. Use Gauss's lemma to evaluate the Legendre symbol $\left(\frac{6}{11}\right)$. By Gauss's lemma, $\left(\frac{6}{11}\right) = (-1)^\omega$, where ω is the number of integers in the set

$$\{ 1 \cdot 6, \ 2 \cdot 6, \ 3 \cdot 6, \ 4 \cdot 6, \ 5 \cdot 6 \}$$

whose least residues modulo 11 are negative. Clearly,

$$(6, 12, 18, 24, 30) \bmod 11 \equiv (6, 1, 7, 2, 8) \equiv (-5, 1, -4, 2, -3) \pmod{11}$$

So there are 3 least residues that are negative. Thus, $\omega = 3$. Therefore, $\left(\dfrac{6}{11}\right) = (-1)^3 = -1$. Consequently, the quadratic congruence $x^2 \equiv 6 \pmod{11}$ is not solvable.

Remark 1.7.3. Gauss's lemma is similar to Euler's criterion in the following ways:

(1) Gauss's lemma provides a method for direct evaluation of the Legendre symbol;

(2) It has more significance as a theoretical tool than as a computational tool.

Gauss's lemma provides, among many others, a means for deciding whether or not 2 is a quadratic residue modulo an odd prime p.

Theorem 1.7.7. If p is an odd prime, then

$$\left(\frac{2}{p}\right) = (-1)^{(p^2-1)/8} = \begin{cases} 1, & \text{if } p \equiv \pm 1 \pmod{8} \\ -1, & \text{if } p \equiv \pm 3 \pmod{8}. \end{cases} \tag{1.150}$$

Proof. By Gauss's lemma, we know that if ω is the number of least positive residues of the integers

$$1 \cdot 2, \ 2 \cdot 2, \ldots, \ \frac{p-1}{2} \cdot 2$$

that are greater than $p/2$, then $\left(\dfrac{2}{p}\right) = (-1)^{\omega}$. Let $k \in \mathbb{Z}$ with $1 \leq k \leq (p-1)/2$. Then $2k < p/2$ if and only if $k < p/4$; so $[p/4]$ of the integers $1 \cdot 2, \ 2 \cdot 2, \ldots, \frac{p-1}{2} \cdot 2$ are less than $p/2$. So there are $\omega = (p-1)/2 - [p/4]$ integers greater than $p/2$. Therefore, by Gauss's lemma, we have

$$\left(\frac{2}{p}\right) = (-1)^{\frac{p-1}{2} - \left[\frac{p}{4}\right]}.$$

For the first equality, it suffices to show that

$$\frac{p-1}{2} - \left[\frac{p}{4}\right] \equiv \frac{p^2-1}{8} \pmod{2}.$$

If $p \equiv 1 \pmod{8}$, then $p = 8k + 1$ for some $k \in \mathbb{Z}$, from which

$$\frac{p-1}{2} - \left[\frac{p}{4}\right] = \frac{(8k+1)-1}{2} - \left[\frac{8k+1}{4}\right] = 4k - 2k = 2k \equiv 0 \pmod{2},$$

and

$$\frac{p^2 - 1}{8} = \frac{(8k+1)^2 - 1}{8} = \frac{64k^2 + 16k}{8} = 8k^2 + 2k \equiv 0 \ (\text{mod } 2),$$

so the desired congruence holds for $p \equiv 1 \ (\text{mod } 8)$. The cases for $p \equiv -1, \pm 3 \ (\text{mod } 8)$ are similar. This completes the proof for the first equality of the theorem. Note that the cases above yield

$$\frac{p^2 - 1}{8} = \begin{cases} \text{even,} & \text{if } p \equiv \pm 1 \ (\text{mod } 8) \\ \text{odd,} & \text{if } p \equiv \pm 3 \ (\text{mod } 8) \end{cases}$$

which implies

$$(-1)^{(p^2-1)/8} = \begin{cases} 1, & \text{if } p \equiv \pm 1 \ (\text{mod } 8) \\ -1, & \text{if } p \equiv \pm 3 \ (\text{mod } 8) \end{cases}$$

This completes the second equality of the theorem.   □

Example 1.7.8. Evaluate $\left(\dfrac{2}{7}\right)$ and $\left(\dfrac{2}{53}\right)$.

(1) By Theorem 1.7.7, we have $\left(\dfrac{2}{7}\right) = 1$, since $7 \equiv -1 \ (\text{mod } 8)$. Consequently, the quadratic congruence $x^2 \equiv 2 \ (\text{mod } 7)$ is solvable.

(2) By Theorem 1.7.7, we have $\left(\dfrac{2}{53}\right) = -1$, since $53 \equiv -3 \ (\text{mod } 8)$. Consequently, the quadratic congruence $x^2 \equiv 2 \ (\text{mod } 53)$ is not solvable.

Using Lemma 1.7.1, Gauss proved the following theorem, which is one of the great results of mathematics:

Theorem 1.7.8 (Quadratic reciprocity law). If p and q are distinct odd primes, then

(1) $\left(\dfrac{p}{q}\right) = \left(\dfrac{q}{p}\right)$ if one of $p, q \equiv 1 \ (\text{mod } 4)$;

(2) $\left(\dfrac{p}{q}\right) = -\left(\dfrac{q}{p}\right)$ if both $p, q \equiv 3 \ (\text{mod } 4)$.

Remark 1.7.4. This theorem may be stated equivalently in the form

$$\left(\frac{p}{q}\right)\left(\frac{q}{p}\right) = (-1)^{(p-1)(q-1)/4}. \qquad (1.151)$$

Proof. We first observe that, by Gauss's lemma, $\left(\dfrac{p}{q}\right) = 1^\omega$, where ω is the number of lattice points (x, y) (that is, pairs of integers) satisfying

$0 < x < q/2$ and $-q/2 < px - qy < 0$. These inequalities give $y < (px/q) + 1/2 < (p+1)/2$. Hence, since y is an integer, we see ω is the number of lattice points in the rectangle R defined by $0 < x < q/2$, $0 < y < p/2$, satisfying $-q/2 < px - qy < 0$ (see Figure 1.5). Similarly, $\left(\dfrac{q}{p}\right) = 1^\mu$, where μ is the

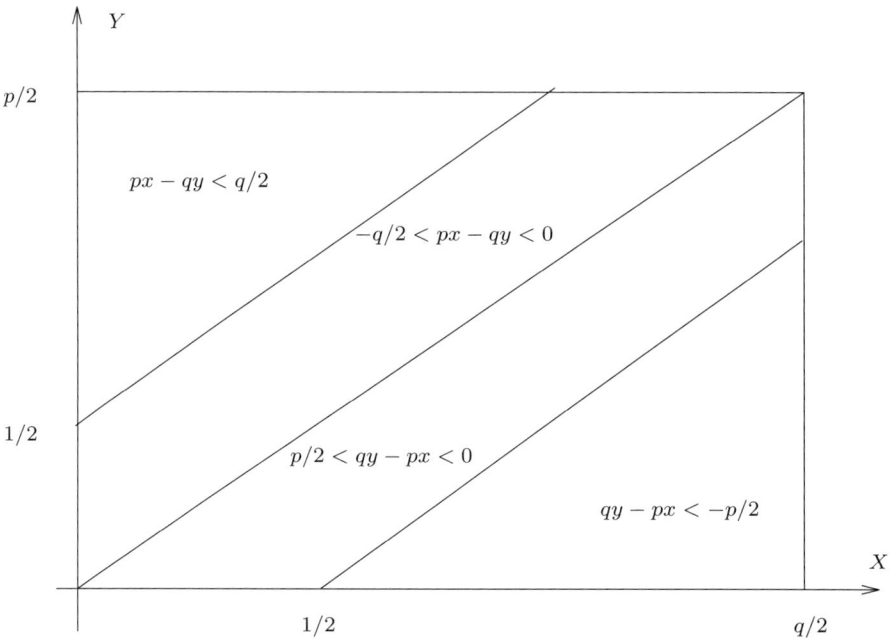

Figure 1.5. Proof of the quadratic reciprocity law

number of lattice points in R satisfying $-p/2 < qx - py < 0$. Now it suffices to prove that $(p-1)(q-1)/4 - (\omega + \mu)$ is even. But $(p-1)(q-1)/4$ is just the number of lattice points in R satisfying that $px - qy \le q/2$ or $qy - px \le -p/2$. The regions in R defined by these inequalities are disjoint and they contain the same number of lattice points, since the substitution

$$
\begin{aligned}
x &= (q+1)/2 - x', \\
y &= (p+1)/2 - y'
\end{aligned}
$$

furnishes a one-to-one correspondence between them. The theorem follows.
□

Remark 1.7.5. The Quadratic Reciprocity Law was one of Gauss's major contributions. For those who consider number theory "the Queen of Mathematics", this is one of the jewels in her crown. Since Gauss's time, over 150 proofs of it have been published; Gauss himself published not less than six different proofs. Among the eminent mathematicians who contributed to the proofs are Cauchy, Jacobi, Dirichlet, Eisenstein, Kronecker and Dedekind.

Combining all the above results for Legendre symbols, we get the following set of formulas for evaluating Legendre symbols:

$$\left(\frac{a}{p}\right) \equiv a^{(p-1)/2} \pmod{p} \tag{1.152}$$

$$\left(\frac{1}{p}\right) = 1 \tag{1.153}$$

$$\left(\frac{-1}{p}\right) = (-1)^{(p-1)/2} \tag{1.154}$$

$$a \equiv b \pmod{p} \implies \left(\frac{a}{p}\right) = \left(\frac{b}{p}\right) \tag{1.155}$$

$$\left(\frac{a_1 a_2 \cdots a_k}{p}\right) = \left(\frac{a_1}{p}\right)\left(\frac{a_2}{p}\right)\cdots\left(\frac{a_k}{p}\right) \tag{1.156}$$

$$\left(\frac{ab^2}{p}\right) = \left(\frac{a}{p}\right), \text{ for } p \nmid b \tag{1.157}$$

$$\left(\frac{2}{p}\right) = (-1)^{(p^2-1)/8} \tag{1.158}$$

$$\left(\frac{p}{q}\right) = (-1)^{(p-1)(q-1)/4}\left(\frac{q}{p}\right) \tag{1.159}$$

Example 1.7.9. Evaluate the Legendre symbol $\left(\dfrac{33}{83}\right)$.

$$\left(\frac{33}{83}\right) = \left(\frac{-50}{83}\right) \qquad\qquad \text{by (1.155)}$$

$$= \left(\frac{-2}{83}\right)\left(\frac{5^2}{83}\right) \qquad\qquad \text{by (1.156)}$$

$$= \left(\frac{-2}{83}\right) \qquad\qquad \text{by (1.157)}$$

$$= -\left(\frac{2}{83}\right) \qquad\qquad \text{by (1.154)}$$

$$= 1 \qquad\qquad \text{by (1.158)}$$

It follows that the quadratic congruence $33 \equiv x^2 \pmod{83}$ is soluble.

Example 1.7.10. Evaluate the Legendre symbol $\left(\dfrac{46}{997}\right)$.

$$\left(\frac{46}{997}\right) = \left(\frac{2}{997}\right)\left(\frac{23}{997}\right) \qquad\qquad \text{by} \quad (1.156)$$

$$= -\left(\frac{23}{997}\right) \qquad\qquad \text{by} \quad (1.158)$$

$$= -\left(\frac{997}{23}\right) \qquad\qquad \text{by} \quad (1.159)$$

$$= -\left(\frac{8}{23}\right) \qquad\qquad \text{by} \quad (1.155)$$

$$= -\left(\frac{2^2 \cdot 2}{23}\right) \qquad\qquad \text{by} \quad (1.156)$$

$$= -\left(\frac{2}{23}\right) \qquad\qquad \text{by} \quad (1.157)$$

$$= -1 \qquad\qquad \text{by} \quad (1.158)$$

It follows that the quadratic congruence $46 \equiv x^2 \pmod{997}$ is not soluble.

Gauss's quadratic reciprocity law enables us to evaluate the values of Legendre symbols $\left(\dfrac{a}{p}\right)$ very quickly provided a is a prime or a product of primes, and p is an odd prime. However, when a is a composite, we must factor it into its prime factorization form in order to use Gauss's quadratic reciprocity law. Unfortunately, there is no efficient algorithm so far for prime factorization (see Chapter 3 for more information). One way to overcome the difficulty of factoring a is to introduce the following Jacobi symbol (in honor of the German mathematician Carl Gustav Jacobi (1804–1851), which is a natural generalization of the Legendre symbol:

Definition 1.7.6. Let a be an integer and $n > 1$ an odd positive integer. If $n = p_1^{\alpha_1} p_2^{\alpha_2} \cdots p_k^{\alpha_k}$, then the *Jacobi symbol*, $\left(\dfrac{a}{n}\right)$, is defined by

$$\left(\frac{a}{n}\right) = \left(\frac{a}{p_1}\right)^{\alpha_1} \left(\frac{a}{p_2}\right)^{\alpha_2} \cdots \left(\frac{a}{p_k}\right)^{\alpha_k}, \qquad (1.160)$$

where $\left(\dfrac{a}{p_i}\right)$ for $i = 1, 2, \ldots, k$ is the Legendre symbol for the odd prime p_i. If n is an odd prime, the Jacobi symbol is *just* the Legendre symbol.

The Jacobi symbol has some similar properties to the Legendre symbol, as shown in the following theorem.

Theorem 1.7.9. Let m and n be any positive odd composites, and $\gcd(a, n) = \gcd(b, n) = 1$. Then

(1) If $a \equiv b \pmod{n}$, then $\left(\dfrac{a}{n}\right) = \left(\dfrac{b}{n}\right)$;

(2) $\left(\dfrac{a}{n}\right)\left(\dfrac{b}{n}\right) = \left(\dfrac{ab}{n}\right)$;

(3) If $\gcd(m, n) = 1$, then $\left(\dfrac{a}{mn}\right)\left(\dfrac{a}{m}\right) = \left(\dfrac{a}{n}\right)$;

(4) $\left(\dfrac{-1}{n}\right) = (-1)^{(n-1)/2}$;

(5) $\left(\dfrac{2}{n}\right) = (-1)^{(n^2-1)/8}$;

(6) If $\gcd(m, n) = 1$, then $\left(\dfrac{m}{n}\right)\left(\dfrac{n}{m}\right) = (-1)^{(m-1)(n-1)/4}$.

Remark 1.7.6. It should be noted that the Jacobi symbol $\left(\dfrac{a}{n}\right) = 1$ does not imply that a is a quadratic residue modulo n. Indeed a is a quadratic residue modulo n if and only if a is a quadratic residue modulo p for each prime divisor p of n. For example, the Jacobi symbol $\left(\dfrac{2}{3599}\right) = 1$, but the quadratic congruence $x^2 \equiv 2 \pmod{3599}$ is actually not soluble. This is the significant difference between the Legendre symbol and the Jacobi symbol. However, $\left(\dfrac{a}{n}\right) = -1$ does imply that a is a quadratic non-residue modulo n. For example, the Jacobi symbol

$$\left(\frac{6}{35}\right) = \left(\frac{6}{5}\right)\left(\frac{6}{7}\right) = \left(\frac{1}{5}\right)\left(\frac{-1}{7}\right) = -1,$$

and so we can conclude that 6 is a quadratic non-residue modulo 35. In short, we have

$$\left.\begin{array}{l} \left(\dfrac{a}{p}\right) = \begin{cases} 1, & a \equiv x^2 \pmod{p} \text{ is soluble} \\ -1, & a \equiv x^2 \pmod{p} \text{ is not soluble} \end{cases} \\[4ex] \left(\dfrac{a}{n}\right) = \begin{cases} 1, & a \equiv x^2 \pmod{n} \text{ may or may not be soluble} \\ -1, & a \equiv x^2 \pmod{n} \text{ is not soluble} \end{cases} \end{array}\right\} \quad (1.161)$$

Combining all the above results for Jacobi symbols, we get the following set of formulas for evaluating Jacobi symbols:

$$\left(\frac{1}{n}\right) = 1 \tag{1.162}$$

$$\left(\frac{-1}{n}\right) = (-1)^{(n-1)/2} \tag{1.163}$$

$$a \equiv b \ (\text{mod } p) \implies \left(\frac{a}{n}\right) = \left(\frac{b}{n}\right) \tag{1.164}$$

$$\left(\frac{a_1 a_2 \cdots a_k}{n}\right) = \left(\frac{a_1}{n}\right)\left(\frac{a_2}{n}\right)\cdots\left(\frac{a_k}{n}\right) \tag{1.165}$$

$$\left(\frac{ab^2}{n}\right) = \left(\frac{a}{n}\right), \ \text{for } \gcd(b,n) = 1 \tag{1.166}$$

$$\left(\frac{2}{n}\right) = (-1)^{(n^2-1)/8} \tag{1.167}$$

$$\left(\frac{m}{n}\right) = (-1)^{(m-1)(m-1)/4}\left(\frac{n}{m}\right) \tag{1.168}$$

Example 1.7.11. Evaluate the Jacobi symbol $\left(\dfrac{286}{563}\right)$.

$$\left(\frac{286}{563}\right) = \left(\frac{2}{563}\right)\left(\frac{143}{563}\right) \qquad \text{by} \quad (1.165)$$

$$= -\left(\frac{143}{563}\right) \qquad \text{by} \quad (1.167)$$

$$= \left(\frac{563}{143}\right) \qquad \text{by} \quad (1.168)$$

$$= \left(\frac{-3^2}{143}\right) \qquad \text{by} \quad (1.154)$$

$$= -\left(\frac{3^2}{143}\right) \qquad \text{by} \quad (1.163)$$

$$= -1 \qquad \text{by} \quad (1.166)$$

It follows that the quadratic congruence $286 \equiv x^2 \ (\text{mod } 563)$ is not soluble.

Example 1.7.12. Evaluate the Jacobi symbol $\left(\dfrac{1009}{2307}\right)$.

$$\left(\frac{1009}{2307}\right) = \left(\frac{2307}{1009}\right) \qquad \text{by} \quad (1.168)$$

$$= \left(\frac{289}{1009}\right) \qquad \text{by} \quad (1.164)$$

$$= \left(\frac{17^2}{1009}\right) \qquad \text{by} \quad (1.165)$$

$$= 1 \qquad \text{by} \quad (1.166)$$

Although the Jacobi symbol $\left(\dfrac{1009}{2307}\right) = 1$, we still cannot determine whether or not the quadratic congruence $1009 \equiv x^2 \pmod{2307}$ is soluble.

Remark 1.7.7. Jacobi symbols can be used to facilitate the calculation of Legendre symbols. In fact, Legendre symbols can be eventually calculated by Jacobi symbols. That is, the Legendre symbol can be calculated as if it were a Jacobi symbol. For example, consider the Legendre symbol $\left(\dfrac{335}{2999}\right)$, where $335 = 5 \cdot 67$ is not a prime (of course, 2999 is prime, otherwise, it is not a Legendre symbol). To evaluate this Legendre symbol, we first regard it as a Jacobi symbol and evaluate it as if it were a Jacobi symbol (note that once it is regarded as a Jacobi symbol, it does not matter whether or not 335 is prime; it even does not matter whether or not 2999 is prime, but anyway, it is a Legendre symbol).

$$\left(\frac{335}{2999}\right) = -\left(\frac{2999}{335}\right) = -\left(\frac{-16}{335}\right) = -\left(\frac{-1 \cdot 4^2}{335}\right) = -\left(\frac{-1}{335}\right) = 1.$$

Since 2999 is prime, $\left(\dfrac{335}{2999}\right)$ is a Legendre symbol, and so 355 is a quadratic residue modulo 2999.

Example 1.7.13. In Table 1.10, we list the elements in $(\mathbb{Z}/21\mathbb{Z})^*$ and their Jacobi symbols. Incidentally, exactly half of the Legendre and Jacobi symbols

$a \in (\mathbb{Z}/21\mathbb{Z})^*$	1	2	4	5	8	10	11	13	16	17	19	20
a^2 mod 21	1	4	16	4	1	16	16	1	4	16	4	1
$\left(\dfrac{a}{3}\right)$	1	-1	1	-1	-1	1	-1	1	1	-1	1	-1
$\left(\dfrac{a}{7}\right)$	1	1	1	-1	1	-1	1	-1	1	-1	-1	-1
$\left(\dfrac{a}{21}\right)$	1	-1	1	1	-1	-1	-1	-1	1	1	-1	1

Table 1.10. Jacobi Symbols for $a \in (\mathbb{Z}/21\mathbb{Z})^*$

$\left(\dfrac{a}{3}\right)$, $\left(\dfrac{a}{7}\right)$ and $\left(\dfrac{a}{21}\right)$ are equal to 1 and half equal to -1. Also for those Jacobi symbols $\left(\dfrac{a}{21}\right) = 1$, exactly half of the a's are indeed quadratic residues, whereas the other half are not. (Note that a is a quadratic residue of 21 if and only if it is a quadratic residue of both 3 and 7.) That is,

$$\left(\frac{a}{3}\right) = \begin{cases} 1, & \text{for } a \in \{1, 4, 10, 13, 16, 19\} = Q_3 \\ -1, & \text{for } a \in \{2, 5, 8, 11, 17, 20\} = \overline{Q}_3 \end{cases}$$

$$\left(\frac{a}{7}\right) = \begin{cases} 1, & \text{for } a \in \{1, 2, 4, 8, 11, 16\} = Q_7 \\ -1, & \text{for } a \in \{5, 10, 13, 17, 19, 20\} = \overline{Q}_7 \end{cases}$$

$$\left(\frac{a}{21}\right) = \begin{cases} 1, & \text{for } a \in \{1, 4, 5, 16, 17, 20\} \begin{cases} a \in \{1, 4, 16\} = Q_{21} \\ a \in \{5, 17, 20\} \subset \overline{Q}_{21} \end{cases} \\ \\ -1, & \text{for } a \in \{2, 8, 10, 11, 13, 19\} \subset \overline{Q}_{21}. \end{cases}$$

Problems for Section 1.7

Problem 1.7.1. Find the x in $2x^2 + 3x + 1 \equiv 0 \pmod 7$ and $2x^2 + 3x + 1 \equiv 0 \pmod{101}$.

Problem 1.7.2. Compute the values for the Legendre symbol:

$$\left(\frac{1234}{4567}\right), \quad \left(\frac{1356}{2467}\right).$$

Problem 1.7.3. Which of the following congruences have solution? If they have, then how many do they have?

$$x^2 \equiv 2 \pmod{61}, \ x^2 \equiv -2 \pmod{61},$$

$$x^2 \equiv 2 \pmod{59}, \ x^2 \equiv -2 \pmod{59},$$

$$x^2 \equiv -1 \pmod{61}, \ x^2 \equiv -1 \pmod{59},$$

$$x^2 \equiv 5 \pmod{229}, \ x^2 \equiv -5 \pmod{229},$$

$$x^2 \equiv 10 \pmod{127}, \ x^2 \equiv 11 \pmod{61}.$$

Problem 1.7.4. Let p be a prime and $\gcd(a, p) = \gcd(b, p) = 1$. Prove that if $x^2 \equiv a \pmod p$, and $x^2 \equiv b \pmod p$ are not soluble, then $x^2 \equiv ab \pmod p$ is soluble.

Problem 1.7.5. Prove that if p is a prime of the form $4k + 1$ then the sum of the quadratic residue modulo p in the interval $[1, p)$ is $p(p - 1)/4$.

Problem 1.7.6. Prove that if r is the quadratic residue modulo $n > 2$, then

$$r^{\phi(n)/2} \equiv 1 \pmod n.$$

Problem 1.7.7. Let p, q be twin primes. Prove that there are infinitely many a such that $p \mid (a^2 - q)$ if and only if there are infinitely many b such that $q \mid (b^2 - p)$.

Problem 1.7.8. Prove that if $\gcd(a, p) = 1$ and p an odd prime, then

$$\sum_{n=1}^{p} \left(\frac{n^2 + a}{p} \right) = -1.$$

Problem 1.7.9. Prove that if $\gcd(a, p) = \gcd(b, p) = 1$ and p an odd prime, then

$$\sum_{n=1}^{p} \left(\frac{an + b}{p} \right) = -1.$$

Problem 1.7.10. Let p be an odd prime, and let $N_{++}(p)$ be the number of n, $1 \leq n < p - 2$ such that

$$\left(\frac{n}{p} \right) = \left(\frac{n+1}{p} \right) = 1.$$

Prove that

$$N_{++}(p) = \left(\frac{p - \left(\frac{-1}{p} \right) - 4}{4} \right).$$

1.8 Primitive Roots and Power Residues

Definition 1.8.1. Let n be a positive integer and a an integer such that $\gcd(a, n) = 1$. Then the *order* of a modulo n, denoted by $\mathrm{ord}_n(a)$ or by $\mathrm{ord}(a, n)$, is the smallest integer r such that $a^r \equiv 1 \pmod{n}$.

Remark 1.8.1. The terminology "the order of a modulo n" is the modern algebraic term from group theory (the theory of groups, rings and fields will be formally introduced in Section 1.2). The older terminology "a belongs to the exponent r" is the classical term from number theory used by Gauss.

Example 1.8.1. In Table 1.11, values of $a^i \bmod 11$ for $i = 1, 2, \ldots, 10$ are given. From Table 1.11, we get

$\mathrm{ord}_{11}(1) = 1,$

$\mathrm{ord}_{11}(2) = \mathrm{ord}_{11}(6) = \mathrm{ord}_{11}(7) = \mathrm{ord}_{11}(8) = 10,$

$\mathrm{ord}_{11}(3) = \mathrm{ord}_{11}(4) = \mathrm{ord}_{11}(5) = \mathrm{ord}_{11}(9) = 5,$

$\mathrm{ord}_{11}(10) = 2.$

a	a^2	a^3	a^4	a^5	a^6	a^7	a^8	a^9	a^{10}
1	1	1	1	1	1	1	1	1	1
2	4	8	5	10	9	7	3	6	1
3	9	5	4	1	3	9	5	4	1
4	5	9	3	1	4	5	9	3	1
5	3	4	9	1	5	3	4	9	1
6	3	7	9	10	5	8	4	2	1
7	5	2	3	10	4	6	9	8	1
8	9	6	4	10	3	2	5	7	1
9	4	3	5	1	9	4	3	5	1
10	1	10	1	10	1	10	1	10	1

Table 1.11. Values of a^i mod 11, for $1 \leq i < 11$

We list in the following theorem some useful properties of the order of an integer a modulo n.

Theorem 1.8.1. Let n be a positive integer, $\gcd(a, n) = 1$, and $r = \mathrm{ord}_n(a)$. Then

(1) If $a^m \equiv 1 \pmod{n}$, where m is a positive integer, then $r \mid m$;

(2) $r \mid \phi(n)$;

(3) For integers s and t, $a^s \equiv a^t \pmod{n}$ if and only if $s \equiv t \pmod{n}$;

(4) No two of the integers $a, a^2, a^3, \ldots, a^r$ are congruent modulo r;

(5) If m is a positive integer, then the order of a^m modulo n is $\dfrac{r}{\gcd(r, m)}$;

(6) The order of a^m modulo n is r if and only if $\gcd(m, r) = 1$.

Definition 1.8.2. Let n be a positive integer and a an integer such that $\gcd(a, n) = 1$. If the order of an integer a modulo n is $\phi(n)$, that is, $\mathrm{order}(a, n) = \phi(n)$, then a is called a *primitive root* of n.

Example 1.8.2. Determine whether or not 7 is a primitive root of 45. First note that $\gcd(7, 45) = 1$. Now observe that

$$7^1 \equiv 7 \pmod{45} \qquad\qquad 7^2 \equiv 4 \pmod{45}$$
$$7^3 \equiv 28 \pmod{45} \qquad\qquad 7^4 \equiv 16 \pmod{45}$$
$$7^5 \equiv 22 \pmod{45} \qquad\qquad 7^6 \equiv 19 \pmod{45}$$
$$7^7 \equiv 43 \pmod{45} \qquad\qquad 7^8 \equiv 31 \pmod{45}$$
$$7^9 \equiv 37 \pmod{45} \qquad\qquad 7^{10} \equiv 34 \pmod{45}$$
$$7^{11} \equiv 13 \pmod{47} \qquad\qquad 7^{12} \equiv 1 \pmod{45}.$$

Thus, $\operatorname{ord}_{45}(7) = 12$. However, $\phi(45) = 24$. That is, $\operatorname{ord}_{45}(7) \neq \phi(45)$. Therefore, 7 is not a primitive root of 45.

Example 1.8.3. Determine whether or not 7 is a primitive root of 46. First note that $\gcd(7, 46) = 1$. Now observe that

$$7^1 \equiv 7 \pmod{46} \qquad\qquad 7^2 \equiv 3 \pmod{46}$$
$$7^3 \equiv 21 \pmod{46} \qquad\qquad 7^4 \equiv 9 \pmod{46}$$
$$7^5 \equiv 17 \pmod{46} \qquad\qquad 7^6 \equiv 27 \pmod{46}$$
$$7^7 \equiv 5 \pmod{46} \qquad\qquad 7^8 \equiv 35 \pmod{46}$$
$$7^9 \equiv 15 \pmod{46} \qquad\qquad 7^{10} \equiv 13 \pmod{46}$$
$$7^{11} \equiv 45 \pmod{46} \qquad\qquad 7^{12} \equiv 39 \pmod{46}$$
$$7^{13} \equiv 43 \pmod{46} \qquad\qquad 7^{14} \equiv 25 \pmod{46}$$
$$7^{15} \equiv 37 \pmod{46} \qquad\qquad 7^{16} \equiv 29 \pmod{46}$$
$$7^{17} \equiv 19 \pmod{46} \qquad\qquad 7^{18} \equiv 41 \pmod{46}$$
$$7^{19} \equiv 11 \pmod{46} \qquad\qquad 7^{20} \equiv 31 \pmod{46}$$
$$7^{21} \equiv 33 \pmod{46} \qquad\qquad 7^{22} \equiv 1 \pmod{46}.$$

Thus, $\operatorname{ord}_{46}(7) = 22$. Note also that $\phi(46) = 22$. That is, $\operatorname{ord}_{46}(7) = \phi(46) = 22$. Therefore 7 is a primitive root of 46.

Theorem 1.8.2 (Primitive roots as residue system). Suppose $\gcd(g, n) = 1$. If g is a primitive root modulo n, then the set of integers $\{g, g^2, g^3, \ldots, g^{\phi(n)}\}$ is a reduced system of residues modulo n.

Example 1.8.4. Let $n = 34$. Then there are $\phi(\phi(34)) = 8$ primitive roots of 34, namely, $3, 5, 7, 11, 23, 27, 29, 31$. Now let $g = 5$ such that $\gcd(g, n) = \gcd(5, 34) = 1$. Then

$$\{g, g^2, \ldots, g^{\phi(n)}\}$$
$$= \{5, 5^2, 5^3, 5^4, 5^5, 5^6, 5^7, 5^8, 5^9, 5^{10}, 5^{11}, 5^{12}, 5^{13}, 5^{14}, 5^{15}, 5^{16}\} \bmod 34$$
$$= \{5, 25, 23, 13, 31, 19, 27, 33, 29, 9, 11, 21, 3, 15, 7, 1\}$$
$$= \{1, 3, 5, 7, 9, 11, 13, 15, 19, 21, 23, 25, 27, 29, 33, 31\}$$

which forms a reduced system of residues modulo 34. We can of course choose $g = 23$ such that $\gcd(g, n) = \gcd(23, 34) = 1$. Then we have

$$\{g, g^2, \ldots, g^{\phi(n)}\}$$
$$= \{23, 23^2, 23^3, 23^4, 23^5, 23^6, 23^7, 23^8, 23^9, 23^{10}, 23^{11}, 23^{12}, 23^{13}, 23^{14},$$
$$\qquad 23^{15}, 23^{16}\} \bmod 34$$
$$= \{23, 19, 29, 21, 7, 25, 31, 33, 11, 15, 5, 13, 27, 9, 3, 1\}$$
$$= \{1, 3, 5, 7, 9, 11, 13, 15, 19, 21, 23, 25, 27, 29, 33, 31\}$$

which again forms a reduced system of residues modulo 34.

Theorem 1.8.3. If p is a prime number, then there exist $\phi(p-1)$ (incongruent) primitive roots modulo p.

Example 1.8.5. Let $p = 47$, then there are $\phi(47-1) = 22$ primitive roots modulo 47, namely,

$$5 \quad 10 \quad 11 \quad 13 \quad 15 \quad 19 \quad 20 \quad 22 \quad 23 \quad 26 \quad 29$$
$$30 \quad 31 \quad 33 \quad 35 \quad 38 \quad 39 \quad 40 \quad 41 \quad 43 \quad 44 \quad 45$$

Note that no method is known for predicting what will be the smallest primitive root of a given prime p, nor is there much known about the distribution of the $\phi(p-1)$ primitive roots among the least residues modulo p.

Corollary 1.8.1. If n has a primitive root, then there are $\phi(\phi(n))$ (incongruent) primitive roots modulo n.

Example 1.8.6. Let $n = 46$, then there are $\phi(\phi(46)) = 10$ primitive roots modulo 46, namely,

$$5 \quad 7 \quad 11 \quad 15 \quad 17 \quad 19 \quad 21 \quad 33 \quad 37 \quad 43$$

Note that not all moduli n have primitive roots; in Table 1.12 we give the smallest primitive root g for $2 \le n \le 911$ that has primitive roots.

The following theorem establishes conditions for moduli to have primitive roots:

Theorem 1.8.4. An integer $n > 1$ has a primitive root modulo n if and only if

$$n = 2, 4, p^\alpha, \text{ or } 2p^\alpha, \tag{1.169}$$

where p is an odd prime and α is a positive integer.

Corollary 1.8.2. If $n = 2^\alpha$ with $\alpha \ge 3$, or $n = 2^\alpha p_1^{\alpha_1} \cdots p_k^{\alpha_k}$ with $\alpha \ge 2$ or $k \ge 2$, then there are no primitive roots modulo n.

Example 1.8.7. For $n = 16 = 2^4$, since it is of the form $n = 2^\alpha$ with $\alpha \ge 3$, there are no primitive roots modulo 16.

Although we know which numbers possess primitive roots, it is not a simple matter to find these roots. Except for trial and error methods, very few general techniques are known. Artin in 1927 made the following conjecture:

Conjecture 1.8.1. Let $N_a(x)$ be the number of primes less than x of which a is a primitive root, and suppose a is not a square and is not equal to -1, 0 or 1. Then

$$N_a(x) \sim A \frac{x}{\ln x}, \tag{1.170}$$

where A depends only on a.

n	g	n	g	n	g	n	g	n	g	n	g	n	g	n	g	n	g	n	g
2	1	3	2	4	3	5	2	6	5	7	3	9	2	10	3	11	2	13	2
14	3	17	3	18	5	19	2	22	7	23	5	25	2	26	7	27	2	29	2
31	3	34	3	37	2	38	3	41	6	43	3	46	5	47	5	49	3	50	3
53	2	54	5	58	3	59	2	61	2	62	3	67	2	71	7	73	5	74	5
79	3	81	2	82	7	83	2	86	3	89	3	94	5	97	5	98	3	101	2
103	5	106	3	107	2	109	6	113	3	118	11	121	2	122	7	125	2	127	3
131	2	134	7	137	3	139	2	142	7	146	5	149	2	151	6	157	5	158	3
162	5	163	2	166	5	167	5	169	2	173	2	178	3	179	2	181	2	193	5
194	5	197	2	199	3	202	3	206	5	211	2	214	5	218	11	223	3	226	3
227	2	229	6	233	3	239	7	241	7	242	7	243	2	250	3	251	6	254	3
257	3	262	17	263	5	269	2	271	6	274	3	277	5	278	3	281	3	283	3
289	3	293	2	298	3	302	7	307	5	311	17	313	10	314	5	317	2	326	3
331	3	334	5	337	10	338	7	343	3	346	3	347	2	349	2	353	3	358	7
359	7	361	2	362	21	367	6	373	2	379	2	382	19	383	5	386	5	389	2
394	3	397	5	398	3	401	3	409	21	419	2	421	2	422	3	431	7	433	5
439	15	443	2	446	3	449	3	454	5	457	13	458	7	461	2	463	3	466	3
467	2	478	7	479	13	482	7	486	5	487	3	491	2	499	7	502	11	503	5
509	2	514	3	521	3	523	2	526	5	529	5	538	3	541	2	542	15	547	2
554	5	557	2	562	3	563	2	566	3	569	3	571	3	577	5	578	3	586	3
587	2	593	3	599	7	601	7	607	3	613	2	614	5	617	3	619	2	622	17
625	2	626	15	631	3	634	3	641	3	643	11	647	5	653	2	659	2	661	2
662	3	673	5	674	15	677	2	683	5	686	3	691	3	694	5	698	7	701	2
706	3	709	2	718	7	719	11	722	3	727	5	729	2	733	6	734	11	739	3
743	5	746	5	751	3	757	2	758	3	761	6	766	5	769	11	773	2	778	3
787	2	794	5	797	2	802	3	809	3	811	3	818	21	821	2	823	3	827	2
829	2	838	11	839	11	841	2	842	23	853	2	857	3	859	2	862	7	863	5
866	5	877	2	878	15	881	3	883	2	886	5	887	5	898	3	907	2	911	17

Table 1.12. Primitive roots g modulo n (if any) for $1 \leq n \leq 911$

Hooley in 1967 showed that if the extended Riemann hypothesis is true then so is Artin's conjecture. It is also interesting to note that before the age of computers Jacobi in 1839 listed all solutions $\{a, b\}$ of the congruences $g^a \equiv b \pmod{p}$ where $1 \leq a < p$, $1 \leq b < p$, g is the least positive primitive root of p and $p < 1000$.

Another very important problem concerning the primitive roots of p is the estimate of the lower bound of the least positive primitive root of p. Let p be a prime and $g(p)$ the least positive primitive root of p. The Chinese mathematician Yuan Wang [253] showed in 1959 that

(1) $g(p) = \mathcal{O}(p^{1/4+\epsilon})$;

(2) $g(p) = \mathcal{O}((\log p)^8)$, if the Generalized Riemann Hypothesis (GRH) is true.

Wang's second result was improved to $g(p) = \mathcal{O}((\log p)^6)$ by Victor Shoup [225] in 1992.

The concept of *index* of an integer modulo n was first introduced by Gauss in his *Disquisitiones Arithmeticae*. Given an integer n, if n has primitive root g, then the set

$$\{g, g^2, g^3, \ldots, g^{\phi(n)}\} \tag{1.171}$$

forms a reduced system of residues modulo n; g is a generator of the cyclic group of the reduced residues modulo n. (Clearly, the group $(\mathbb{Z}/n\mathbb{Z})^*$ is cyclic if $n = 2, 4, p^\alpha$, or $2p^\alpha$, for p odd prime and α positive integer.) Hence, if $\gcd(a, n) = 1$, then a can be expressed in the form:

$$a \equiv g^k \pmod{n} \tag{1.172}$$

for a suitable k with $1 \leq k \leq \phi(n)$. This motivates our following definition, which is an analogue of the real base logarithm function.

Definition 1.8.3. Let g be a primitive root of n. If $\gcd(a, n) = 1$, then the smallest positive integer k such that $a \equiv g^k \pmod{n}$ is called the *index* of a to the base g modulo n and is denoted by $\mathrm{ind}_{g,n}(a)$, or simply by $\mathrm{ind}_g a$.

Clearly, by definition, we have

$$a \equiv g^{\mathrm{ind}_g a} \pmod{n}. \tag{1.173}$$

The function $\mathrm{ind}_g a$ is sometimes called the *discrete logarithm* and is denoted by $\log_g a$ so that

$$a \equiv g^{\log_g a} \pmod{n}. \tag{1.174}$$

Generally, the discrete logarithm is a computationally intractable problem; no efficient algorithm has been found for computing discrete logarithms and hence it has important applications in public-key cryptography.

Theorem 1.8.5 (Index theorem). If g is a primitive root modulo n, then $g^x \equiv g^y \pmod{n}$ if and only if $x \equiv y \pmod{\phi(n)}$.

Proof. Suppose that $x \equiv y \pmod{\phi(n)}$. Then, $x = y + k\phi(n)$ for some integer k. Therefore,

$$
\begin{aligned}
g^x &\equiv g^{y+k\phi(n)} \pmod{n} \\
&\equiv g^y \cdot (g^{\phi(n)})^k \pmod{n} \\
&\equiv g^y \cdot 1^k \pmod{n} \\
&\equiv g^y \pmod{n}.
\end{aligned}
$$

The proof of the "only if" part of the theorem is left as an exercise. $\quad\square$

The properties of the function $\mathrm{ind}_g a$ are very similar to those of the conventional real base logarithm function, as the following theorems indicate:

Theorem 1.8.6. Let g be a primitive root modulo the prime p, and $\gcd(a, p) = 1$. Then $g^k \equiv a \pmod{p}$ if and only if

$$k \equiv \mathrm{ind}_g \, a \pmod{p - 1}. \tag{1.175}$$

Theorem 1.8.7. Let n be a positive integer with primitive root g, and $\gcd(a, n) = \gcd(b, n) = 1$. Then

(1) $\mathrm{ind}_g 1 \equiv 0 \pmod{\phi(n)}$;

(2) $\mathrm{ind}_g(ab) \equiv \mathrm{ind}_g a + \mathrm{ind}_g b \pmod{\phi(n)}$;

(3) $\mathrm{ind}_g a^k \equiv k \cdot \mathrm{ind}_g a \pmod{\phi(n)}$, if k is a positive integer.

Example 1.8.8. Compute the index of 15 base 6 modulo 109, that is, $6^{\mathrm{ind}_6 15} \bmod 109 = 15$. To find the index, we just successively perform the computation $6^k \pmod{109}$ for $k = 1, 2, 3, \ldots$ until we find a suitable k such that $6^k \pmod{109} = 15$:

$$6^1 \equiv 6 \pmod{109} \qquad\qquad 6^2 \equiv 36 \pmod{109}$$
$$6^3 \equiv 107 \pmod{109} \qquad\qquad 6^4 \equiv 97 \pmod{109}$$
$$6^5 \equiv 37 \pmod{109} \qquad\qquad 6^6 \equiv 4 \pmod{109}$$
$$6^7 \equiv 24 \pmod{109} \qquad\qquad 6^8 \equiv 35 \pmod{109}$$
$$6^9 \equiv 101 \pmod{109} \qquad\qquad 6^{10} \equiv 61 \pmod{109}$$
$$6^{11} \equiv 39 \pmod{109} \qquad\qquad 6^{12} \equiv 16 \pmod{109}$$
$$6^{13} \equiv 96 \pmod{109} \qquad\qquad 6^{14} \equiv 31 \pmod{109}$$
$$6^{15} \equiv 77 \pmod{109} \qquad\qquad 6^{16} \equiv 26 \pmod{109}$$
$$6^{17} \equiv 47 \pmod{109} \qquad\qquad 6^{18} \equiv 64 \pmod{109}$$
$$6^{19} \equiv 57 \pmod{109} \qquad\qquad 6^{20} \equiv 15 \pmod{109}.$$

Since $k = 20$ is the smallest positive integer such that $6^{20} \equiv 15 \pmod{109}$, $\mathrm{ind}_6 15 \pmod{109} = 20$.

In what follows, we shall study the congruences of the form $x^k \equiv a \pmod{n}$, where n is an integer with primitive roots and $\gcd(a, n) = 1$. First of all, we present a definition, which is the generalization of quadratic residues.

Definition 1.8.4. Let a, n and k be positive integers with $k \geq 2$. Suppose $\gcd(a, n) = 1$, then a is called a kth (higher) power residue of n if there is an x such that

$$x^k \equiv a \pmod{n}. \tag{1.176}$$

The set of all kth (higher) power residues is denoted by $K(k)_n$. If the congruence has no solution, then a is called a kth (higher) power non-residue of n. The set of such a is denoted by $\overline{K(k)}_n$. For example, $K(9)_{126}$ would denote the set of the 9th power residues of 126, whereas $\overline{K(5)}_{31}$ the set of the 5th power non-residue of 31.

Theorem 1.8.8 (kth power theorem). Let n be a positive integer having a primitive root, and suppose $\gcd(a,n) = 1$. Then the congruence (1.176) has a solution if and only if

$$a^{\phi(n)/\gcd(k,\phi(n))} \equiv 1 \ (\text{mod } n). \tag{1.177}$$

If (1.176) is soluble, then it has exactly $\gcd(k, \phi(n))$ incongruent solutions.

Proof. Let x be a solution of $x^k \equiv a \ (\text{mod } n)$. Since $\gcd(a,n) = 1$, $\gcd(x,n) = 1$. Then

$$
\begin{aligned}
a^{\phi(n)/\gcd(k,\phi(n))} &\equiv (x^k)^{\phi(n)/\gcd(k,\phi(n))} \\
&\equiv (x^{\phi(n)})^{k/\gcd(k,\phi(n))} \\
&\equiv 1^{k/\gcd(k,\phi(n))} \\
&\equiv 1 \ (\text{mod } n).
\end{aligned}
$$

Conversely, if $a^{\phi(n)/\gcd(k,\phi(n))} \equiv 1 \ (\text{mod } n)$, then $r^{(\text{ind}_r a)\phi(n)/\gcd(k,\phi(n))} \equiv 1 \ (\text{mod } n)$. Since $\text{ord}_n r = \phi(n)$, $\phi(n) \mid (\text{ind}_r a)\phi(n)/\gcd(k,\phi(n))$, and hence $\gcd(k,\phi(n)) \mid \text{ind}_r a$ because $(\text{ind}_r a)/\gcd(k,\phi(n))$ must be an integer. Therefore, there are $\gcd(k,\phi(n))$ incongruent solutions to $k(\text{ind}_r x) \equiv (\text{ind}_r a) \ (\text{mod } \phi(n))$ and hence $\gcd(k,\phi(n))$ incongruent solutions to $x^k \equiv a \ (\text{mod } n)$. □

If n is a prime number, say, p, then we have

Corollary 1.8.3. Suppose p is prime and $\gcd(a,p) = 1$. Then a is a kth power residue of p if and only if

$$a^{(p-1)/\gcd(k,(p-1))} \equiv 1 \ (\text{mod } p). \tag{1.178}$$

Example 1.8.9. Determine whether or not 5 is a sixth power of 31, that is, decide whether or not the congruence

$$x^6 \equiv 5 \ (\text{mod } 31)$$

has a solution. First of all, we compute

$$5^{(31-1)/\gcd(6,31-1)} \equiv 25 \not\equiv 1 \ (\text{mod } 31)$$

since 31 is prime. By Corollary 1.8.3, 5 is not a sixth power of 31. That is, $5 \notin K(6)_{31}$. However,

$$5^{(31-1)/\gcd(7,31-1)} \equiv 1 \ (\text{mod } 31).$$

So, 5 is a seventh power of 31. That is, $5 \in K(7)_{31}$.

Now let us introduce a new symbol $\left(\dfrac{a}{p}\right)_k$, the kth power residue symbol, analogous to the Legendre symbol for quadratic residues.

Definition 1.8.5. Let p be an odd prime, $k > 1$, $k \mid p - 1$ and $q = \dfrac{p-1}{k}$. Then the symbol

$$\left(\frac{\alpha}{p}\right)_k = \alpha^q \bmod p \tag{1.179}$$

is called the kth power residue symbol modulo p, where $\alpha^q \bmod p$ represents the absolute smallest residue of α^q modulo p. (The complete set of the absolute smallest residues modulo p are: $-(p-1)/2, \ldots, -1, 0, 1, \ldots, (p-1/2)$).

Theorem 1.8.9. Let $\left(\dfrac{\alpha}{p}\right)_k$ be the kth power residue symbol. Then

(1) $p \mid a \implies \left(\dfrac{a}{p}\right)_k = 0$;

(2) $a \equiv a_1 \pmod{p} \implies \left(\dfrac{a}{p}\right)_k = \left(\dfrac{a_1}{p}\right)_k$;

(3) For $a_1, a_2 \in \mathbb{Z} \implies \left(\dfrac{a_1 a_2}{p}\right)_k \equiv \left(\dfrac{a_1}{p}\right)_k \left(\dfrac{a_2}{p}\right)_k$;

(4) $\operatorname{ind}_g a \equiv b \pmod{k}, 0 \le b < k \implies \left(\dfrac{a}{p}\right)_k \equiv g^{aq} \pmod{p}$;

(5) a is the kth power residue of $p \iff \left(\dfrac{a}{p}\right)_k = 1$;

(6) $n = p_1^{\alpha_1} p_2^{\alpha_2} \cdots p_l^{\alpha_l} \implies \left(\dfrac{n}{p}\right)_k = \left(\dfrac{p_1}{p}\right)_k^{\alpha_1} \left(\dfrac{p_2}{p}\right)_k^{\alpha_2} \cdots \left(\dfrac{p_l}{p}\right)_k^{\alpha_l}$.

Example 1.8.10. Let $p = 19$, $k = 3$ and $q = 6$. Then

$$\left(\frac{-1}{19}\right)_3 = \left(\frac{1}{19}\right)_3 = 1.$$

$$\left(\frac{2}{19}\right)_3 = 7.$$

$$\left(\frac{3}{19}\right)_3 = \left(\frac{-16}{19}\right)_3 \equiv \left(\frac{-1}{19}\right)_3 \left(\frac{16}{19}\right)_3 \equiv \left(\frac{-1}{19}\right)_3 \left(\frac{2}{19}\right)_3^4 = \left(\frac{2}{19}\right)_3 = 7.$$

$$\left(\frac{5}{19}\right)_3 = \left(\frac{24}{19}\right)_3 \equiv \left(\frac{2}{19}\right)_3^3 \left(\frac{3}{19}\right)_3 = \left(\frac{3}{19}\right)_3 = 7.$$

$$\left(\frac{7}{19}\right)_3 = \left(\frac{45}{19}\right)_3 \equiv \left(\frac{3}{19}\right)_3^2 \left(\frac{5}{19}\right)_3 = 7^3 \equiv 1.$$

$$\left(\frac{11}{19}\right)_3 = \left(\frac{30}{19}\right)_3 \equiv \left(\frac{2}{19}\right)_3 \left(\frac{3}{19}\right)_3 \left(\frac{5}{19}\right)_3 = 7^3 \equiv 1.$$

$$\left(\frac{13}{19}\right)_3 = \left(\frac{32}{19}\right)_3 \equiv \left(\frac{2}{19}\right)_3 = -8.$$

$$\left(\frac{17}{19}\right)_3 = \left(\frac{-2}{19}\right)_3 \equiv \left(\frac{-1}{19}\right)_3 \left(\frac{2}{19}\right)_3 = 7.$$

All the above congruences are modular 19.

Problems for Section 1.8

Problem 1.8.1. Find the primitive roots for primes $3, 5, 7, 11, 13, 17, 23$.

Problem 1.8.2. Prove $a^2 \equiv 1 \pmod{p}$ if and only if $a \equiv -1 \pmod{p}$.

Problem 1.8.3. Show that the numbers $1^k, 2^k, 3^k, \ldots, (p-1)^k$ form a reduced residue system modulo p if and only if $\gcd(k, p-1) = 1$.

Problem 1.8.4. Prove that if g and g' are primitive roots modulo an odd prime p, then gg' is not a primitive root modulo p.

Problem 1.8.5. Let g be a primitive root modulo a prime p. Show that

$$(p-1)! \equiv g \cdot g^2 \cdot g^3 \cdots g^{p-1} \equiv g^{p(p-1)/2} \pmod{p}.$$

Use this to prove the Wilson theorem:

$$(p-1)! \equiv -1 \pmod{p}.$$

Problem 1.8.6. Prove that if a and $n > 1$ be any integers such that $a^{n-1} \equiv 1 \pmod{n}$, but $a^d \not\equiv 1 \pmod{n}$ for every proper divisor d of $n-1$, then n is a prime.

Problem 1.8.7. For any positive integer n, prove that the arithmetic progression

$$n+1, 2n+1, 3n+1, \ldots$$

contains infinitely many primes.

Problem 1.8.8. Show that if $n > 1$, then $n \nmid (2^n - 1)$.

Problem 1.8.9. Determine how many solutions each of the following congruence have?

$$x^{12} \equiv 16 \pmod{17}, \ x^{48} \equiv 9 \pmod{17},$$

$$x^{20} \equiv 13 \pmod{17}, \ x^{11} \equiv 9 \pmod{17}.$$

Problem 1.8.10. (Victor Shoup) Let $g(p)$ be the least positive primitive root modulo a prime p. Show that $g(p) = \mathcal{O}((\log p)^6)$.

1.9 Arithmetic of Elliptic Curves

The study of elliptic curves is intimately connected with the the study of Diophantine equations. The theory of Diophantine equations is a branch of number theory which deals with the solution of polynomial equations in either integers or rational numbers. As a solvable polynomial equation always has a corresponding geometrical diagram (e.g., curves or even surfaces). thus to find the integer or rational solution to a polynomial equation is equivalent to find the integer or rational points on the corresponding geometrical diagram, this leads naturally to *Diophantine geometry*, a subject dealing with the integer or rational points on curves or surfaces represented by polynomial equations. For example, in analytic geometry, the linear equation

$$f(x, y) = ax + by + c \qquad (1.180)$$

represents a straight line. The points (x, y) in the plane whose coordinates x and y are integers are called *lattice points*. Solving the linear equation in integers is therefore equivalent to determine those lattice points that lie on the line; The integer points on this line give the solutions to the linear Diophantine equation $ax + by + c = 0$. The general form of the integral solutions for the equation shows that if (x_0, y_0) is a solution, then there are lattice points on the line:

$$x_0, x_0 \pm b, x_0 \pm 2b, \ldots . \qquad (1.181)$$

If the polynomial equation is

$$f(x, y) = x^2 + y^2 - 1 \qquad (1.182)$$

then its associate algebraic curve is the unit circle. The solution (x, y) for which x and y are rational correspond to the Pythagorean triples $x^2 + y^2 = 1$. In general, a polynomial $f(x, y)$ of degree 2

$$ax^2 + bxy + cy^2 + dx + ey + f = 0 \qquad (1.183)$$

gives either an ellipse, a parabola, or a hyperbola, depending on the values of the coefficients. If $f(x, y)$ is a cubic polynomial in (x, y), then the locus of points satisfying $f(x, y) = 0$ is a cubic curve. A general cubic equation in two variables is of the form

$$ax^3 + bx^2y + cxy^2 + dy^3 + ex^2 + fxy + gy^2 + hx + iy + j = 0. \qquad (1.184)$$

Again, we are only interested in the integer solutions of the Diophantine equations, or equivalently, the integer points on the curves of the equations.

The above discussions leads us very naturally to *Diophantine geometry*, a subject dealing with the integer or rational points on algebraic curves or even surfaces of Diophantine equations (a straight line is a special case of algebraic curves).

Definition 1.9.1. A *rational number*, as everybody knows, is a quotient of two integers. A point in the (x, y)-plane is called a *rational point* if both its coordinates are rational numbers. A line is a *rational line* if the equation of the line can be written with rational numbers; that is, the equation is of the form

$$ax + by + c = 0 \tag{1.185}$$

where a, b, c are rational numbers.

Definition 1.9.2. Let

$$ax^2 + bxy + cy^2 + dx + ey + f = 0. \tag{1.186}$$

be a *conic*. Then the conic is rational if we can write its equation with rational numbers.

We have already noted that the point of intersection of two rational lines is rational point. But what about the intersection of a rational line with a rational conic? Will it be true that the points of intersection are rational? In general, they are not. In fact, the two points of intersection are rational if and only if the roots of the quadratic equation are rational. However, if one of the points is rational, then so is the other.

There is a very general method to test, in a finite number of steps, whether or not a given rational conic has a rational point, due to Legendre. The method consists of determining whether a certain congruence can be satisfied.

Theorem 1.9.1 (Legendre). For the Diophantine equation

$$ax^2 + by^2 = cz^2, \tag{1.187}$$

there is an integer n, depending on a, b, c, such that the equation has a solution in integers, not all zero, if and only if the congruence

$$ax^2 + by^2 \equiv cz^2 \ (\mathrm{mod} \ n) \tag{1.188}$$

has a solution in integers relatively prime to n.

An elliptic curve is an algebraic curve given by a *cubic Diophantine equation*

$$y^2 = x^3 + ax + b. \tag{1.189}$$

More general cubics in x and y can be reduced to this form, known as Weierstrass normal form, by rational transformations.

Example 1.9.1. Two examples of elliptic curves are shown in Figure 1.6 (from left to right). The graph on the left is the graph of a *single* equation, namely $E_1: y^2 = x^3 - 4x + 2$; even though it breaks apart into two pieces, we refer to it as a *single* curve. The graph on the right is given by the

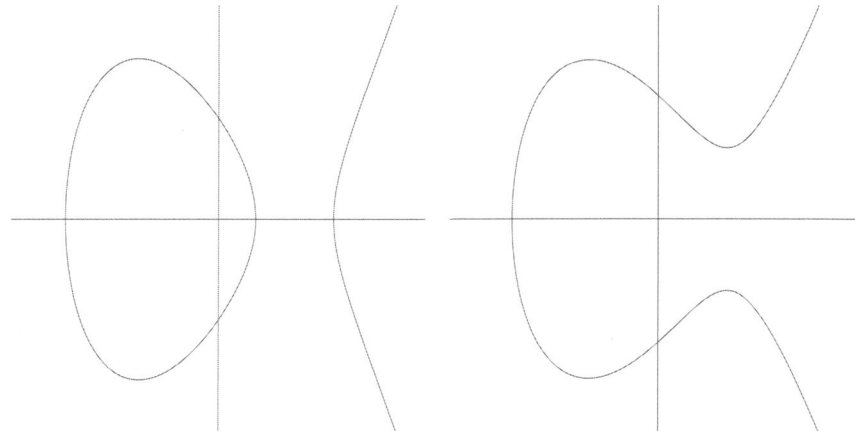

Figure 1.6. Two examples of elliptic curves

equation $E_2 : y^2 = x^3 - 3x + 3$. Note that an elliptic curve is not an *ellipse*; a more accurate name for an elliptic curve, in terms of *algebraic geometry*, is an *Abelian variety of dimension one*. It should also be noted that *quadratic* polynomial equations are fairly well understood by mathematicians today, but cubic equations still pose enough difficulties to be topics of current research.

Definition 1.9.3. An elliptic curve $E : y^2 = x^3 + ax + b$ is called non-singular if its discriminant

$$\Delta(E) = -16(4a^3 + 27b^2) \neq 0. \tag{1.190}$$

Remark 1.9.1. By elliptic curve, we always mean that the cubic curve is non-singular. A cubic curve, such as $y^2 = x^3 - 3x + 2$ for which $\Delta = -16(4(-3)^3 + 27 \cdot 2^2) = 0$, is actually not an elliptic curve; such a cubic curve with $\Delta(E) = 0$ is called a *singular curve*. It can be shown that a cubic curve $E : y^2 = x^3 + ax + b$ is singular if and only if $\Delta(E) = 0$.

Definition 1.9.4. Let $\mathcal{K}$ be a field. Then the *characteristic* of the field $\mathcal{K}$ is 0 if

$$\underbrace{1 \oplus 1 \oplus \cdots \oplus 1}_{n \text{ summands}}$$

is never equal to 0 for any $n > 1$. Otherwise, the *characteristic* of the field $\mathcal{K}$ is the least positive integer n such that

$$\sum_{i=1}^{n} 1 = 0.$$

Example 1.9.2. The fields $\mathbb{Z}$, $\mathbb{Q}$, $\mathbb{R}$ and $\mathbb{C}$ all have characteristic 0, whereas the field $\mathbb{Z}/p\mathbb{Z}$ is of characteristic p, where p is prime.

Definition 1.9.5. Let $\mathcal{K}$ be a field (either the field $\mathbb{Q}$, $\mathbb{R}$, $\mathbb{C}$, or the finite field $\mathbb{F}_q$ with $q = p^\alpha$ elements), and $x^3 + ax + b$ with $a, b \in \mathcal{K}$ be a cubic polynomial. Then

(1) If $\mathcal{K}$ is a field of characteristic $\neq 2, 3$, then an *elliptic curve* over $\mathcal{K}$ is the set of points (x, y) with $x, y \in \mathcal{K}$ that satisfy the following cubic Diophantine equation:

$$E: \quad y^2 = x^3 + ax + b, \tag{1.191}$$

(where the cubic on the right-hand side has no multiple roots) together with a single element, denoted by $\mathcal{O}_E = (\infty, \infty)$, called the *point at infinity*.

(2) If $\mathcal{K}$ is a field of characteristic 2, then an *elliptic curve* over $\mathcal{K}$ is the set of points (x, y) with $x, y \in \mathcal{K}$ that satisfy one of the following cubic Diophantine equations:

$$\left. \begin{array}{ll} E: & y^2 + cy = x^3 + ax + b, \\[4pt] E: & y^2 + xy = x^3 + ax^2 + b, \end{array} \right\} \tag{1.192}$$

(here we do not care whether or not the cubic on the right-hand side has multiple roots) together with a *point at infinity* $\mathcal{O}_E$.

(3) If $\mathcal{K}$ is a field of characteristic 3, then an *elliptic curve* over $\mathcal{K}$ is the set of points (x, y) with $x, y \in \mathcal{K}$ that satisfy the cubic Diophantine equation:

$$E: \quad y^2 = x^3 + ax^2 + bx + c, \tag{1.193}$$

(where the cubic on the right-hand side has no multiple roots) together with a *point at infinity* $\mathcal{O}_E$.

In practice, we are actually more interested in the elliptic curves modulo a positive integer N.

Definition 1.9.6. Let N be a positive integer with $\gcd(n, 6) = 1$. An *elliptic curve* over $\mathbb{Z}/n\mathbb{Z}$ is given by the following cubic Diophantine equation:

$$E: \quad y^2 = x^3 + ax + b, \tag{1.194}$$

where $a, b \in \mathbb{Z}$ and $\gcd(N, \ 4a^3 + 27b^2) = 1$. The set of points on E is the set of solutions in $(\mathbb{Z}/n\mathbb{Z})^2$ to equation (1.194), together with a *point at infinity* $\mathcal{O}_E$.

Remark 1.9.2. The subject of elliptic curves is one of the jewels of 19th century mathematics, originated by Abel, Gauss, Jacobi and Legendre. Contrary to popular opinion, an elliptic curve (i.e., a non-singular cubic curve) is not an ellipse; as Niven, Zuckerman and Montgomery remarked, it is natural to express the arc length of an ellipse as an integral involving the square root of a quadratic polynomial. By making a rational change of variables, this may be reduced to an integral involving the square root of a cubic polynomial. In general, an integral involving the square root of a quadratic or cubic polynomial is called an *elliptic integral*. So, the word *elliptic* actually came from the theory of elliptic integrals of the form

$$\int R(x, y)dx \tag{1.195}$$

where $R(x, y)$ is a rational function in x and y, and y^2 is a polynomial in x of degree 3 or 4 having no repeated roots. Such integrals were intensively studied in the 18th and 19th centuries. It is interesting to note that elliptic integrals serve as a motivation for the theory of elliptic functions, whilst elliptic functions parametrize elliptic curves. It is not our intention here to explain fully the theory of elliptic integrals and elliptic functions; interested readers are recommended to consult some more advanced texts such as Lang [128], McKean and Moll [142], and Silverman [227] for more information.

The geometric interpretation of addition of points on an elliptic curve is quite straightforward. Suppose E is an elliptic curve as shown in Figure 1.7. A straight line L connecting points P and Q intersects the elliptic curve at a third point R, and the point $P \oplus Q$ is the reflection of R in the X-axis.

As can be seen from Figure 1.7, an elliptic curve can have many rational points; any straight line connecting two of them intersects a third. The point at infinity $\mathcal{O}_E$ is the third point of intersection of any two points (not necessarily distinct) of a vertical line with the elliptic curve E. This makes it possible to generate all rational points out of just a few.

The above observations lead naturally to the following geometric composition law of elliptic curves [227].

Proposition 1.9.1 (Geometric composition law (See 1.7)). Let $P, Q \in E$, L the line connecting P and Q (tangent line to E if $P = Q$), and R the third point of intersection of L with E. Let L' be the line connecting R and $\mathcal{O}_E$ (the point at infinity). Then $P \oplus Q$ is the point such that L' intersects E at R, $\mathcal{O}_E$ and $P \oplus Q$.

The points on an elliptic curve form an Abelian group with addition of points as the binary operation on the group.

Theorem 1.9.2 (Group laws on elliptic curves). The geometric composition laws of elliptic curves have the following group-theoretic properties:

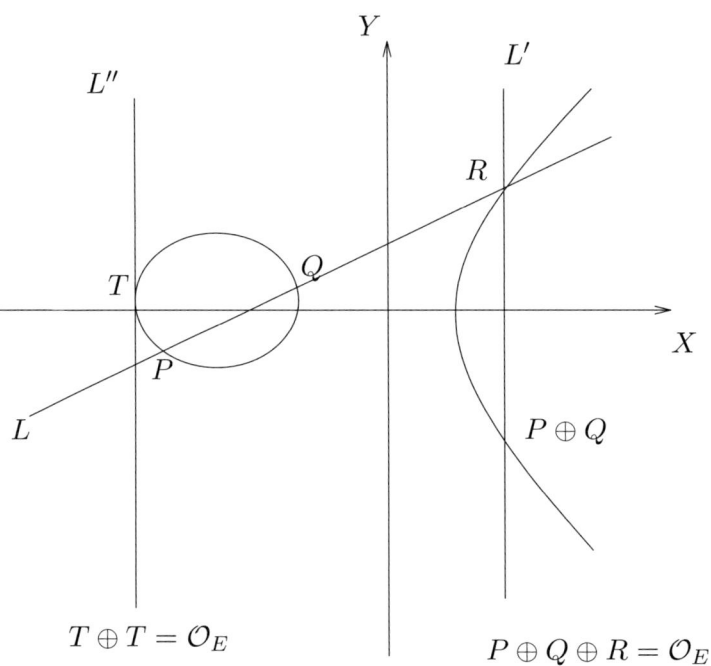

Figure 1.7. Geometric composition laws of an elliptic curve

(1) If a line L intersects E at the (not necessary distinct) points P, Q, R, then
$$(P \oplus Q) \oplus R = \mathcal{O}_E.$$

(2) $P \oplus \mathcal{O}_E = P, \quad \forall P \in E.$

(3) $P \oplus Q = Q \oplus P, \quad \forall P, Q \in E.$

(4) Let $P \in E$, then there is a point of E, denoted $\ominus P$, such that
$$P \oplus (\ominus P) = \mathcal{O}_E.$$

(5) Let $P, Q, R \in E$, then
$$(P \oplus Q) \oplus R = P \oplus (Q \oplus R).$$

In other words, the composition law makes E into an Abelian group with identity element $\mathcal{O}_E$. We further have

(6) Let E be defined over a field $\mathcal{K}$, then

$$E(\mathcal{K}) = \{(x, y) \in \mathcal{K}^2 : y^2 = x^3 + ax + b\} \cup \{\mathcal{O}_E\}.$$

is a subgroup of E.

Example 1.9.3. Let $E(\mathbb{Q})$ be the set of rational points on E. Then $E(\mathbb{Q})$ with the addition operation defined on it forms an Abelian group.

We shall now introduce the important concept of the order of a point on E.

Definition 1.9.7. Let P be an element of the set $E(\mathbb{Q})$. Then P is said to have *order k* if

$$kP = \underbrace{P \oplus P \oplus \cdots \oplus P}_{k \text{ summands}} = \mathcal{O}_E \qquad (1.196)$$

with $k'P \neq \mathcal{O}_E$ for all $1 < k' < k$ (that is, k is the smallest integer such that $kP = \mathcal{O}_E$). If such a k exists, then P is said to have *finite order*, otherwise, it has *infinite order*.

Example 1.9.4. Let $P = (3, 2)$ be a point on the elliptic curve $E : y^2 = x^3 - 2x - 3$ over $\mathbb{Z}/7\mathbb{Z}$ (see Example 1.9.9). Since $10P = \mathcal{O}_E$ and $kP \neq \mathcal{O}_E$ for $k < 10$, P has order 10.

Example 1.9.5. Let $P = (-2, 3)$ be a point on the elliptic curve $E : y^2 = x^3 + 17$ over $\mathbb{Q}$ (see Example 1.9.10). Then P apparently has infinite order.

Now let us move on to the problem as to *how many points (rational or integral) are there on an elliptic curve?* First let us look at an example:

Example 1.9.6. Let E be the elliptic curve $y^2 = x^3 + 3x$ over $\mathbb{F}_5$, then

$$\mathcal{O}_E, \ (0,0), \ (1,2), \ (1,3), \ (2,2), \ (2,3), \ (3,1), \ (3,4), \ (4,1), \ (4,4)$$

are the 10 points on E. However, the elliptic curve $y^2 = 3x^3 + 2x$ over $\mathbb{F}_5$ has only two points:

$$\mathcal{O}_E, \ (0,0).$$

How many points are there on an elliptic curve $E : y^2 = x^3 + ax + b$ over $\mathbb{F}_p$? The following theorem answers this question:

Theorem 1.9.3. Let $|E(\mathbb{F}_p)|$ with p prime be the number of points on $E : y^2 = x^3 + ax + b$ over $\mathbb{F}_p$. Then

$$|E(\mathbb{F}_p)| = 1 + p + \sum_{x \in \mathbb{F}_p} \left(\frac{x^3 + ax + b}{p} \right) = 1 + p + \epsilon \qquad (1.197)$$

points on E over $\mathbb{F}_p$, including the point at infinity $\mathcal{O}_E$, where $\left(\dfrac{x^3 + ax + b}{p}\right)$ is the Legendre symbol.

The quantity ϵ in (1.197) is constrained in the following theorem, due to Hasse (1898–1979) in 1933:

Theorem 1.9.4 (Hasse).
$$|\epsilon| \leq 2\sqrt{p}. \tag{1.198}$$

That is,
$$1 + p - 2\sqrt{p} \leq |E(\mathbb{F}_p)| \leq 1 + p + 2\sqrt{p}. \tag{1.199}$$

Example 1.9.7. Let $p = 5$, then $|\epsilon| \leq 4$. Hence, $1 + 5 - 4 \leq |E(\mathbb{F}_5)| \leq 1 + 5 + 4$, that is, we have between 2 and 10 points on an elliptic curve over $\mathbb{F}_5$. In fact, all the possibilities occur in the elliptic curves given in Table 1.13.

Elliptic curve	Number of points	Elliptic curve	Number of points
$y^2 = x^3 + 2x$	2	$y^2 = x^3 + 4x + 2$	3
$y^2 = x^3 + x$	4	$y^2 = x^3 + 3x + 2$	5
$y^2 = x^3 + 1$	6	$y^2 = x^3 + 2x + 1$	7
$y^2 = x^3 + 4x$	8	$y^2 = x^3 + x + 1$	9
$y^2 = x^3 + 3x$	10		

Table 1.13. Number of points on elliptic curves over $\mathbb{F}_5$

A more general question is: How many rational points are there on an elliptic curve $E : y^2 = x^3 + ax + b$ over $\mathbb{Q}$? Louis Joel Mordell (1888–1972) solved this problem in 1922:

Theorem 1.9.5 (Mordell's finite basis theorem). Suppose that the cubic polynomial $f(x, y)$ has rational coefficients, and that the equation $f(x, y) = 0$ defines an elliptic curve E. Then the group $E(\mathbb{Q})$ of rational points on E is a finitely generated Abelian group.

In elementary language, this says that on any elliptic curve that contains a rational point, there exists a finite collection of rational points such that all other rational points can be generated by using the chord-and-tangent method. From a group-theoretic point of view, Mordell's theorem tells us that we can produce all of the rational points on E by starting from some finite set and using the group laws. It should be noted that for some cubic curves, we have tools to find this generating set, but unfortunately, there is no general method (i.e., algorithm) guaranteed to work for all cubic curves.

The fact that the Abelian group is finitely generated means that it consists of a finite "torsion subgroup" E_{tors}, consisting of the rational points of finite order, plus the subgroup generated by a finite number of points of infinite order:

$$E(\mathbb{Q}) \simeq E_{tors} \oplus \mathbb{Z}^r.$$

The number r of generators needed for the infinite part is called the *rank* of $E(\mathbb{Q})$; it is zero if and only if the entire group of rational points is finite. The study of the rank r and other features of the group of points on an elliptic curve over $\mathbb{Q}$ are related to many interesting problems in number theory and arithmetic algebraic geometry. One of such problems is the Birch and Swinerton-Dyer conjecture (BSD conjecture); we shall make a further study on BSD conjecture in Section 2.2 of Chapter 2.

The most important and fundamental operation on an elliptic curve is the addition of points on the curve. To perform the addition of points on elliptic curves systematically, we need an algebraic formula. The following gives us a convenient computation formula.

Theorem 1.9.6 (Algebraic computation law). Let $P_1 = (x_1, y_1)$, $P_2 = (x_2, y_2)$ be points on the elliptic curve:

$$E: \; y^2 = x^3 + ax + b, \tag{1.200}$$

then $P_3 = (x_3, y_3) = P_1 \oplus P_2$ on E may be computed by

$$P_1 \oplus P_2 = \begin{cases} \mathcal{O}_E, & \text{if } x_1 = x_2 \ \& \ y_1 = -y_2 \\ (x_3, y_3), & \text{otherwise.} \end{cases} \tag{1.201}$$

where

$$(x_3, y_3) = (\lambda^2 - x_1 - x_2, \; \lambda(x_1 - x_3) - y_1) \tag{1.202}$$

and

$$\lambda = \begin{cases} \dfrac{(3x_1^2 + a)}{2y_1}, & \text{if } P_1 = P_2, \\[2mm] \dfrac{(y_2 - y_1)}{(x_2 - x_1)}, & \text{otherwise.} \end{cases} \tag{1.203}$$

Example 1.9.8. Let E be the elliptic curve $y^2 = x^3 + 17$ over $\mathbb{Q}$, and let $P_1 = (x_1, y_1) = (-2, 3)$ and $P_2 = (x_2, y_2) = (1/4, \ 33/8)$ be two points on E. To find the third point P_3 on E, we perform the following computation:

$$\lambda = \frac{y_2 - y_1}{x_2 - x_1} = \frac{1}{2},$$
$$x_3 = \lambda^2 - x_1 - x_2 = 2,$$
$$y_3 = \lambda(x_1 - x_3) - y_1 = -5.$$

So $P_3 = P_1 \oplus P_2 = (x_3, \ y_3) = (2, -5)$.

Example 1.9.9. Let $P = (3,2)$ be a point on the elliptic curve $E : y^2 = x^3 - 2x - 3$ over $\mathbb{Z}/7\mathbb{Z}$. Compute

$$10P = \underbrace{P \oplus P \oplus \cdots \oplus P}_{10 \text{ summands}} \quad (\bmod\ 7).$$

According to (1.202), we have:

$$2P = P \oplus P = (3,2) \oplus (3,2) = (2,6),$$

$$3P = P \oplus 2P = (3,2) \oplus (2,6) = (4,2),$$

$$4P = P \oplus 3P = (3,2) \oplus (4,2) = (0,5),$$

$$5P = P \oplus 4P = (3,2) \oplus (0,5) = (5,0),$$

$$6P = P \oplus 5P = (3,2) \oplus (5,0) = (0,2),$$

$$7P = P \oplus 6P = (3,2) \oplus (0,2) = (4,5),$$

$$8P = P \oplus 7P = (3,2) \oplus (4,5) = (2,1),$$

$$9P = P \oplus 8P = (3,2) \oplus (2,1) = (3,5),$$

$$10P = P \oplus 9P = (3,2) \oplus (3,5) = \mathcal{O}_E.$$

Example 1.9.10. Let $E : y^2 = x^3 + 17$ be the elliptic curve over $\mathbb{Q}$ and $P = (-2,3)$ a point on E. Then

$P = (-2,3),$

$2P = (8, -23),$

$3P = \left(\frac{19}{25}, \frac{522}{125}\right),$

$4P = \left(\frac{752}{529}, \frac{-54239}{12167}\right),$

$5P = \left(\frac{174598}{32761}, \frac{76943337}{5929741}\right),$

$6P = \left(\frac{-4471631}{3027600}, \frac{-19554357097}{5268024000}\right),$

$7P = \left(\frac{12870778678}{76545001}, \frac{1460185427995887}{669692213749}\right),$

$8P = \left(\frac{-3705032916448}{1556248765009}, \frac{3635193007425360001}{1941415665602432473}\right),$

$9P = \left(\frac{1508016107720305}{1146705139411225}, \frac{-185877155243117444 0537502}{38830916270562191567875}\right),$

$10P = \left(\frac{2621479238320017368}{21550466484219504001}, \frac{41250808450252350540981 3257257}{10004260991388455752541474 3999}\right),$

$11P = \left(\frac{98386489129108787338247 8}{4557708224535761198560 81}, \frac{-16005818393035651701390378886 10254293}{30769453204705350935032590551 7943271}\right),$

$12P = \left(\frac{17277017794597335695799625921}{46306885438389913760299536 00}, \frac{2616325792251321558429704062367454 696426719}{31511447812142672670439205364233763321 6000}\right).$

Suppose now we are interested in measuring the *size* (or the *height of point on elliptic curve*) of points on an elliptic curve E. One way to do this is to look at the numerator and denominator of the x-coordinates. If we write the coordinates of kP as

$$kP = \left(\frac{A_k}{B_k}, \frac{C_k}{D_k} \right), \tag{1.204}$$

we may define the height of these points as follows

$$H(kP) = \max(|A_k|, |B_k|). \tag{1.205}$$

It is interesting to note that for large k, the height of kP looks like:

$$D(H(kP)) \approx 0.1974k^2, \tag{1.206}$$

$$H(kP) \approx 10^{0.1974k^2}$$

$$\approx (1.574)^{k^2} \tag{1.207}$$

where $D(H(kP))$ denotes the number of digits in $H(kP)$.

Remark 1.9.3. To provide greater flexibility, we may also consider the following form of elliptic curves:

$$E : y^2 = x^3 + ax^2 + bx + c. \tag{1.208}$$

In order for E to be an elliptic curve, it is necessary and sufficient that

$$\Delta(E) = a^2b^2 - 4a^3c - 4b^3 + 18abc - 27c^2 \neq 0. \tag{1.209}$$

Thus,

$$P_3(x_3, y_3) = P_1(x_1, y_1) \oplus P_2(x_2, y_2),$$

on E may be computed by

$$(x_3, y_3) = (\lambda^2 - a - x_1 - x_2, \ \lambda(x_1 - x_3) - y_1) \tag{1.210}$$

where

$$\lambda = \begin{cases} (3x_1^2 + 2a + b)/2y_1, & \text{if } P_1 = P_2 \\ (y_2 - y_1)/(x_2 - x_1), & \text{otherwise.} \end{cases} \tag{1.211}$$

Problems for Section 1.9

Problem 1.9.1. Describe an algorithm to find a point on an elliptic curve $E : y^2 = x^3 + ax + b$ over $\mathbb{Q}$. Use your algorithm to find a point on the $E : y^2 = x^3 - 13x + 21$ over $\mathbb{Q}$.

Problem 1.9.2. Find all the rational points on the elliptic curve $y^2 = x^3 - x$.

Problem 1.9.3. Find all the rational points on the elliptic curve $y^2 = x^3 + 4x$.

Problem 1.9.4. Describe an algorithm to find the order of a point on an elliptic curve $E : y^2 = x^3 + ax + b$ over $\mathbb{Q}$. Let $P = (2, 4)$ be a point on $E : y^2 = x^3 - 13x + 21$ over $\mathbb{Q}$. Use your algorithm to find the order of the point on E.

Problem 1.9.5. Find all the torsion points of the elliptic curve $E : y^2 = x^3 - 13x + 21$ over $\mathbb{Q}$.

Problem 1.9.6. Find the point of infinite order on the elliptic curve $E : y^2 = x^3 - 2x$ over $\mathbb{Q}$.

Problem 1.9.7. Determine the number of points of the elliptic curve $E : y^2 = x^3 - 1$ for all odd primes up to 100.

Problem 1.9.8. Let $P = (0, 0)$ be a point on the elliptic curve $E : y^2 = x^3 + x^2 + 2x$. Compute $100P$ and $200P$.

Problem 1.9.9. Derive addition formula for rational points on the elliptic curve
$$E : y^2 = x^3 + ax^2 + bx + c.$$

Problem 1.9.10. Show that $P = (9/4, 29/8)$ is a point on the elliptic curve $E : y^2 = x^3 - x + 4$.

Problem 1.9.11. Let $P = (1, 1)$ be a point on the elliptic curve $E : y^2 = x^3 - 6x + 6$ over $\mathbb{Z}_{4247}$. Compute $100P$ on $E(\mathbb{Z}_{4727})$.

Problem 1.9.12. Let n be a positive integers greater than 1, and P a point on an elliptic curve $E(\mathbb{Z}_{4727})$. Prove that there are some integers s and t such that $sP = tP$.

1.10 Chapter Notes and Further Reading

This chapter is mainly concerned with the elementary theory of numbers, including Euclid's algorithm, continued fractions, the Chinese remainder theorem, Diophantine equations, arithmetic functions, quadratic and power residues, and the arithmetic of elliptic curves. It also includes some algebraic topics such as groups, rings, fields, polynomials, and algebraic numbers. The theory of numbers is one of the most beautiful and pure parts of mathematics, and there are many books in this field. For a quick, concise and friendly introduction to the subject, see Backer [13], Davenport [64], and Silverman [228]. Kato, et al. [117] is also to be recommended. For a comprehensive treatment of the subject, see the most authoritative text by Hardy and Wright [97], as well as Cohen [49], Hua [105], Ireland and Rosen [109], Niven, Zuckerman and Montgomery [165], Rose [204], Rosen [206], and Stillwell [241].

Elliptic curves are used throughout the book for primality testing in Chapter 2, integer factorization in Chapter 3 and cryptography in Chapter 4. In Section 1.9 of this chapter we have only introduced the necessary background information about elliptic curves; serious readers who wish to know more about elliptic curves are recommended to consult e.g., Husemöller [108], Silverman and Tate [226], Silverman [227] and Washington [254]. Number theory and algebra are two sister subjects, and in fact many of the concepts and results in number theory can be described in the language of modern algebra, particularly groups, rings and fields. We have introduced the basic concepts of groups, rings, fields polynomials and algebraic numbers in Section 1.2 of this chapter; readers who want to know more information about abstract algebra are recommended to consult e.g., Herstein ([99] and [100]), and Rotman ([210] and [209]).

More references in number theory and abstract algebra/algebraic geometry can be found in the Bibliography section at the end of the book, see, e.g., Anderson and Bell [7], Andrews [8], Apostol [9], Childs [47], Dickson [69], Dirichlet [72], Euclid [75], Gauss [81], Hardy [96], Jackson [110], Koblitz ([121], [122] and [123]), Mollin ([151] and [152]), Motwani and Raghavan [161], Nathanson [163], Ore [169], Redmond [192], Ribenboim ([193] and [195]), and Yan [268].

2. Primality Testing and Prime Generation

It would be interesting to know, for example, what the situation is with the determination if a number is a prime, and in general how much we can reduce the number of steps from the method of simply trying for finite combinatorial problems.

KURT GÖDEL (1906–1978)

2.1 Computing with Numbers and Curves

Computational number theory is a new branch of number theory. One of the main goals of computational number theory is to use theories and techniques from computing science to solve problems in various branches of number theory such as elementary number theory, algebraic number theory, analytic number theory, combinatorial number theory, probabilistic number theory, arithmetic algebraic geometry, to name just a few; see Figure 2.1 for relationships among number theory, computational number theory and computing science. The main driving force for computational number theory is, however, cryptography, particularly public-key cryptography, as many of the cryptographic schemes and protocols rely their security on the intractability of some computational number-theoretic problems. For example, the most famous and widely used public-key cryptosystem RSA is based on the simple number-theoretic fact that it should be easy to find large prime number but it should be very difficult to factor a large composite number. By *easy* we mean that it can be done in (deterministic) polynomial time (denoted by $\mathcal{P}$), and by *difficult* we mean that it cannot be done in polynomial time; for example, it can only be done in non-deterministic polynomial-time ($\mathcal{NP}$) or in exponential-time ($\mathcal{EXP}$). More precisely, we define:

(1) An algorithm/problem is of polynomial-time complexity if its required running time is bounded by

$$\mathcal{O}((\log n)^k) \qquad (2.1)$$

where $\log n$ is the length of the input n and k a constant.

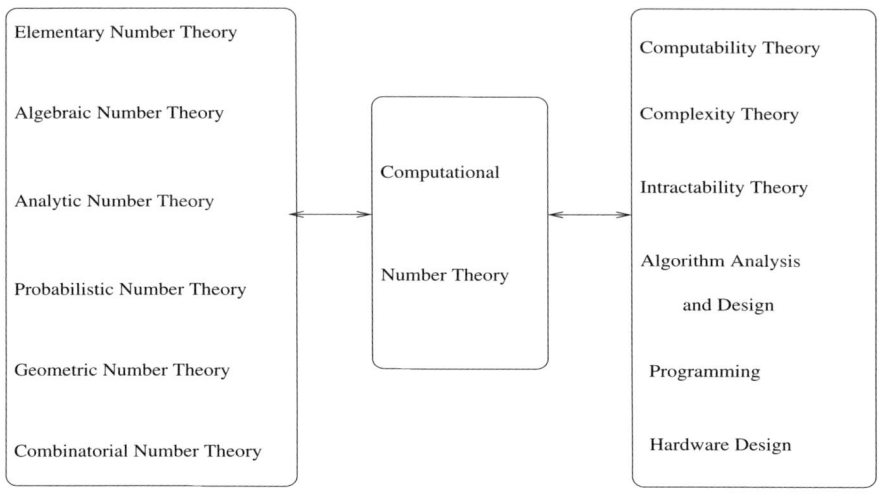

Figure 2.1. Computational number theory

(2) An algorithm/program is of exponential-time complexity if its required running time is bounded by

$$\mathcal{O}(n^{\epsilon}) \tag{2.2}$$

where $\epsilon < 1$ is a small positive real number[1].

(3) An algorithm/problem is of superpolynomial-time complexity if its running time is bounded by

$$\mathcal{O}((\log n)^{c \log n}). \tag{2.3}$$

Note that we normally do not regard superpolynomial as polynomial.

(4) An algorithm/problem is of subexponential-time complexity if its running time is bounded by

$$\mathcal{O}(\exp(c(\log n)^{\alpha}(\log \log n)^{1-\alpha})). \tag{2.4}$$

The relationship among the four different complexity classes can be described as follows:

$$\text{Polynomial} \subset \text{Superpolynomial} \subset \text{Subexponential} \subset \text{Exponential}. \tag{2.5}$$

The polynomial-time mentioned above is specifically for *deterministic* polynomial-time, denoted by $\mathcal{P}$. There are also many other type of polynomial-times, listed below are just some of them.

---

[1] Note that $\mathcal{O}((\log n)^{12})$ can be regarded as polynomial complexity, but $\mathcal{O}(n^{0.1})$ can not, it is in fact exponential complexity, because $\mathcal{O}(n^{0.1}) = \mathcal{O}(2^{0.1 \log n})$.

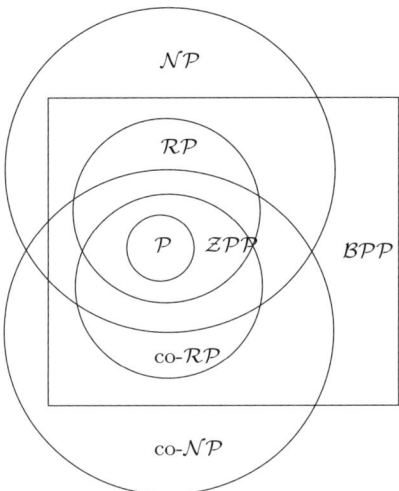

Figure 2.2. Common complexity classes

(1) Randomized Polynomial $\mathcal{RP}$: Problems can be solved in *expected* polynomial time with *one-sided* probabilistic error.

(2) Randomized Polynomial $\mathcal{BPP}$: Problems can be solved in *expected* polynomial time with *two-sided* probabilistic error.

(3) Zero-error Probabilistic Polynomial $\mathcal{ZPP}$: Problems can be solved in *expected* polynomial time with *zero* probabilistic error.

(4) Non-deterministic Polynomial $\mathcal{NP}$: Problems can be solved in *non-deterministic* polynomial time.

It is generally believed that

$$\mathcal{P} \subseteq \mathcal{ZPP} \subseteq \mathcal{RP} \subseteq \left(\begin{array}{c} \mathcal{BPP} \\ \mathcal{NP} \end{array} \right) \subseteq \mathcal{EXP}.$$

It is not known whether any of the inclusions in the above hierarchy is proper, and more importantly, we do not know if $\mathcal{P} = \mathcal{NP}$. The problem whether or not $\mathcal{P} = \mathcal{NP}$ is one of the most important unsolved problems in both mathematics and computer science, and also one of the seven Millennium Prize Problems[2].

---

[2] On 24 May 2000, the Clay Mathematics Institute of Cambridge, Massachusetts announced seven Millennium Prize Problems (Birch and Swinnerton-Dyer Conjecture, Hodge Conjecture, Navier-Stokes Equations, P vs NP Problem, Poincare Conjecture, Riemann Hypothesis, Yang Mills Theory), each with $1 million for a solution. It is interesting to note that computational number theory has a direct connection to three of the seven problems, namely, the Birch and Swinnerton-Dyer Conjecture for elliptic curves, P vs NP Problem, and the Riemann Hypothesis.

In practice, only those problems in $\mathcal{P}$ are tractable; anything beyond $\mathcal{P}$ is intractable although they are still computable. Thus, from a computational point of view, we have two main problems: computable problems and uncomputable problems. Within computable problems, there are also two types of problems: tractable problems and intractable problems.

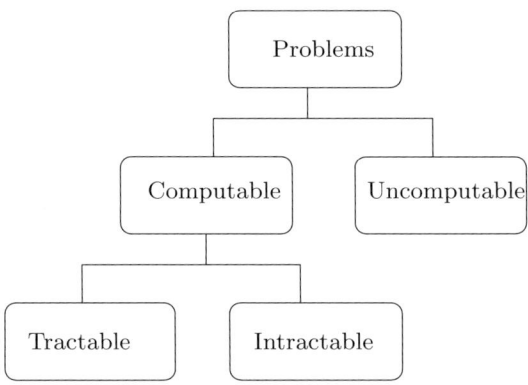

Figure 2.3. Tractable/intractable problems

(1) The primality testing problem (PTP): Given a positive integer greater than 1, determine whether or not it is prime. This problem can be solved by the AKS algorithm [5] deterministically and unconditionally in $\mathcal{O}(\log n)^{12}$ (improvements are possible). That is, PTP is tractable.

(2) The integer factorization problem (IFP): Given a positive integer $n > 1$, find a factor $1 < a < n$ of n such that $n = ab$ for $1 < b < n$. We shall be mainly concerned with the most difficult case: given a large positive composite integer $n > 1$, typically with more than 200 digits, find p and q:

$$n = p \cdot q,$$

where p and q are primes, each with about 100 digits. This is the problem that the most famous RSA cryptosystem is based on. The fastest factoring algorithm, Number Field Sieve, takes time

$$\mathcal{O}(\exp((64/9)^{1/3}(\log n)^{1/3}(\log \log n)^{2/3}))$$

to factor a general integer and

$$\mathcal{O}(\exp((32/9)^{1/3}(\log n)^{1/3}(\log \log n)^{2/3}))$$

for a special integer.

(3) The discrete logarithm problem (DLP) over the finite field $\mathbb{F}_p$ (we can also say for the multiplicative group $\mathbb{F}_p^*$): Let α and β be nonzero elements of a finite field $\mathbb{F}_p$. Then the DLP over $\mathbb{F}_p$ is to determine, from the data (p, α, β), the integer k, such that

$$\alpha = \beta^k \in \mathbb{F}_p, \quad \text{or} \quad \alpha \equiv \beta^k \pmod{p}. \tag{2.6}$$

We write

$$k = \log_\beta \alpha \in \mathbb{F}_p, \quad \text{or} \quad k \equiv \log_\beta \alpha \pmod{p}. \tag{2.7}$$

The DLP is another difficult problem, at least as difficult as the IFP, which the US government's digital signature algorithm/standard (DSA/DSS) is based on.

(4) The elliptic curve discrete logarithm problem (ECDLP): Let E be an elliptic curve over the finite field $\mathbb{F}_p$, say, given by a Weierstrass equation

$$E : \ y^2 \equiv x^3 + ax + b \pmod{p}, \tag{2.8}$$

S and T the two points in the elliptic curve group $E(\mathbb{F}_p)$. Then the ECDLP is to find the integer k (assuming that such an integer k exists)

$$k = \log_T S \in \mathbb{Z}, \quad \text{or} \quad k \equiv \log_T S \pmod{p} \tag{2.9}$$

such that

$$S = kT \in E(\mathbb{F}_p), \quad \text{or} \quad S \equiv kT \pmod{p}. \tag{2.10}$$

The ECDLP is an even more difficult problem than the DLP, for which the elliptic curve digital signature algorithm (ECDSA) is based on. Clearly, the ECDLP is the generalization of DLP, which extends the the multiplicative group $\mathbb{F}_p^*$ to the elliptic curve group $E(\mathbb{F}_p)$.

No polynomial-time algorithm[3] is known for IFP, DLP or the ECDLP, nor can anyone prove that no such algorithm exists. However, we do have subexponential-time algorithms, namely, the index calculus[4] for IFP and DLP. Unfortunately, no subexponential-time algorithm is known for the ECDLP. At present, the only known algorithm, Xedni calculus (see [229] and [269]), is of exponential-time complexity.

The following two number theoretic operations are at the heart of efficient primality testing and integer factorization, and of much of modern cryptography:

---

[3] We do have quantum polynomial-time algorithms for IFP and DLP, although we still do not have a quantum computer; a practically useful quantum computer can be very difficult to build.

[4] Index calculus belongs to a wide range of algorithms, including CFRAC (Continued FRACtion method), QS (Quadratic Sieve) and NFS (Number Field Sieve) for IFP; some types or variants of index calculus are also applicable to solve DLP.

(1) modular exponentiation: find y such that $y \equiv a^k \pmod{n}$, where $a, k, n \in \mathbb{Z}$.

(2) elliptic curve point addition: Find Q such that $Q \equiv kP \pmod{n}$, where $k, n \in \mathbb{Z}$ and P, Q are the points on ellitpic curve $y^2 \equiv x^3 + ax + b \pmod{n}$.

In what follows we shall introduce algorithms for *fast* modular exponentiation and elliptic curve point addition.

Theorem 2.1.1. Suppose we want to compute $x^e \bmod n$ with $x, e, n \in \mathbb{N}$. Suppose moreover that the binary form of e is as follows:

$$e = \beta_k 2^k + \beta_{k-1} 2^{k-1} + \cdots + \beta_1 2^1 + \beta_0 2^0, \qquad (2.11)$$

where each β_i $(i = 0, 1, 2, \ldots k)$ is either 0 or 1. Then we have

$$x^e = x^{\beta_k 2^k + \beta_{k-1} 2^{k-1} + \cdots + \beta_1 2^1 + \beta_0 2^0} = \prod_{i=0}^{k} x^{\beta_i 2^i} = \prod_{i=0}^{k} \left(x^{2^i} \right)^{\beta_i}.$$

Furthermore, by the exponentiation law

$$x^{2^{i+1}} = (x^{2^i})^2, \qquad (2.12)$$

and so the final value of the exponentiation can be obtained by *repeated squaring and multiplication* operations.

Example 2.1.1. Suppose we wish to compute a^{100}; we first write $100_{10} = 1100100_2 := e_6 e_5 e_4 e_3 e_2 e_1 e_0$, and then compute

$$a^{100} = (((((((a)^2 \cdot a)^2)^2)^2 \cdot a)^2)^2 \implies a, \ a^3, \ a^6, \ a^{12}, \ a^{24}, \ a^{25}, \ a^{50}, \ a^{100}.$$

Note that for each e_i, if $e_i = 1$, we perform a *squaring* and a *multiplication* operation (except "$e_6 = 1$", for which we just write down a, as indicated in the first bracket), otherwise, we perform only a *squaring* operation. That is,

e_6	1	a	initialization
e_5	1	$(a)^2 \cdot a$	squaring and multiplication
e_4	0	$((a)^2 \cdot a)^2$	squaring
e_3	0	$(((a)^2 \cdot a)^2)^2$	squaring
e_2	1	$((((a)^2 \cdot a)^2)^2)^2 \cdot a$	squaring and multiplication
e_1	0	$(((((a)^2 \cdot a)^2)^2)^2 \cdot a)^2$	squaring
e_0	0	$((((((a)^2 \cdot a)^2)^2)^2 \cdot a)^2)^2$	squaring

$$\| \atop a^{100}$$

We are now in a position to introduce a fast algorithm for modular exponentiations (note that we can simply remove the "mod n" operation if we only wish to compute the exponentiation $c = x^e$):

Algorithm 2.1.1 (Fast modular exponentiation $x^e \bmod n$). This algorithm will compute the modular exponentiation

$$c = x^e \bmod n,$$

where $x, e, n \in \mathbb{N}$ with $n > 1$. It requires at most $2 \log e$ multiplication and $2 \log e$ divisions (divisions are only needed for modular operations; they can be removed if only $c = x^e$ are required to be computed).

[1] [Precomputation] Let

$$e_{\beta-1} e_{\beta-2} \cdots e_2 e_1 e_0 \tag{2.13}$$

be the binary representation of e (i.e., e has β bits). For example, for $562 = 1000110010$, we have $\beta = 10$ and

1	0	0	0	1	1	0	0	1	0
↑	↑	↑	↑	↑	↑	↑	↑	↑	↑
e_9	e_8	e_7	e_6	e_5	e_4	e_3	e_2	e_1	e_0

[2] [Initialization] Set $c \leftarrow 1$.

[3] [Modular Exponentiation] Compute $c = x^e \bmod n$ in the following way:

> for i from $\beta - 2$ down to 0 do
> $\quad c \leftarrow c^2 \bmod n$ (squaring)
> $\quad$ if $e_i = 1$ then
> $\quad\quad c \leftarrow c \cdot x \bmod n$ (multiplication)

[4] [Exit] Print c and terminate the algorithm.

Theorem 2.1.2. Let x, e and n be positive integers with $n > 1$. Then the modular exponentiation $x^e \bmod n$ can be computed in $\mathcal{O}(\log e)$ arithmetic operations and $\mathcal{O}\left((\log e)(\log n)^2\right)$ bit operations. That is,

$$\text{Time}(x^e \bmod n) = \mathcal{O}_A(\log e) = \mathcal{O}_B\left((\log e)(\log n)^2\right). \tag{2.14}$$

Proof. We first find the least positive residues of $x, x^2, x^4, \ldots, x^{2^k}$ modulo n, where $2^k \le e < 2^{k+1}$, by successively squaring and reducing modulo n. This requires a total of $\mathcal{O}\left((\log e)(\log n)^2\right)$ bit operations, since we perform $\mathcal{O}(\log e)$ squarings modulo n, each requiring $\mathcal{O}(\log n)^2$ bit operations. Next, we multiply together the least positive residues of the integers x^{2^i} corresponding to the binary bits of e which are equal to 1, and reduce modulo n. This also requires $\mathcal{O}\left((\log e)(\log n)^2\right)$ bit operations, since there are at most $\mathcal{O}(\log e)$ multiplications, each requiring $\mathcal{O}(\log n)^2$ bit operations. Therefore, a total of $\mathcal{O}\left((\log e)(\log n)^2\right)$ bit operations are needed to find the least positive residue of $x^e \bmod n$. $\square$

Example 2.1.2. Use the above algorithm to compute $7^{9007} \bmod 561$ (here $x = 7$, $e = 9007$ and $m = 561$). By writing e in the binary form $e = e_{\beta-1}e_{\beta-2}\cdots e_1 e_0$, we have

$$9007 = 10001100101111 = e_{13}e_{12}\cdots e_1 e_0.$$

Now we just perform the following computations as described in Algorithm 2.1.1:

$$
\begin{aligned}
&c \leftarrow 1 \\
&x \leftarrow 7 \\
&n \leftarrow 561 \\
&\text{for } i \text{ from } \beta - 2 \text{ down to } 0 \text{ do} \\
&\quad c \leftarrow c^2 \bmod n \\
&\quad \text{if } e_i = 1 \text{ then } c \leftarrow c \cdot x \bmod n \\
&\text{print } c; \text{ (now } c = x^e \bmod n \text{)}
\end{aligned}
$$

The values of (i, e_i, c) at each loop for i from 13 down to 0 are as follows:

13	12	11	10	9	8	7	6	5	4	3	2	1	0
1	0	0	0	1	1	0	0	1	0	1	1	1	1
7	49	157	526	160	241	298	166	469	49	538	337	46	226

So, at the end of the computation, the final result $c = 7^{9007} \bmod 561 = 226$ will be returned. It is clear that at most $2\log_2 9007$ multiplications and $2\log_2 9007$ divisions will be needed for the computation. In fact, only 22 multiplications and 22 divisions will be needed for this computation task.

Remark 2.1.1. The above fast exponentiation algorithm is about half as good as the best; more efficient algorithms are known. For example, Brickell, et. al. [40] developed a more efficient algorithm, using precomputed values to reduce the number of multiplications needed. Their algorithm allows the computation of g^n in time $\mathcal{O}(\log n/\log\log n)$. They also showed that their method can be parallelized, to compute powers in time $\mathcal{O}(\log\log n)$ with $\mathcal{O}(\log n/\log\log n)$ processors.

Now we move on to the fast computation of point additions on elliptic curves:

$$kP = \underbrace{P \oplus P \oplus \cdots \oplus P}_{k \text{ summands}} \qquad (2.15)$$

where $P = (x, y)$ is a point on an elliptic curve $E : y^2 = x^3 + ax + b$, and k a very large positive integer. Since the computation of kP is so fundamental in all elliptic curve related computations and applications, it is desirable that such computations are carried out as fast as possible. The basic idea of the fast computation of kP is as follows:

[1] Compute $2^i P$, for $i = 0, 1, 2, \ldots, \beta - 1$, with $\beta = \lfloor 1.442 \ln k + 1 \rfloor$.

[2] Add together suitable multiples of P, determined by the binary expansion of k, to get kP.

For example, to compute kP where $k = 232792560$, we first compute:

$$\beta = \lfloor 1.442 \ln k + 1 \rfloor = 28,$$

then compute $2^i P$, for $i = 0, 1, 2, \ldots, 27$ as follows:

P	$2P$	$2^2 P$	$2^3 P$	$2^4 P$	$\cdots$	$2^{25} P$	$2^{26} P$	$2^{27} P$
	$\parallel$	$\parallel$	$\parallel$	$\parallel$		$\parallel$	$\parallel$	$\parallel$
	$2(2P)$	$2(2^2 P)$	$2(2^3 P)$	$\cdots$		$2(2^{24} P)$	$2(2^{25} P)$	$2(2^{26} P)$

By the binary expansion of k

$$k = 232792560_{10} = 11011110000000010000111110000_2 := e_{27} e_{26} e_{25} \cdots e_2 e_1 e_0$$

we add only those multiples that correspond to 1:

1	1	1	1	1	1	1	1	1	1	1	1
$\updownarrow$	$\updownarrow$	$\updownarrow$	$\updownarrow$	$\updownarrow$	$\updownarrow$	$\updownarrow$	$\updownarrow$	$\updownarrow$	$\updownarrow$	$\updownarrow$	$\updownarrow$
2^{27}	2^{26}	2^{24}	2^{23}	2^{22}	2^{21}	2^{13}	2^8	2^7	2^6	2^5	2^4

and ignore those multiples that correspond to 0:

$$2^{25}, 2^{20}, 2^{19}, 2^{18}, 2^{17}, 2^{16}, 2^{15}, 2^{14}, 2^{12}, 2^{11}, 2^{10}, 2^9, 2^3, 2^2, 2^1, 2^0.$$

Thus, we finally have:

$$\begin{aligned} kP &= 2^{27}P \oplus 2^{26}P \oplus 2^{24}P \oplus 2^{23}P \oplus 2^{22}P \oplus 2^{21}P \\ &\quad \oplus 2^{13}P \oplus 2^8 P \oplus 2^7 P \oplus 2^6 P \oplus 2^5 P \oplus 2^4 P. \\ &= 232792560P. \end{aligned}$$

Remarkably enough, the idea of *repeated squaring* for fast exponentiations can be used almost directly for fast group operations (i.e., fast point additions) on elliptic curves. The idea of fast group additions is as follows: Let $e_{\beta-1} e_{\beta-2} \cdots e_1 e_0$ be the binary representation of k. Then for i starting from $e_{\beta-1}$ down to e_0 ($e_{\beta-1}$ is always 1 and used for initialization), check whether or not $e_i = 1$. If $e_i = 1$, then perform a doubling and an addition group operation; otherwise, just perform a doubling operation. For example, to compute $89P$, since $89 = 1011001$, we have:

e_6	1	P	initialization
e_5	0	$2P$	doubling
e_4	1	$2(2P) + P$	doubling and addition
e_3	1	$2(2(2P) + P) + P$	doubling and addition
e_2	0	$2(2(2(2P) + P) + P)$	doubling
e_1	0	$2(2(2(2P) + P) + P))$	doubling
e_0	1	$2(2(2(2(2P) + P) + P))) + P$	doubling and addition

$$\parallel$$
$$89P$$

The following algorithm implements this idea of *repeated doubling and addition* for computing kP.

Algorithm 2.1.2 (Fast group operations kP on elliptic curves).
This algorithm computes kP, where k is a large integer, and P is assumed to be a point on an elliptic curve $E : y^2 = x^3 + ax + b$ (note that we do not actually do the additions for the coordinates of P in this algorithm).

[1] Write k in the binary expansion form $k = e_{\beta-1}e_{\beta-2}\cdots e_2e_1e_0$, where each e_i is either 1 or 0. (Assume k has β bits.)

[2] Set $c \leftarrow 0$

[3] Compute kP:

$$\text{for } i \text{ from } \beta - 2 \text{ down to } 0 \text{ do}$$
$$c \leftarrow 2c \text{ (doubling)}$$
$$\text{if } e_i = 1 \text{ then } c \leftarrow c + P \text{ (addition)}$$

[4] Print c; (now $c = kP$)

Example 2.1.3. Use Algorithm 2.1.2 to compute $105P$. Let

$$k = 105 = 1101001 := e_6e_5e_4e_3e_2e_1e_0.$$

At the initial stage of the algorithm, we set $c = 0$. Now we perform the following computation steps according to Algorithm 2.1.2:

$$e_6 = 1:\ c \leftarrow P + 2c \implies c \leftarrow P \qquad\qquad\qquad\qquad \implies c = P$$
$$e_5 = 1:\ c \leftarrow P + 2c \implies c \leftarrow P + 2P \qquad\qquad\quad \implies c = 3P$$
$$e_4 = 0:\ c \leftarrow 2c \qquad \implies c \leftarrow 2(P + 2P) \qquad\qquad \implies c = 6P$$
$$e_3 = 1:\ c \leftarrow P + 2c \implies c \leftarrow P + 2(2(P + 2P)) \qquad \implies c = 13P$$
$$e_2 = 0:\ c \leftarrow 2c \qquad \implies c \leftarrow 2(P + 2(2(P + 2P))) \qquad \implies c = 26P$$
$$e_1 = 0:\ c \leftarrow 2c \qquad \implies c \leftarrow 2(2(P + 2(2(P + 2P)))) \qquad \implies c = 52P$$
$$e_0 = 1:\ c \leftarrow P + 2c \implies c \leftarrow P + 2(2(2(P + 2(2(P + 2P))))) \implies c = 105P.$$

That is, $P + 2(2(2(P + 2(2(P + 2P)))))) = 105P$.

Example 2.1.4. Suppose we wish to compute $kP \bmod 1997$, where $k = 9007 = 10001100101111_2$. The computation can be summarized in the following table which shows the values of (i, e_i, c) for each execution of the "for" loop in Algorithm 2.1.2 (plus an additional modular operation "mod 1997" at the end of each loop):

13	12	11	10	9	8	7	6	$\cdots$	2	1	0
1	0	0	0	1	1	0	0	$\cdots$	1	1	1
P	2P	4P	8P	17P	35P	70P	140P	$\cdots$	254P	509P	1019P

The final result of the computation is $c \equiv 1019P \pmod{1997}$.

Note that Algorithm 2.1.2 does not actually calculate the coordinates (x, y) of kP on an elliptic curve over $\mathbb{Q}$ or over $\mathbb{Z}/n\mathbb{Z}$. To make Algorithm 2.1.2 a practically useful algorithm for point additions on an elliptic curve E, we must incorporate the actual coordinate addition $P_3(x_3, y_3) = P_1(x_1, y_1) + P_2(x_2, y_2)$ on E into the algorithm. To do this, we use the following formulas to compute x_3 and y_3 for P_3:

$$\left\{ \begin{array}{l} x_3 = \lambda^2 - x_1 - x_2, \\ y_3 = \lambda(x_1 - x_3) - y_1 \end{array} \right.$$

where

$$\lambda = \left\{ \begin{array}{ll} \dfrac{3x_1^2 + a}{2y_1} & \text{if } P_1 = P_2 \\[2mm] \dfrac{y_1 - y_2}{x_1 - x_2} & \text{otherwise.} \end{array} \right.$$

Algorithm 2.1.3 (Fast group operations kP on elliptic curves).
This algorithm will compute the point kP mod n, where $k \in \mathbb{Z}^+$ and P is an initial point (x, y) on an elliptic curve $E : y^2 = x^3 + ax + b$ over $\mathbb{Z}/n\mathbb{Z}$; if we require E over $\mathbb{Q}$, just compute kP, rather than kP mod n. Let the initial point $P = (x_1, y_1)$, and the result point $P = (x_c, y_c)$.

[1] (Precomputation) Write k in the following binary expansion form $k = e_{\beta-1}e_{\beta-2}\cdots e_1e_0$. (Suppose k has β bits).

[2] (Initialization) Initialize the values for a, x_1 and y_1. Let $(x_c, y_c) = (x_1, y_1)$; this is exactly the computation task for e_1 (e_1 always equals 1).

[3] (Doublings and Additions) Computing kP mod n.

        for i from $\beta - 2$ down to 0 do
            $m_1 \leftarrow 3x_c^2 + a \bmod n$
            $m_2 \leftarrow 2y_c \bmod n$
            $M \leftarrow m_1/m_2 \bmod n$
            $x_3 \leftarrow M^2 - 2x_c \bmod n$
            $x_3 \leftarrow M^2 - 2x_c \bmod n$
            $y_3 \leftarrow M(x_c - x_3) - y_c \bmod n$
            $x_c \leftarrow x_3$
            $y_c \leftarrow y_3$
            if $e_i = 1$
               then $c \leftarrow 2c + P$
                    $m_1 \leftarrow y_c - y_1 \bmod n$
                    $m_2 \leftarrow x_c - x_1 \bmod n$
                    $M \leftarrow m_1/m_2 \bmod n$
                    $x_3 \leftarrow M^2 - x_1 - x_c \bmod n$
                    $y_3 \leftarrow M(x_1 - x_3) - y_1 \bmod n$
                    $x_c \leftarrow x_3$
                    $y_c \leftarrow y_3$
            else $c \leftarrow 2c$

[4] Print c (now $c = kP \bmod n$) and terminate the algorithm. (Note that this algorithm will stop whenever $m_1/m_2 \equiv \mathcal{O}_E \pmod{n}$, that is, it will stop whenever a modular inverse does not exit at any step of the computation).

Example 2.1.5. Let
$$E : \; y^2 = x^3 - x - 1$$
be an elliptic curve over $\mathbb{Z}/1098413\mathbb{Z}$ and $P = (0,1)$ be a point on E. Then using Algorithm 2.1.3, we get:

$$92P = (159895, 673343),$$
$$7892P = (371451, 650785),$$
$$10319P = (258834, 557124).$$

Problems for Section 2.1

Problem 2.1.1. Show that the four arithmetic operations $+, -, \times, \div$ can be performed in polynomial-time.

Problem 2.1.2. Show that matrix multiplication can be performed in polynomial-time.

Problem 2.1.3. Give an efficient algorithm to convert a given β-bit binary integer to a decimal representation.

Problem 2.1.4. Prove or disprove that $\mathcal{P} = \mathcal{NP}$.

Problem 2.1.5. Show that if $\mathcal{P} = \mathcal{NP}$, then the integer factorization problem can be solved in $\mathcal{P}$.

Problem 2.1.6. Let $\mathcal{EXP}$ be deterministic exponential-time and $\mathcal{NEXP}$ non-deterministic exponential-time. Prove that if $\mathcal{EXP} \neq \mathcal{NEXP}$, then $\mathcal{P} \neq \mathcal{NP}$.

Problem 2.1.7. Show that $\mathcal{ZPP} = \mathcal{RP} \cap \mathrm{co}\mathcal{RP}$.

Problem 2.1.8. Show that if $\mathcal{NP} \subseteq \mathcal{BPP}$, then $\mathcal{NP} = \mathcal{RP}$.

Problem 2.1.9. Show that $\mathcal{BPP} \subseteq \mathcal{PSPACE}$.

Problem 2.1.10. Give an efficient modular exponentiation algorithm for computing $y \equiv x^k \pmod{n}$ that examines the bits of k from right to left rather than from left to right. Use your algorithm to compute $y = 3^{257}$ and $y \equiv 3^{257} \pmod{103}$.

Problem 2.1.11. Let E be an elliptic curve $y^2 = x^3 - a - 1$ over $\mathbb{Z}_{1098413}$, and $P = (0,1)$ a point on E. Compute kP for $k = 2, 3, 8, 20, 31, 45, 92, 261, 513, 875$. Find the smallest k such that $kP = (467314, 689129)$.

Problem 2.1.12. In 1985 René Schoof [217] proposed a polynomial-time algorithm for computing the order of a point of elliptic curve over $\mathbb{F}_p$. Describe Schoof's algorithm in full and give a computational complexity analysis of the algorithm.

2.2 Riemann ζ and Dirichlet L Functions

Two types of number-theoretic functions, the Riemann ζ-function and the Dirichlet L-function, are studied in this section, as they are fundamental to many computational number-theoretic problems, including the primality testing and integer factorization problems.

Definition 2.2.1. Let x be a positive real number greater than 1. Then $\pi(x)$, the *prime counting function* is defined as follows:

$$\pi(x) = \sum_{\substack{p \leq x \\ p \in \text{Primes}}} 1. \tag{2.16}$$

That is, $\pi(x)$ is the number of primes less than or equal to x.

Although the distribution of primes among the integers is very irregular, the prime counting function $\pi(x)$ is surprisingly well behaved (see Table 2.1). It can be easily seen from Table 2.1 that the approximation $x/\ln x$ gives reasonably accurate estimates of $\pi(x)$. In fact, the study of this approximation leads to the following *famous* theorem of number theory, and indeed of all mathematics.

Theorem 2.2.1 (Prime Number Theorem). $\pi(x)$ is asymptotic to $\dfrac{x}{\ln x}$. That is,

$$\lim_{x \to \infty} \frac{\pi(x)}{x/\ln x} = 1. \tag{2.17}$$

The Prime Number Theorem was postulated by Gauss (1777–1855) in 1792 on numerical evidence; it is known that Gauss constructed by hand a table of all primes up to three million, and investigated the number of primes occurring in each group of 1000. Note also that it was conjectured by Legendre (1752–1833) before Gauss, in a different form, but of course both Legendre and Gauss were unable to prove the theorem. Around 1850 Chebyshev (1821–1894) showed that

$$0.92129 \frac{x}{\ln x} < \pi(x) < 1.1056 \frac{x}{\ln x} \tag{2.18}$$

for large x. Chebyshev's result was further refined by Sylvester (1814–1897) in 1892 to

$$0.95695 \frac{x}{\ln x} < \pi(x) < 1.04423 \frac{x}{\ln x} \tag{2.19}$$

x	$\pi(x)$	$\dfrac{x}{\ln x}$	$\dfrac{\pi(x)}{x/\ln x}$
10^1	4	$4.3\cdots$	$0.93\cdots$
10^2	25	$21.7\cdots$	$1.152\cdots$
10^3	168	$144.8\cdots$	$1.16\cdots$
10^4	1229	$1085.7\cdots$	$1.13\cdots$
10^5	9592	$8685.8\cdots$	$1.131\cdots$
10^6	78498	$72382.5\cdots$	$1.084\cdots$
10^7	664579	$620420.5\cdots$	$1.071\cdots$
10^8	5761455	$5428680.9\cdots$	$1.061\cdots$
10^9	50847534	$48254942.5\cdots$	$1.053\cdots$
10^{10}	455052511	$434294481.9\cdots$	$1.047\cdots$
10^{11}	4118054813	$3948131653.7\cdots$	$1.043\cdots$
10^{12}	37607912018	$36191206825.3\cdots$	$1.039\cdots$
10^{13}	346065536839	$334072678387.1\cdots$	$1.035\cdots$
10^{14}	3204941750802	$3102103442166.0\cdots$	$1.033\cdots$
10^{15}	29844570422669	$28952965460216.8\cdots$	$1.030\cdots$
10^{16}	279238341033925	$271434051189532.4\cdots$	$1.028\cdots$
10^{17}	2625557157654233	$2554673422960304.8\cdots$	$1.027\cdots$
10^{18}	24739954287740860	$24127471216847323.8\cdots$	$1.025\cdots$
10^{19}	234057667276344607	$228576043106974646.1\cdots$	$1.023\cdots$
10^{20}	2220819602560918840	$2171472409516259138.2\cdots$	$1.022\cdots$
10^{21}	21127269486018731928	$20680689614440563221.4\cdots$	$1.021\cdots$
10^{22}	201467286689315906290	$197406582683296285295.9\cdots$	$1.020\cdots$
10^{23}	1925320391606818006727		

Table 2.1. Approximations to $\pi(x)$ by $x/\ln x$

for every sufficiently large x. It can be seen that Chebyshev came rather close to the Prime Number Theorem; however, the complete proof of the prime number theorem had to wait for about 50 years more. Bernhard Riemann (1826–1866) astounded the mathematical world in 1859 by writing a six-page memoir on $\pi(x)$ entitled *Über die Anzahl der Primzahlen unter einer gegebenen Grösse* (*On the Number of Primes Less Than a Given Magnitude*; see Figure 2.4 for the first two pages of the paper) which is now regarded as one of the greatest classics of mathematics. In this remarkable paper, which was incidentally the only paper he ever wrote on Number Theory, Riemann related the study of prime numbers to the properties of various functions of a *complex* number. In particular, he studied the ζ-function

$$\zeta(s) = \sum_{n=1}^{\infty} \frac{1}{n^s} \qquad (2.20)$$

where $s = \sigma + it$ is a complex number, and made various conjectures about its behavior. It is clear that the series $\zeta(s)$ converges absolutely for $\sigma > 1$, where $s = \sigma + it$ with σ and t real, and indeed that it converges uniformly for

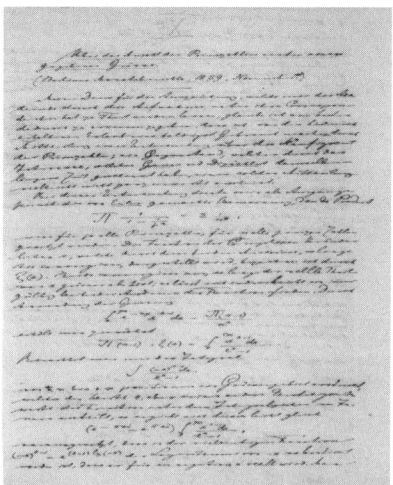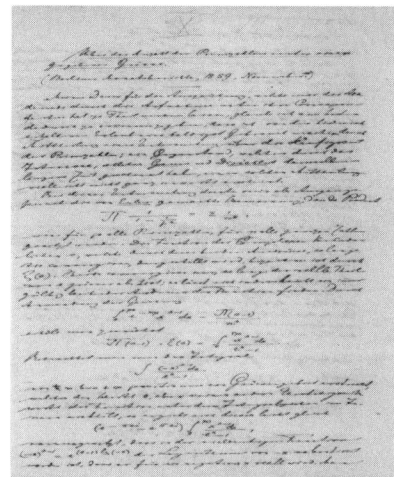

Figure 2.4. The first two pages of Riemann's paper

$\sigma > 1 + \delta$ for any $\delta > 0$. Euler actually studied the zeta function earlier, but only considered it for real values of s. The most interesting thing about the Riemann ζ-function is the *distribution* of the zeros of the ζ-function, since it is intimately connected with the distribution of the prime numbers. It has been known that

(1) The ζ-function has no zeros in the half-plane $\sigma > 1$. (Since by Euler's product, if $\sigma > 1$, then $\zeta(s) \neq 0$.)

(2) The ζ-function has no zeros on the line $\sigma = 1$. (Since for any real value of t, $\zeta(1 + it) \neq 0$.)

Therefore, there are only three possible types of zeros of $\zeta(s)$:

(1) Zeros lying outside the critical strip $0 < \sigma < 1$: These are the zeros at the points
$$-2, \ -4, \ -6, \ -8, \ -10, \ \cdots .$$
These zeros are the only zeros of $\zeta(s)$ outside the critical strip and are called *trivial zeros* of $\zeta(s)$. They are also called *real zeros* of $\zeta(s)$, since the zeros $-2, -4, \ldots$ are certainly real, and no other zeros are real.

(2) Zeros lying in the critical strip $0 < \sigma < 1$: These zeros are called *non-trivial zeros* of $\zeta(s)$; there are infinitely many such nontrivial zeros. Note that the nontrivial zeros are *not* real, and hence they are sometimes called *complex zeros*. Note also that these zeros are symmetric about the real axis (so that if s_0 is a zero, so is $\bar{s}_0$, where the bar denotes the complex

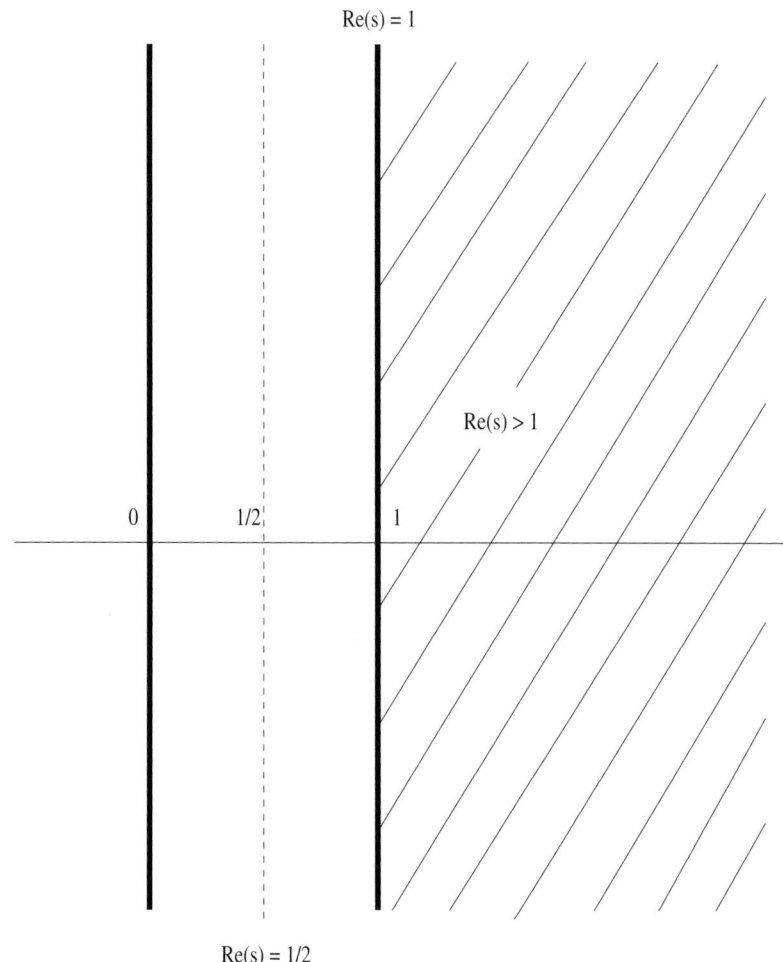

Figure 2.5. The complex plane of the Riemann ζ-function

conjugate) and the critical line $\sigma = \dfrac{1}{2}$ so that if $\dfrac{3}{4} + it$ were a zero, then $\dfrac{1}{4} + it$ would also be a zero).

(3) Zeros lying on the critical line $\sigma = \dfrac{1}{2}$: These are the zeros at $\dfrac{1}{2} + it$. These zeros are, of course, nontrivial (complex) zeros (because they all lie in the critical strip). There are infinitely many such nontrivial zeros lying on the critical line.

Riemann made the somewhat startling conjecture about the distribution of the nontrivial zeros of $\zeta(s)$ in his famous memoir, namely that

Conjecture 2.2.1 (Riemann Hypothesis (RH)). All the nontrivial (complex) zeros ρ of $\zeta(s)$ lying in the critical strip $0 < \sigma < 1$ must lie on the critical line $\sigma = \dfrac{1}{2}$, that is, $\rho = \dfrac{1}{2} + it$, where ρ denotes a nontrivial zero of $\zeta(s)$.

Remark 2.2.1. The Riemann Hypothesis may be true; if it is true, then it can be diagrammatically shown as in the left picture of Figure 2.6. The Riemann Hypothesis may also be false; if it is false, then it can be diagrammatically shown as in the right picture of Figure 2.6. At present, no one knows whether or not the Riemann Hypothesis is true.

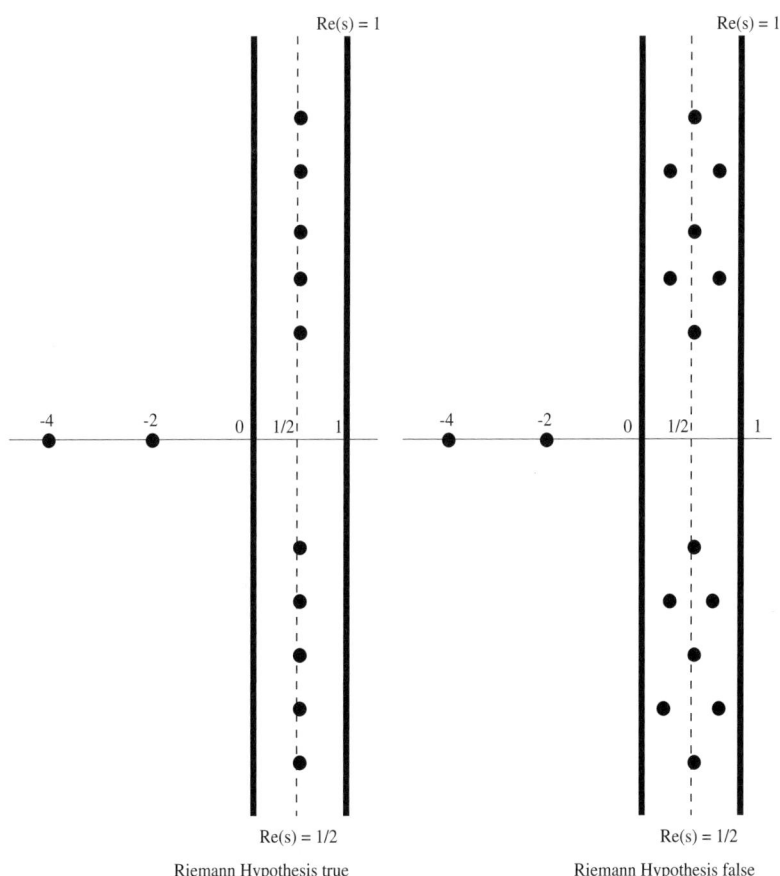

Figure 2.6. True of false of the Riemann hypothesis

Remark 2.2.2. The Riemann Hypothesis has never been proved or disproved; in fact, finding a proof or a counter-example is generally regarded as one of most difficult and important unsolved problems in all of mathematics. There is, however, a lot of numerical evidence to support the conjecture; as we move away from the real axis, the first thirty nontrivial zeros ρ_n (where ρ_n denotes the nth nontrivial zero) of $\zeta(s)$ are given in Table 2.2 (all figures here are given to six decimal digits). In fact, as we move further and fur-

n	t_n	n	t_n	n	t_n
1	14.134725	2	21.022040	3	25.010857
4	30.424876	5	32.935062	6	37.586178
7	40.918719	8	43.327073	9	48.005151
10	49.773832	11	52.970321	12	56.446248
13	59.347044	14	60.831779	15	65.112544
16	67.079811	17	69.546402	18	72.067158
19	75.704691	20	77.144840	21	79.337375
22	82.910381	23	84.735479	24	87.425275
25	88.809111	26	92.491899	27	94.651344
28	95.874634	29	98.831194	30	101.317851

Table 2.2. The first thirty nontrivial zeros of $\zeta(s)$

ther away from the real axis, the first 1500000001 nontrivial zeros of $\zeta(s)$ in the critical strip have been calculated; all these zeros lie on the critical line $\mathrm{Re}(s) = \dfrac{1}{2}$ and have imaginary part with $0 < t < 545439823.215$. That is, $\rho_n = \dfrac{1}{2} + it_n$ with $n = 1, 2, \ldots, 1500000001$ and $0 < t_n < 545439823.215$. In spite of this, there are several distinguished number theorists who believe the Riemann Hypothesis to be false, and that the presence of the first 1500000001 nontrivial zeros of $\zeta(s)$ on the critical line $\mathrm{Re}(s) = \dfrac{1}{2}$ is just a coincidence! does not indicate the behaviour of $\zeta(s)$ for every large t. The current status of knowledge of this conjecture is:

(1) The ζ-function has infinitely many zeros lying on the critical line $\mathrm{Re}(s) = \dfrac{1}{2}$.

(2) A positive proportion of the zeroes of $\zeta(s)$ in the critical strip $0 < \mathrm{Re}(s) < 1$ lie on the critical line $\mathrm{Re}(s) = \dfrac{1}{2}$ (thanks to Selberg).

(3) It is not known whether there are any nontrivial zeros which are not simple; certainly, none has ever been found.

Remark 2.2.3. The Riemann Hypothesis (RH) is fundamental to the Prime Number Theorem (PNT). For example, if this conjecture is true, then there is a refinement of the Prime Number Theorem

$$\pi(x) = \int_2^x \frac{dt}{\ln t} + \mathcal{O}\left(xe^{-c\sqrt{\ln x}}\right) \tag{2.21}$$

to the effect that

$$\pi(x) = \int_2^x \frac{dt}{\ln t} + \mathcal{O}\left(\sqrt{x}\ln x\right). \tag{2.22}$$

Remark 2.2.4. The knowledge of a large zero-free region for $\zeta(s)$ is important in the proof of the PNT and better estimates of the various functions connected with the distribution of prime numbers; the larger the region, the better the estimates of differences $|\pi(x) - \mathrm{Li}(x)|$ and $|\psi(x) - x|$, appearing in the PNT. If we assume RH, we then immediately have a good zero-free region and hence the proof of PNT becomes considerably easier (see picture on the right in Figure 2.7). De la Vallée-Poussin constructed in 1896 a zero-free region in the critical strip (see the picture on the left in Figure 2.7). This zero-free region is not as good as that given by the RH, but it turns out to be good enough for the purpose of proving the PNT.

In a celebrated memoir published in 1837, when studying the infinitude of primes in the arithmetic progression $kn + h$, Peter Dirichlet (1805–1859) introduced new arithmetic functions $\chi(n)$, called *Dirichlet characters* modulo k. These are multiplicative functions that have period k and vanish only on numbers not relatively prime to k. Clearly, there are $\phi(k)$ Dirichlet characters modulo k. In terms of Dirichlet characters, Dirichlet also introduced functions analogous to the Riemann ζ-function $\zeta(s)$ (note that Riemann actually studied the ζ-function in 1859, 22 years later than that Dirichlet studied the L-functions), called Dirichlet L-functions. These functions are defined by infinite series of the form:

$$L(s, \chi) = \sum_{n=1}^{\infty} \frac{\chi(n)}{n^s}, \tag{2.23}$$

where $\chi(n)$ is a Dirichlet character modulo k and s is a real number greater than 1 (or a complex number with real part greater than 1). Dirichlet's work on L-functions led naturally to the description of a more general class of functions defined by infinite series of the form:

$$F(s) = \sum_{n=1}^{\infty} \frac{f(n)}{n^s}, \tag{2.24}$$

where $f(n)$ is a given arithmetic function. This is called a *Dirichlet series* with coefficients $f(n)$, and the function $F(s)$ is called a *generating function*

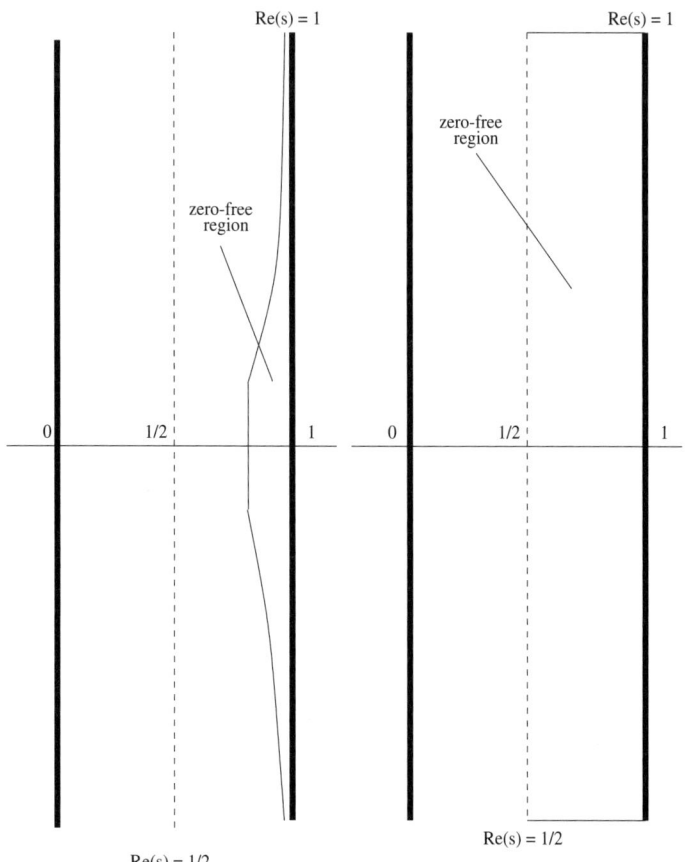

Figure 2.7. Zero-free region for $\zeta(s)$

of $f(n)$. For example, the simplest possible Dirichlet series is the Riemann ζ-function $\zeta(s)$, which generates the constant function $f(n) = 1$ for all n,

$$\zeta(s) = \sum_{n=1}^{\infty} \frac{1}{n^s}. \tag{2.25}$$

The square of the ζ-function generates the divisor function $\tau(n)$,

$$\zeta(s)^2 = \sum_{n=1}^{\infty} \frac{\tau(n)}{n^s}, \tag{2.26}$$

and the reciprocal of the ζ-function generates the Möbius function $\mu(n)$,

$$\zeta(s)^{-1} = \sum_{n=1}^{\infty} \frac{\mu(n)}{n^s}. \tag{2.27}$$

The study of L-functions is an active area of contemporary mathematical research, but it is not our purpose to explain here the theory and applications of Dirichlet L-functions in detail; we shall only use the basic concepts of Dirichlet L-functions to formulate the *Generalized Riemann Hypothesis*. Later, we shall study the L-functions attached to elliptic curves E and form the Birch and Swinerton-Dyer Conjecture in terms of the L-functions for E.

Conjecture 2.2.2 (Generalized Riemann Hypothesis (GRH)).
All the nontrivial zeros of the Dirichlet L-functions in the critical strip $0 < \mathrm{Re}(s) < 1$ must lie on the critical line $\mathrm{Re}(s) = \dfrac{1}{2}$.

Clearly, the Generalized (or Extended) Riemann Hypothesis (GRH) generalizes the (plain) Riemann Hypothesis (RH) to Dirichlet L-functions. There are again many consequences of the generalized Riemann hypothesis. For example, if this conjecture is true, then the primality testing problem is in $\mathcal{P}$. Of course, recent research shows that to prove primality testing problem to be in $\mathcal{P}$, it is not necessary to assume GRH.

Just the same as that the Riemann ζ-function has the Euler product expansion:

$$\zeta(s) = \prod_p \left(1 - \frac{1}{p^s}\right)^{-1}$$

when $\sigma > 1$, the Dirichlet L-functions for elliptic curve E also have the similar product expansion:

$$L(E, s) = \prod_p (1 - a_p p^{-s} + p^{1-2s})^{-1},$$

where $a_p = 1 + p + N_p$ and $N_p = \#E(\mathbb{F}_p)$, the number of points of the elliptic curve E over $\mathbb{F}_p$.

Analog to the Riemann Hypothesis, the famous Birch and Swinnerton-Dyer Conjecture can be formulated in its current form:

Conjecture 2.2.3 (The Birch and Swinnerton-Dyer Conjecture).
For an elliptic curve E, the Taylor expansion of the L-function $L(E, s)$, $s \in \mathbb{C}$, around $s = 1$ is

$$L(E, s) = c(s - 1)^r + \text{ higher order terms}$$

where r is the rank of $E(\mathbb{Q})$ and $c \neq 0$.

If this conjecture is true, we get an immediate corollary that

$$L(E, s) = 0 \Longleftrightarrow Q(Q) = \infty.$$

Same as the Riemann Hypothesis and the $\mathcal{P}$ Versus $\mathcal{NP}$ Problem, the Birch and Swinnerton-Dyer Conjecture is also one of the seven Millennium Prize Problems. Up to date, only partial results about the Birch and Swinnerton-Dyer Conjecture have been obtained, e.g., in 1977, Coates and Wiles proved that if $L(E, 1) \neq 0$ then $r = 0$. The entire problem, however, remains open.

Problems for Section 2.2

Problem 2.2.1. Prove that if $\{a_n\}$ is any sequence of integers with $\gcd(a_n, a_m)$, $n \neq m$, then there exist infinitely many primes.

Problem 2.2.2. Prove that a positive integer n is prime if and only of $\sigma(n) = n + 1$.

Problem 2.2.3. Prove that if p and q are primes, then $pq + 1$ is a square if and only if p and q are twin primes

Problem 2.2.4. Show that

$$\mathrm{Li}(x) = \int_2^x \frac{\mathrm{d}t}{\ln t}$$

is asymptotically equal to $x / \ln x$. That is,

$$\mathrm{Li}(x) \sim \frac{x}{\ln x}.$$

Problem 2.2.5. Let p_n be the nth prime. Show that

$$\lim_{n \to \infty} \frac{p_{n+1}}{p_n} = 1.$$

Problem 2.2.6. Prove that there exist real numbers a and b such that for all n,

$$n^{an} < \prod_{i=1}^n p_i < n^{bn}$$

with p_i the ith prime.

Problem 2.2.7. Prove for prime p,

$$x^p + y^p \equiv (x + y)^p \pmod{p}.$$

Problem 2.2.8. Prove that

$$\pi(2x) - \pi(x) \sim \pi(x).$$

Problem 2.2.9. Prove that

$$\frac{1}{\zeta(s)} = \sum_{n=1}^{\infty} \frac{\nu(n)}{n^s}$$

where $\nu(n)$ is the Möbius function.

Problem 2.2.10. Are there infinitely many primes p of the form $p = n^2 + 1$? Up to date, the best result is due to Iwaniec who proved in 1978 that there are infinitely many integers n such that $n^2 + 1$ is either a prime or a product of two primes.

Problem 2.2.11. Prove that

$$\zeta(s) = \sum_{n=1}^{\infty} \frac{1}{n^s} = \prod_{p} \frac{1}{1 - p^{-s}}$$

if and only if every positive integer $n > 1$ has a unique prime factorization.

Problem 2.2.12. Prove or disprove the Birch and Swinerton-Dyer Conjecture. (This problem has been opened for nearly 50 years, and is one of the seven Millennium Prize Problems, with one million US dollar prize attached to it.)

Problem 2.2.13. Show that if the Birch and Swinerton-Dyer Conjecture is true, then

$$L(E, 1) = 0 \iff E(Q) = \infty.$$

Problem 2.2.14. Show that, independent of the Birch and Swinerton-Dyer Conjecture, if $L(E, s) \neq 0$, then the group of solutions $E(\mathbb{Q})$ is finite.

2.3 Rigorous Primality Tests

This section introduces some simple, but rigorous primality tests.

First, we introduce a simple number-theoretic fact.

Theorem 2.3.1. Let $n > 1$. If n has no prime factor less than or equal to $\lfloor \sqrt{n} \rfloor$, then n is prime.

Thus the simplest possible primality test of n is by trial divisions of all possible prime factors of n up to $\lfloor \sqrt{n} \rfloor$ as follows (the sieve of Eratosthenes for finding, or a table containing prime numbers up to $\sqrt{n}$ is used in this test).

Primality test by trial divisions:

$$\text{Test}(p_i) \overset{\text{def}}{=} p_1, p_2, \ldots, p_k \leq \lfloor\sqrt{n}\rfloor, \quad p_i \nmid n$$
$$\Uparrow \tag{2.28}$$
$$\text{Eratosthenes Sieve}$$

Thus, if n passes $\text{Test}(p_i)$, then n is prime:

$$n \text{ passes } \text{Test}(p_i) \implies n \in \text{Primes}. \tag{2.29}$$

Example 2.3.1. To test whether or not 3271 is prime, we only needs to test the primes up to $\lfloor\sqrt{3271}\rfloor = 57$. That is, we will only need to do at most 16 trial divisions as follows:

$$\frac{3271}{2}, \frac{3271}{3}, \frac{3271}{5}, \frac{3271}{7}, \ldots, \frac{3271}{47}, \frac{3271}{53}.$$

As none of these division gives a zero remainder, so 3271 is a prime number. However, for $n = 3273$, we would normally expect to do the following trial divisions:

$$\frac{3273}{2}, \frac{3273}{3}, \frac{3273}{5}, \frac{3273}{7}, \ldots, \frac{3273}{47}, \frac{3273}{53}.$$

but fortunately we do not need to do all these trial divisions as 3273 is a composite, in fact, when we do the trial division 3273/3, it gives a zero remainder, so we conclude immediately that 3273 is a composite number.

This test, although easy to implement, is not practically useful for large numbers since it needs $\mathcal{O}\left(2^{(\log n)/2}\right)$ bit operations.

In 1876 (although it was published in 1891), Lucas discovered a type of converse of the Fermat little theorem, based on the use of primitive roots.

Theorem 2.3.2 (Lucas' converse of Fermat's little theorem, 1876). Let $n > 1$. Assume that there exists a primitive root of n, i.e., an integer a such that

(1) $a^{n-1} \equiv 1 \pmod{n}$,

(2) $a^{(n-1)/p} \not\equiv 1 \pmod{n}$, for each prime divisor p of $n - 1$.

Then n is prime.

Proof. Since $a^{n-1} \equiv 1 \pmod{n}$, Part (a) of Theorem 1.8.1 (see Chapter 1) tells us that $\text{ord}_n(a) \mid (n - 1)$. We will show that $\text{ord}_n(a) = n - 1$. Suppose $\text{ord}_n(a) \neq n - 1$. Since $\text{ord}_n(a) \mid (n - 1)$, there is an integer k satisfying $n - 1 = k \cdot \text{ord}_n(a)$. Since $\text{ord}_n(a) \neq n - 1$, we know that $k > 1$. Let p be a prime factor of k. Then

$$x^{n-1}/q = x^{k/q \cdot \text{ord}_n(a)} = \left(x^{\text{ord}_n(a)}\right)^{k/q} \equiv 1 \pmod{n}.$$

However, this contradicts the hypothesis of the theorem, so we must have $\mathrm{ord}_n(a) = n - 1$. Now, since $\mathrm{ord}_n(a) \leq \phi(n)$ and $\phi(n) \leq n - 1$, it follows that $\phi(n) = n - 1$. So finally by Part (b) of Theorem 1.5.7, n must be prime. $\square$

Lucas' theorem can be converted to rigorous primality test as follows:

Primality test based on primitive roots:

$$\text{Test}(a) \quad \stackrel{\text{def}}{=} \quad a^{n-1} \equiv 1 \pmod{n},$$
$$a^{(n-1)/p} \not\equiv 1 \pmod{n}, \forall p \mid (n-1). \tag{2.30}$$

If n passes the test, then n is prime:

$$n \text{ passes Test}(a) \implies n \in \text{Primes.} \tag{2.31}$$

Primality test based on primitive roots is also called $n - 1$ primality test, as it is based on the prime factorization of $n - 1$.

Example 2.3.2. Let $n = 2011$, then $2011 - 1 = 2 \cdot 3 \cdot 5 \cdot 67$. Note first 3 is a primitive root (in fact, the smallest) primitive root of 2011, since $\mathrm{order}(3, 2011) = \phi(2011) = 2010$. So, we have

$$3^{2011-1} \equiv 1 \pmod{2011},$$

$$3^{(2011-1)/2} \equiv -1 \not\equiv 1 \pmod{2011},$$

$$3^{(2011-1)/3} \equiv 205 \not\equiv 1 \pmod{2011},$$

$$3^{(2011-1)/5} \equiv 1328 \not\equiv 1 \pmod{2011},$$

$$3^{((2011-1)/67} \equiv 1116 \not\equiv 1 \pmod{2011}.$$

Thus, by Theorem 2.3.2, 2011 must be prime.

Remark 2.3.1. In practice, primitive roots tend to be small integers and can be quickly found (although there are some primes with arbitrary large smallest primitive roots), and the computation for $a^{n-1} \equiv 1 \pmod{n}$ and $a^{(n-1)/p} \not\equiv 1 \pmod{n}$ can also be performed very efficiently by Algorithm 2.1.1. However, to determine if n is prime, the above test requires the prime factorization of $n - 1$, a problem of almost the same size as that of factoring n, a problem that is much harder than the primality testing of n.

Note that Theorem 2.3.2 is actually equivalent to the following theorem:

Theorem 2.3.3. If there is an integer a for which the order of a modulo n is equal to $\phi(n)$ and $\phi(n) = n - 1$, then n is prime. That is, if

$$\mathrm{ord}_n(a) = \phi(n) = n - 1, \tag{2.32}$$

or

$$\mathbb{Z}_n^+ = \mathbb{Z}_n^*, \tag{2.33}$$

then n is prime.

Thus, we have the following primality test.

Primality test based on $\text{ord}_n(a)$

$$\text{Test}(a) \stackrel{\text{def}}{=} \text{ord}_n(a) = \phi(n) = n - 1 \tag{2.34}$$

Thus, if n passes the test, then n is prime:

$$n \text{ passes Test}(a) \implies n \in \text{Primes}. \tag{2.35}$$

Example 2.3.3. Let $n = 3779$. We find e.g., that the integer $a = 19$ with $\gcd(19, 3779) = 1$ satisfies

(1) $\text{ord}_{3779}(19) = 3778$,

(2) $\phi(3779) = 3778$.

That is, $\text{ord}_{3779}(19) = \phi(3779) = 3778$. Thus by Theorem 2.3.3, 3779 is prime.

Remark 2.3.2. It is not a simple matter to find the order of an element a modulo n, $\text{ord}_n(a)$, if n is large. In fact, if $\text{ord}_n(a)$ can be calculated efficiently, the primality and prime factorization of n can be easily determined. At present, the best known method for computing $\text{ord}_n(a)$ requires to factor n.

Remark 2.3.3. If we know the value of $\phi(n)$, we can immediately determine whether or not n is prime, since by Part (2) of Theorem 1.5.7 we know that n is prime if and only if $\phi(n) = n - 1$. Of course, this method is not practically useful, since to determine the primality of n, we need to find $\phi(n)$, but to find $\phi(n)$, we need to factor n, a problem even harder than the primality testing of n.

Remark 2.3.4. The difficulty in applying Theorem 2.3.3 for primality testing lies in finding the order of an integer a modulo n, which is computationally intractable. As we will show later, the finding of the order of an integer a modulo n can be efficiently done on a quantum computer.

It is possible to use different bases a_i (rather than a single base a) for different prime factors p_i of $n - 1$ in Theorem 2.3.2:

Theorem 2.3.4. If for each prime p_i of $n - 1$ there exists an integer a_i such that

(1) $a_i^{n-1} \equiv 1 \pmod{n}$,

(2) $a_i^{(n-1)/p_i} \not\equiv 1 \pmod{n}$.

Then n is prime.

Proof. Suppose that $n - 1 = \prod_{i=1}^{k} p_i^{\alpha_i}$, with $\alpha_i > 0$, for $i = 1, 2, \ldots, k$. Let also $r_i = \mathrm{ord}_n(a_i)$. Then $r_i \mid (n-1)$ and $r_i \nmid (n-1)/p_i$ gives that $p_i^{\alpha_i} \mid r_i$. But for each i, we have $r_i \mid \phi(n)$ and hence $p_i^{\alpha_i} \mid \phi(n)$. This gives $(n-1) \mid \phi(n)$, so n must be prime. $\qquad\square$

Example 2.3.4. Let $n = 997$, then $n - 1 = 2^2 \cdot 3 \cdot 83$. We choose three different bases $5, 7, 11$ for the prime factors $2, 3, 83$, respectively, and get

$$5^{997-1} \equiv 1 \ (\mathrm{mod} \ 997)$$

$$5^{(997-1)/2} \equiv -1 \ \not\equiv 1 \ (\mathrm{mod} \ 997)$$

$$7^{(997-1)/3} \equiv 304 \ \not\equiv 1 \ (\mathrm{mod} \ 997)$$

$$11^{(997-1)/83} \equiv 697 \ \not\equiv 1 \ (\mathrm{mod} \ 997)$$

Thus, we can conclude that 997 is prime.

The above tests require the factorization of $n - 1$, a problem even harder than the primality test of n. In 1914, Henry C. Pocklington (1870–1952) showed that it is not necessary to know all the prime factors of $n - 1$; part of them will be sufficient, as indicated in the following theorem.

Theorem 2.3.5 (Pocklington, 1914). Let $n - 1 = mj$, with $m = p_1^{\alpha_1} p_2^{\alpha_2} \cdots p_k^{\alpha_k}$, $m \geq \sqrt{n}$, and $\gcd(m, j) = 1$. If for each prime $p_i, i = 1, 2, \ldots, k$, there exists an integer a such that

(1) $a^{n-1} \equiv 1 \ (\mathrm{mod} \ n)$,

(2) $\gcd(a^{(n-1)/p_i} - 1, n) = 1$.

Then n is prime.

Proof. Let q be any one of the prime factors of n, and $\mathrm{ord}_n(a)$ the order of a modulo n. We have $\mathrm{ord}_n(a) \mid (q - 1)$ and also $\mathrm{ord}_n(a) \mid (n - 1)$, but $\mathrm{ord}_n(a) \nmid (n - 1)/p_i$. Hence, $p_i^{\alpha_i} \mid \mathrm{ord}_n(a)$, Since $\mathrm{ord}_n(a) \mid (q - 1)$, the result thus follows. $\qquad\square$

As already pointed out by Pocklington, the above theorem can lead to a primality test:

Pocklington's Test

$$\mathrm{Test}(a, p_i) \ \overset{\mathrm{def}}{=} \ n - 1 = mj, \ \ m = p_1^{\alpha_1} p_2^{\alpha_2} \cdots p_k^{\alpha_k}, \ m \geq \sqrt{n}, \ \gcd(m, j) = 1$$

$$a^{n-1} \equiv 1 \ (\mathrm{mod} \ n), \gcd(a^{(n-1)/p_i} - 1, n) = 1. \qquad (2.36)$$

Thus, if n passes the test, then n is prime:

$$n \text{ passes } \mathrm{Test}(a, p_i) \implies n \in \mathrm{Primes}. \qquad (2.37)$$

Example 2.3.5. Let also $n = 997$, and $n - 1 = 12 \cdot 83$, $83 > \sqrt{997}$. Choose $a = 3$ for $m = 83$. Then we have

$$3^{997-1} \equiv 1 \pmod{997},$$

$$\gcd(3^{(997-1)/83} - 1, 997) = 1.$$

Thus, we can conclude that 997 is prime.

There are some other rigorous, although inefficient, primality tests, for example, one of them follows directly from the converse of Wilson's theorem (Theorem 1.6.19)

Wilson's Test

$$\text{Test}(n) \quad \overset{\text{def}}{=} \quad n > 1 \text{ odd}, (n - 1)! \equiv -1 \pmod{n} \tag{2.38}$$

Thus, if n passes the test, then n is prime:

$$n \text{ passes Test}(n) \implies n \in \text{Primes}. \tag{2.39}$$

Example 2.3.6. As discussed in Section 1.6, very few primes satisfy the condition, in fact, for $p \leq 5 \cdot 10^8$, there are only three primes, namely, $p = 5, 13, 563$. So the Wilson test is essentially of no use in primality test, not just for its inefficiency.

Pratt's Primality Proving

It is interesting to note, although primality testing is difficult, the verification (proving) of primality is easy, since the primality (as well as the compositeness) of an integer n can be verified very quickly in polynomial-time:

Theorem 2.3.6. If n is composite, it can be proved to be composite in $\mathcal{O}((\log n)^2)$ bit operations.

Proof. If n is composite, there are integers a and b with $1 < a < n$, $1 < b < n$ and $n = ab$. Hence, given the two integers a and b, we multiply a and b, and verify that $n = ab$. This takes $\mathcal{O}((\log n)^2)$ bit operations and proves that n is composite. □

Theorem 2.3.7. If n is prime, then it can be proved to be prime in $\mathcal{O}((\log n)^4)$ bit operations.

Theorem 2.3.7 was discovered by Pratt [190] in 1975; he interpreted the result as showing that every prime has a *succinct primality certification*. The proof can be written as a finite tree whose vertices are labeled by pairs (p, g_p) where p is a prime number and g_p is primitive root modulo p we illustrate the

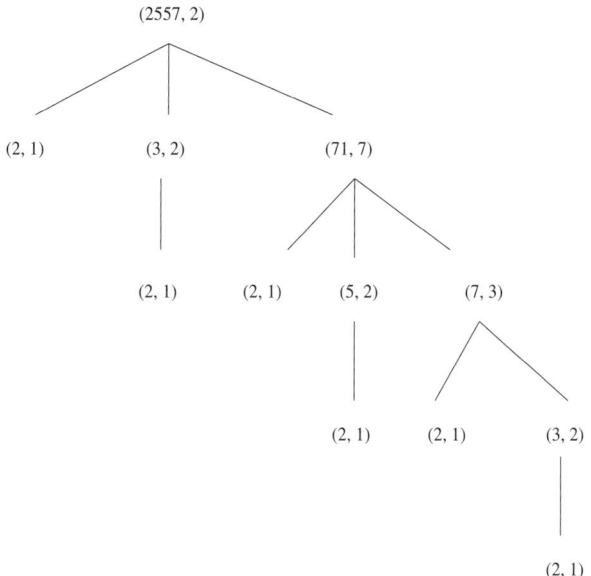

Figure 2.8. Certificate of primality for $n = 2557$

primality proving of prime number 2557 in Figure 2.8. In the top level of the tree, we write $(2557, 2)$ with 2 the primitive root modulo 2557. As $2557 - 1 = 2^2 \cdot 3^2 \cdot 71$, we have in the second level three vertices $(2, 1), (3, 2), (71, 7)$. Since $3, 71 > 2$, we have in the third level the child vertices $(2, 1)$ for $(3, 2)$, $(2, 1)$, $(5, 2)$ and $(7, 3)$ for $(71, 7)$. In the fourth level of the tree, we have $(2, 1)$ for $(5, 2)$, $(2, 1)$ and $(3, 2)$ for $(7, 3)$. Finally, in the fifth level we have $(2, 1)$ for $(3, 2)$. The leaves of the tree now are all labeled $(2, 1)$, completing the certification of the primality of 2557.

Remark 2.3.5. It should be noted that Theorem 2.3.7 cannot be used for finding the short proof of primality, since the factorization of $n - 1$ and the primitive root a of n are required.

Note that for some primes, Pratt's certificate is considerably shorter. For example, if $p = 2^{2^k} + 1$ is a Fermat number with $k \geq 1$, the p is prime if and only if

$$3^{(p-1)/2} \equiv -1 \pmod{p}. \tag{2.40}$$

This result, known as Papin's test, gives a Pratt certificate for Fermat primes. The work in verifying (2.40) is just $\mathcal{O}(p)$, since $2^k - 1 = \lfloor \log_2 p \rfloor - 1$. In fact, as Pomerance [180] showed, every prime p has an $\mathcal{O}(p)$ certificate. More precisely, he proved:

Theorem 2.3.8. For every prime p there is a proof that it is prime which requires for its certification $(5/2 + o(1)) \log_2 p$ multiplications modulo p.

Problems for Section 2.3

Problem 2.3.1. Use the Primality test based on primitive roots defined in (2.30) to prove that 1299709 is prime (choose $a = 6$).

Problem 2.3.2. Use the primality test based on Pocklington's theorem defined in (2.36) to prove that 179424673 is prime.

Problem 2.3.3. Prove that if $p > 1$ is an odd integer and

$$\begin{cases} a^{(p-1)/2} \equiv -1 \ (\mathrm{mod}\ p) \\ a^{(p-1)/2q} \not\equiv -1 \ (\mathrm{mod}\ p), \ \text{for every odd prime } q \mid (p-1) \end{cases} \tag{2.41}$$

Then p is prime. Conversely, if p is an odd prime, then every primitive root a of p satisfy conditions (2.41).

Problem 2.3.4. Prove or disprove that there are infinitely many Wilson primes (i.e., those primes p satisfying the condition $(p-1)! \equiv -1 \ (\mathrm{mod}\ p)$).

Problem 2.3.5. Prove that if

$$\begin{cases} F > \sqrt{n} \\ n - 1 = FR, \ \text{the complete prime factorization of } F \text{ is known} \\ a^{n-1} \equiv 1 \ (\mathrm{mod}\ n) \\ \gcd(a^{(n-1)/q} - 1, \ n) = 1 \text{ for every prime } q \mid F. \end{cases}$$

Then p is prime.

Problem 2.3.6. (Brillhart, Lehmer and Selfridge). Suppose

$$\begin{cases} \sqrt[3]{n} \le F \le \sqrt{n} \\ n - 1 = FR, \ \text{the complete prime factorization of } F \text{ is known} \\ a^{n-1} \equiv 1 \ (\mathrm{mod}\ n) \\ \gcd(a^{(n-1)/q} - 1, \ n) = 1 \text{ for every prime } q \mid F. \end{cases}$$

Represent a as a base F form $n = c_2 F^2 + c_1 F + 1$, where $c_1, c_2 \in [0, F-1]$. Show that p is prime if and only if $c_1^2 - 4c_2$ is not a square.

Problem 2.3.7. (Proth, 1878). Let $n = 2^k m + 1$, with $2 \nmid m$, $2^k > m$, and suppose that $\left(\frac{a}{n}\right) = -1$. Then n is prime if and only if

$$a^{(n-1)/2} \equiv -1 \ (\mathrm{mod}\ n).$$

Problem 2.3.8. Use Proth's result to prove that the following three numbers are prime:

$$
\begin{aligned}
45334718235548594054148161 &= 60 \cdot 2^{76} + 1, \\
6286414261996071708472153 &= 52 \cdot 2^{80} + 1, \\
18878585599102049192211644417 &= 61 \cdot 2^{88} + 1.
\end{aligned}
$$

Problem 2.3.9. (Pocklington, 1914). Let $n = p^k m + 1$, with p prime and $p \nmid m$, Show that if for some a we have

$$
\begin{cases}
a^{n-1} \equiv -1 \ (\mathrm{mod}\ n), \\
\gcd(a^{(n-1)/p} - 1, n) = 1,
\end{cases}
\tag{2.42}
$$

then each prime factor of n is of the form $sp^k + 1$ for some s.

Problem 2.3.10. Use the Pratt tree to prove the primality of 123456791.

2.4 Compositeness and Pseudoprimality Tests

The primality tests discussed in the previous section are rigorous and also easy to implement, they are, however, not practically useful, as they run in exponential-time. In this section, we shall discuss some pseudoprimality tests that run in polynomial-time but with a small error of probability for primality. First, we introduce some compositeness tests.

First notice that by Fermat's little theorem (Theorem 1.6.15), we have

$$
n \in \text{Primes} \implies a^{n-1} \equiv 1 \ (\mathrm{mod}\ n),
$$

but the inverse is not always true

$$
n \in \text{Primes} \overset{?}{\Longleftarrow} a^{n-1} \equiv 1 \ (\mathrm{mod}\ n).
$$

However, the converse (Corollary 1.6.4) is true:

$$
n \in \text{Composite} \overset{\checkmark}{\Longleftarrow} a^{n-1} \not\equiv 1 \ (\mathrm{mod}\ n).
$$

Thus, the converse of the Fermat little theorem can be converted to an efficient compositeness test as follows:

Compositeness Test based on the Converse of Fermat's Little Theorem:

$$\text{Test}(a) \overset{\text{def}}{=} n > 1 \text{ odd}, \gcd(a, n) = 1, a^{n-1} \not\equiv 1 \ (\text{mod } n). \tag{2.43}$$

Thus,

$$n \text{ passes } T(a) \Longrightarrow n \in \text{Composite}. \tag{2.44}$$

Example 2.4.1. Let

$$a = 2,$$
$$n = 1111111111111111111111111111111111111,$$

(n consists of 37 ones) such that $\gcd(2, n) = 1$. Since

$$2^{n-1} \equiv 2127960476395865717726168789645607932 \not\equiv 1 \ (\text{mod } n),$$

we conclude that n is composite. In fact,

$$n = 2028119 \cdot 2212394296770203368013 \cdot 247629013,$$

but to prove the compositeness of n, it is not necessary to know its factors.

Example 2.4.2. Let

$$a = 3,$$
$$n = 9746347772161.$$

Then we have

$$3^{9746347772161-1} \equiv 1 \ (\text{mod } 9746347772161).$$

From this result, we would conclude that 9746347772161 is prime. Unfortunately, it is not; it is in fact composite, since

$$9746347772161 = 7 \cdot 11 \cdot 13 \cdot 17 \cdot 19 \cdot 31 \cdot 37 \cdot 41 \cdot 641.$$

Now let

$$a = 3,$$
$$n = 179443609,$$

we have

$$3^{179443609-1} \equiv 1 \ (\text{mod } 179443609).$$

By this test, we still cannot conclude that 179443609 is prime although it is indeed prime.

Thus, we have

$$a^{n-1} \not\equiv 1 \pmod{n} \overset{100\%}{\Longrightarrow} n \in \text{Composite},$$

$$a^{n-1} \equiv 1 \pmod{n} \overset{?}{\Longrightarrow} n \in \text{Prime},$$

$$a^{n-1} \equiv 1 \pmod{n} \overset{?}{\Longrightarrow} n \in \text{Composite}.$$

So, the above test can prove that a number is composite but can never prove that a number is prime, although it is very efficient. However, the test can be strengthened to a pseudoprimality test, which is the main topic of this section.

Let us re-examine Fermat's little theorem: if b is a positive integer, p a prime and $\gcd(b, p) = 1$, then

$$b^{p-1} \equiv 1 \pmod{p}. \tag{2.45}$$

The converse of Fermat's little theorem is: for some odd positive integer n, if $\gcd(b, n) = 1$ and

$$b^{n-1} \not\equiv 1 \pmod{n}, \tag{2.46}$$

then n is composite. So, if there exists an integer b with $1 < b < n$, $\gcd(b, n) = 1$ and $b^{n-1} \not\equiv 1 \pmod{n}$, then n must be composite. However, as Exercise 2.4.2 shows, n can be prime but can also not be prime when $b^{n-1} \equiv 1 \pmod{n}$. We want to know how sure it is for n to be prime; this leads us naturally to the following important concepts of probable primes and pseudoprimes.

Definition 2.4.1. The positive integer n is a base-b *probable prime* if

$$b^{n-1} \equiv 1 \pmod{n}. \tag{2.47}$$

A base-b probable prime n is called a base-b *pseudoprime* if n is composite. A base-b probable prime and a base-b pseudoprime are also called the base-b *Fermat probable prime* and the base-b *Fermat pseudoprime*, respectively.

Example 2.4.3. Let $n = 1387$, we have $2^{341-1} \equiv 1 \pmod{341}$. Thus 341 is a base-2 probable prime. But since $341 = 11 \cdot 31$ is composite, it is a base-2 pseudoprime. The first few base-2 pseudoprimes are as follows: $341, 561, 645, 1105, 1387, 1729, 1905$.

Note that there are some composite numbers that satisfy (2.47) for every positive integer b such that $\gcd(b, n) = 1$:

Definition 2.4.2. A composite number n that satisfies $b^{n-1} \equiv 1 \pmod{n}$ for every positive integer b such that $\gcd(b, n) = 1$ is called a *Carmichael number*, in honor of the American mathematician Carmichael[5].

---

[5] Robert Carmichael conjectured in 1912 that *there are infinitely many such numbers* that now bear his name. W. Alford, G. Granville and C. Pomerance proved this conjecture in 1992.

Example 2.4.4. The first ten Carmichael numbers are as follows:

$$561, 1105, 1729, 2465, 2821, 6601, 8911, 10585, 15841, 29341.$$

It is usually much harder to show that a given integer (particularly when it is large) is a Carmichael number than to show that it is a base-b pseudoprime, as we can see from the following example.

Example 2.4.5. Show that 561 is a Carmichael number. Note that $561 = 3 \cdot 11 \cdot 17$. Thus $\gcd(b, 561) = 1$ implies that $\gcd(b, 3) = \gcd(b, 11) = \gcd(b, 17) = 1$. To show that $b^{560} \equiv 1 \pmod{561}$ for all b for which $\gcd(b, 3) = \gcd(b, 11) = \gcd(b, 17) = 1$, we use the Chinese Remainder Theorem and Fermat's little theorem, and get

$$b^2 \equiv 1 \pmod{3} \implies b^{560} = (b^2)^{280} \equiv 1 \pmod{3},$$

$$b^{10} \equiv 1 \pmod{11} \implies b^{560} = (b^{10})^{56} \equiv 1 \pmod{11},$$

$$b^{16} \equiv 1 \pmod{17} \implies b^{560} = (b^{16})^{35} \equiv 1 \pmod{17}.$$

Hence $b^{560} \equiv 1 \pmod{561}$ for all b satisfying $\gcd(b, 561) = 1$. Therefore, 561 is a Carmichael number.

Carmichael numbers are characterized by the following property.

Theorem 2.4.1. A composite integer $n > 2$ is a Carmichael number if and only if

$$n = \prod_{i=1}^{k} p_i, \quad k \geq 3$$

for all distinct odd primes p_i such that $\lambda(n) \mid n-1$, or equivalently $p_i - 1 \mid n-1$, for all non-negative integers $i \leq k$.

Fermat's little theorem implies that if n is prime, then n satisfies congruence (2.47) for every a in $(\mathbb{Z}/n\mathbb{Z})^+$. Thus, if we can find an integer $b \in (\mathbb{Z}/n\mathbb{Z})^+$ such that n does not satisfy congruence (2.47), then n is certainly composite. Surprisingly, the converse almost holds, so that this criterion forms an *almost* perfect test for primality. The following is the algorithm for $b = 2$:

Algorithm 2.4.1 (Base-2 Fermat pseudoprimality test). This algorithm will test numbers from 3 up to j, say, $j = 10^{10}$ for primality. If it outputs n is composite, then n is certainly composite. Otherwise, n is almost surely prime.

[1] Initialize the values $i \geq 3$ and $j > i$. Set $n \leftarrow i$.

[2] If $2^n \pmod{n} = 2$, then n is a *base-2 probable prime*, else n is composite.

[3] $n \leftarrow n + 1$. If $n \leq j$ goto [2], else goto [4].

[4] Terminate the execution of the algorithm.

The above base-2 pseudoprimality test is also called the Chinese test, since the Chinese mathematicians had this idea earlier than Fermat. Among the numbers below 2000 that can pass the Chinese test, only six are composites: 341, 561, 645, 1105, 1729 and 1905; all the rest are indeed primes. Further computation shows that such composite numbers seem to be rare. To exhibit quite how rare these are, note that up to 10^{10} there are around 450 million primes, but only about fifteen thousand base-2 pseudoprimes, while up to $2 \cdot 5 \times 10^{10}$ there are over a billion primes, and yet fewer than 22 thousand base-2 pseudoprimes. So, if we were to choose a random number $n < 2 \cdot 5 \times 10^{10}$ for which n divides $2^n - 2$, then there would be a less than a one-in-fifty-thousand chance that our number would be composite. We quote the following comments on the usefulness of the Chinese test from Rosen's book [206]:

> Because most composite integers are not pseudoprimes, it is possible to develop primality tests based on the original Chinese idea, together with extra observations.

Now we are in a position to present an improved version of the pseudoprimality tests, the strong pseudoprimality test, (or just the strong test, for short), due to Gary Miller, John Selfridge and Michael Rabin.

Theorem 2.4.2. Let p be a prime. Then

$$x^2 \equiv 1 \pmod{p} \tag{2.48}$$

if and only if $x \equiv \pm 1 \pmod{p}$.

Proof. First notice that

$$
\begin{aligned}
x^2 \equiv \pm 1 \pmod{p} &\iff (x+1)(x-1) \equiv 0 \pmod{p} \\
&\iff p \mid (x+1)(x-1) \\
&\iff p \mid (x+1) \text{ or } p \mid (x-1) \\
&\iff x+1 \equiv 0 \pmod{p} \text{ or } x-1 \equiv 0 \pmod{p} \\
&\iff x \equiv -1 \pmod{p} \text{ or } x \equiv 1 \pmod{p}.
\end{aligned}
$$

Conversely, if either $x \equiv -1 \pmod{p}$ or $x \equiv 1 \pmod{p}$ holds, then $x^2 \equiv 1 \pmod{p}$. □

Definition 2.4.3. The number x is called a *nontrivial square root* of 1 modulo n if it satisfies (2.48) but $x \not\equiv \pm 1 \pmod{n}$.

Example 2.4.6. The number 6 is a nontrivial square root of 1 modulo 35. since $x^2 = 6^2 \equiv 1 \pmod{35}$, $x = 6 \not\equiv \pm 1 \pmod{35}$.

Corollary 2.4.1. If there exists a nontrivial square root of 1 modulo n, then n is composite.

Example 2.4.7. Show that 1387 is composite. Let $x = 2^{693}$. We have $x^2 = (2^{693})^2 \equiv 1 \pmod{1387}$, but $x = 2^{693} \equiv 512 \not\equiv \pm 1 \pmod{1387}$. So, 2^{693} is a nontrivial square root of 1 modulo 1387. Then by Corollary 2.4.1, 1387 is composite.

Now we are in a position to introduce the strong pseudoprimality test, an improved version of the (Fermat) pseudoprimality test.

Theorem 2.4.3 (Strong pseudoprimality test). Let n be an odd prime number: $n = 1 + 2^j d$, with d odd. Then the b-sequence defined by

$$\{b^d, \ b^{2d}, \ b^{4d}, \ b^{8d}, \ \ldots, \ b^{2^{j-1}d}, \ b^{2^j d}\} \ \bmod n \tag{2.49}$$

has one of the following two forms:

$$(1, \ 1, \ \ldots, \ 1, \quad 1, \ 1, \ \ldots, \ 1), \tag{2.50}$$

$$(?, \ ?, \ \ldots, \ ?, \quad -1, \ 1, \ \ldots, \ 1), \tag{2.51}$$

reduced to modulo n, for any $1 < b < n$. (The question mark "?" denotes a number different from ± 1.)

The correctness of the above theorem relies on Theorem 2.4.2: if n is prime, then the only solutions to $x^2 \equiv 1 \pmod{n}$ are $x \equiv \pm 1$. To use the strong pseudoprimality test on n, we first choose a base b, usually a small prime. Then compute the b-sequence of n; write $n - 1$ as $2^j d$ where d is odd, compute $b^d \bmod n$, the first term of the b-sequence, and then square repeatedly to obtain the b-sequence of $j + 1$ numbers defined in (2.49), all reduced to modulo n. If n is prime, then the b-sequence of n will be of the form of either (2.50) or (2.51). If the b-sequence of n has any one of the following three forms

$$(?, \ \ldots, \ ?, \ 1, \ 1, \ \ldots, \quad 1), \tag{2.52}$$

$$(?, \ \ldots, \ ?, \ ?, \ ?, \ \ldots, -1), \tag{2.53}$$

$$(?, \ \ldots, \ ?, \ ?, \ ?, \ \ldots, \quad ?), \tag{2.54}$$

then n is *certainly* composite. However, a composite can masquerade as a prime for a few choices of base b, but should not be "too many". The above idea leads naturally to a very efficient and also practically useful algorithm for (pseudo)primality testing:

Algorithm 2.4.2 (Strong pseudoprimality test). This algorithm will test n for primality with high probability:

[1] Let n be an odd number, and the base b a random number in the range $1 < b < n$. Find j and d with d odd, so that $n - 1 = 2^j d$.

[2] Set $i \leftarrow 0$ and $y \leftarrow b^d \pmod{n}$.

[3] If $i = 0$ and $y = 1$, or $y = n - 1$, then terminate the algorithm and output "n is probably prime". If $i > 0$ and $y = 1$ goto [5].

[4] $i \leftarrow i + 1$. If $i < j$, set $y \equiv y^2 \pmod{n}$ and return to [3].

[5] Terminate the algorithm and output "n is definitely not prime".

The strong pseudoprimality test is most often called the Miller-Rabin test, in honor of Gary L. Miller and Michael O. Rabin It is also called the Miller-Selfridge-Rabin test, because Selfridge used the test in 1974 before Miller first published the result.

Theorem 2.4.4. The strong pseudoprimality test above runs in time $\mathcal{O}((\log n)^3)$.

Definition 2.4.4. A positive integer n with $n - 1 = d \cdot 2^j$ and d odd, is called a base-b *strong probable prime* if it passes the strong pseudoprimality test described above (i.e., the last term in sequence 2.49 is 1, and the first occurrence of 1 is either the first term or is preceded by -1). A base-b strong probable prime is called a base-b *strong pseudoprime* if it is a composite.

If n is prime and $1 < b < n$, then n passes the test. The converse is usually true, as shown by the following theorem.

Theorem 2.4.5. Let $n > 1$ be an odd composite integer. Then n passes the strong test for at most $(n - 1)/4$ bases b with $1 \le b < n$.

Proof. The proof is rather lengthy, we thus only give a sketch of the proof. First note that if p is an odd prime, and α and q are positive integers then the number of incongruent solutions of the congruence

$$x^{q-1} \equiv 1 \pmod{p^\alpha}$$

is $\gcd(q, p^{\alpha-1}(p - 1))$. Let $n - 1 = d \cdot 2^j$, where d is an odd positive integer and j is a positive integer. For n to be a strong pseudoprime to the base b, either

$$b^d \equiv 1 \pmod{n}$$

or

$$b^{2^i d} \equiv -1 \pmod{n}$$

for some integer i with $0 < i < j - 1$. In either case, we have

$$b^{n-1} \equiv 1 \pmod{n}.$$

Let the standard prime factorization of n be

$$n = p_1^{\alpha_1} p_2^{\alpha_2} \cdots p_k^{\alpha_k}.$$

By the assertion made at the beginning of the proof, we know that there are

$$\gcd\left(n - 1, \ p_i^{\alpha_i}(p_i - 1)\right) = \gcd(n - 1, \ p_i - 1)$$

incongruent solutions to the congruence

$$x^{n-1} \equiv 1 \pmod{p_i^{\alpha_i}}, \qquad i = 1, 2, \ldots, k.$$

Further, by the Chinese remainder theorem, we know that there are exactly

$$\prod_{i=1}^{k} \gcd(n - 1, \ p_i - 1)$$

incongruent solutions to the congruence

$$x^{n-1} \equiv 1 \pmod{n}.$$

To prove the theorem, there are three cases to consider:

(1) the standard prime factorization of n contains a prime power $p_r^{\alpha_r}$ with exponent $\alpha_r \geq 2$;

(2) $n = pq$, with p and q distinct odd primes.

(3) $n = p_1 p_2 \cdots p_k$, with $p_1, p_2, \ldots, p_k$ distinct odd primes.

The second case can actually be included in the third case. We consider here only the first case. Since

$$\frac{p_r - 1}{p_r^{\alpha_r}} = \frac{1}{p_r^{\alpha_r - 1}} - \frac{1}{p_r^{\alpha_r}}$$

$$\leq \frac{2}{9},$$

we have

$$\prod_{i=1}^{k} \gcd(n-1,\ p_i - 1) \quad \leq \quad \prod_{i=1}^{k} (p_i - 1)$$

$$\leq \quad \left(\prod_{\substack{i=1 \\ i \neq r}}^{k} p_i \right) \left(\frac{2}{9} p_r^{\alpha_r} \right)$$

$$\leq \quad \frac{2}{9} n$$

$$\leq \quad \frac{n-1}{4} \qquad \text{for } n \geq 9.$$

Thus, there are at most $(n-1)/4$ integers b, $1 < b < n-1$, for which n is a base-b strong pseudoprime and n can pass the strong test. $\qquad\square$

A probabilistic interpretation of Theorem 2.4.5 is as follows:

Corollary 2.4.2. Let $n > 1$ be an odd composite integer and b be chosen randomly from $\{2, 3, \ldots, n-1\}$. Then the probability that n passes the strong test is less than $1/4$.

From Corollary 2.4.2, we can construct a simple, general purpose, polynomial time primality test which has a positive (but arbitrarily small) probability of giving the wrong answer. Suppose an error probability of ϵ is acceptable. Choose k such that $4^{-k} < \epsilon$, and select $b_1, b_2, \ldots, b_k$ randomly and independently from $\{2, 3, \ldots, n-1\}$. If n fails the strong test on b_i, $i = 1, 2, \ldots, k$, then n is a strong probable prime.

Theorem 2.4.6. The strong test (i.e., Algorithm 2.4.2) requires, for $n - 1 = 2^j d$ with d odd and for k randomly selected bases, at most $k(2 + j) \log n$ steps. If n is prime, then the result is always correct. If n is composite, then the probability that n passes all k tests is at most $1/4^k$.

Proof. The first two statements are obvious, only the last statement requires proof. An error will occur only when the n to be tested is composite and the bases $b_1, b_2, \ldots, b_k$ chosen in this particular run of the algorithm are all non-witnesses. (An integer a is a *witness* to the compositeness of n if it is possible using a to prove that n is composite, otherwise it is a *non-witness*). Since the probability of randomly selecting a non-witness is smaller than $1/4$ (by Corollary 2.4.2), then the probability of independently selecting k non-witnesses is smaller than $1/4^k$. Thus the probability that with any given number n, a particular run of the algorithm will produce an erroneous answer is smaller than $1/4^k$. $\qquad\square$

Finally in this section, we move on to another pseudoprimality test.

Theorem 2.4.7. Let n be a positive integer greater than 1, choose, at random, k integers $b_1, b_2, \ldots, b_k$ with $0 < b_i < n$ and $\gcd(b_i, n) = 1$ and compute

$$b_i^{(n-1)/2} \equiv \left(\frac{b_i}{n}\right) \pmod{n}, \quad \text{for } i = 1, 2, \ldots, k. \tag{2.55}$$

If (2.55) fails to hold for any i, then n is composite. The probability that n is composite but (2.55) holds for every i is less than $1/2^k$.

How many b_i satisfy (2.55)? The following is the answer.

Theorem 2.4.8. Let n be an odd composite number. If n is an Euler pseudoprime, then the number of b_i in $1 \le b_i < n$ with $\gcd(b_i, n) = 1$ is less than or equal to $\phi(n)/2$.

The above theorem leads naturally to the following pseudoprimality test:

Algorithm 2.4.3 (Euler's pseudoprimality test). This algorithm will test n for primality with high probability; an odd composite n that can pass the test is at most $1/2^k$.

[1] Let n be an odd number and k the number of bases b_i. Set $i \leftarrow 1$

[2] Choose at random b_i in $1 < b_i < n - 1$.

[3] If $\gcd(b_i, n) \ne 1$ or $b_i^{(n-1)/2} \not\equiv \left(\frac{b_i}{n}\right) \pmod{n}$, then output "$n$ is composite" and terminate the algorithm.

[4] $i \leftarrow i + 1$. If $i < k$, go to Step [2]. Output "n is probably prime" and terminate the algorithm.

Euler's pseudoprimality test is often called the Solovay-Strassen test, in honor of its inventors Solovay and Strassen [239].

Definition 2.4.5. If the positive integer $n > 1$ passes Euler's pseudoprimality test on base b, then n is the base-b *Euler probable prime*. A Euler probable prime is called the base-b *Euler pseudoprime* if it is composite.

Example 2.4.8. Let $n = 1105 = 5 \cdot 13 \cdot 17$ and $b = 2$. Then we have $b^{(n-1)/2} \bmod n = 2^{(1105-1)/2} \bmod 1105 = 1$ and $\left(\frac{b}{n}\right) = \left(\frac{2}{1105}\right) = 1$. Thus, $b^{(n-1)/2} \equiv \left(\frac{2}{1105}\right) \pmod{n}$. Therefore, 1105 is a base-2 Euler pseudoprime. However, 1105 is not a base-2 strong pseudoprime (why? left as an exercise).

Remark 2.4.1. Since every base-a strong pseudoprime is a base-a Euler pseudoprime, more composites pass Euler's pseudoprimality test than the strong pseudoprimality test, although both require $\mathcal{O}(\log n)^3$ bit operations.

Problems for Section 2.4

Problem 2.4.1. Show that for any positive integer $n > 4$, the following three statements are equivalent:

(1) n is composite,

(2) $(n - 1)! \equiv 0 \pmod{n}$,

(3) $(n - 1)! \not\equiv -1 \pmod{n}$.

Problem 2.4.2. Let

$$
\begin{cases}
p \text{ be an odd prime,} \\
p \nmid b(b^2 - 1), \\
n = \dfrac{b^{2p} - 1}{b^2 - 1}.
\end{cases}
$$

Then n is a based b pseudoprime.

Problem 2.4.3. Prove that there are infinitely many Carmichael numbers.

Problem 2.4.4. Show that if n is an odd composite number, then there exists an integer b such that $\gcd(b, n) = 1$ and n is not a base b Euler pseudoprime.

Problem 2.4.5. Show that if n is a base b strong pseudoprime, then n is a base b Euler pseudoprime.

Problem 2.4.6. Let $n > 1$ be an odd integer. Show that n is prime if and only if

$$
\forall a \in \mathbb{Z}_n^*, a^{(n-1)/2} \equiv \left(\frac{a}{n}\right) \pmod{n}.
$$

Problem 2.4.7. Let $n > 1$ be an odd integer. Show that n is prime if and only if

$$
\begin{aligned}
\forall a \in \mathbb{Z}_n^*, a^{(n-1)/2} &\equiv \pm 1 \pmod{n}, \\
\exists a \in \mathbb{Z}_n^*, a^{(n-1)/2} &\equiv -1 \pmod{n}.
\end{aligned}
$$

Problem 2.4.8. Let R_n be the set

$$\{b \in \mathbb{Z}^* : b^{n-1} \not\equiv 1 \pmod{n} \text{ or } \exists t \geq 0, 1 < \gcd(b^{(n-1)/2^t} - 1, n) < n\}$$

where $n - 1 = 2^k u$, u odd, and $\exists t$ is restricted to $t \leq k$. Prove that for all odd composite integer $n > 9$,

$$
\frac{|\mathbb{Z}_n^* - R_n|}{\phi(n)} \leq \frac{1}{4}.
$$

Problem 2.4.9. Prove or disprove the following conjecture: there is an algorithm to determine all prime numbers less than or equal to n in $\mathcal{O}(\log n)$ bit operations.

2.5 Lucas Pseudoprimality Test

In this subsection, we shall study Lucas sequences and their applications to primality testing.

Let a, b be non-zero integers and $D = a^2 - 4b$. Consider the equation $x^2 - ax + b = 0$; its discriminant is $D = a^2 - 4b$, and α and β are the two roots:

$$\left.\begin{aligned} \alpha &= \frac{a + \sqrt{D}}{2} \\ \beta &= \frac{a - \sqrt{D}}{2}. \end{aligned}\right\} \tag{2.56}$$

So

$$\left.\begin{aligned} \alpha + \beta &= a \\ \alpha - \beta &= \sqrt{D} \\ \alpha\beta &= b. \end{aligned}\right\} \tag{2.57}$$

We define the sequences (U_k) and (V_k) by

$$\left.\begin{aligned} U_k(a, b) &= \frac{\alpha^k - \beta^k}{\alpha - \beta} \\ V_k(a, b) &= \alpha^k + \beta^k. \end{aligned}\right\} \tag{2.58}$$

In particular, $U_0(a, b) = 0, U_1(a, b) = 1$, while $V_0(a, b) = 2, V_1(a, b) = a$. For $k \geq 2$, we also have

$$\left.\begin{aligned} U_k(a, b) &= aU_{k-1} - bU_{k-2} \\ V_k(a, b) &= aV_{k-1} - bV_{k-2}. \end{aligned}\right\} \tag{2.59}$$

The sequences

$$\left.\begin{aligned} U(a, b) &= (U_k(a, b))_{k \geq 0} \\ V(a, b) &= (V_k(a, b))_{k \geq 0} \end{aligned}\right\} \tag{2.60}$$

are called the *Lucas sequences* associated with the pair (a, b). Special cases of Lucas sequences were considered by Fibonacci, Fermat, and Pell, among others. For example, the sequence $U_k(a, b)$, $k = 0, 1$, corresponding to $a = 1$, $b = -1$, was first considered by Fibonacci, and it begins as follows:

$$0, 1, 1, 2, 3, 5, 8, 13, 21, 34, 55, 89, 144, 233, 377, 610, 987, 1597, 2584, 4181, \ldots$$

These are called *Fibonacci numbers*, in honour of the Italian mathematician Fibonacci. The companion sequence to the Fibonacci numbers, still with $a = 1$, $b = -1$, is the sequence of *Lucas numbers*: $V_0 = V_0(1, -1) = 2$, $V_1 = V_1(1, -1) = 1$, and it begins as follows:

$$2, 1, 3, 4, 7, 11, 18, 29, 47, 76, 123, 199, 322, 521, 843, 1364, 2207, 3571,$$

$$5778, 9349, 15127, \ldots$$

If $a = 3$, $b = 2$, then the sequences obtained are

$$U_k(3, 2) = 2^k - 1 :$$

$$0, 1, 3, 7, 15, 31, 63, 127, 255, 511, 1023, 2047, 4095, 8191, 16383, \ldots$$

$$V_k(3, 2) = 2^k + 1 :$$

$$2, 3, 5, 9, 17, 33, 65, 129, 257, 513, 1025, 2049, 4097, 8193, 16385, \ldots$$

for $k \geq 0$. The sequences associated with $a = 2$, $b = -1$ are called *Pell sequences*; they begin as follows:

$$U_k(2, -1) : 0, 1, 2, 5, 12, 29, 70, 169, 408, 985, 2378, 5741, 13860, 33461, 80782, \ldots$$

$$V_k(2, -1) : 2, 2, 6, 14, 34, 82, 198, 478, 1154, 2786, 6726, 16238, 39202, 94642, \ldots$$

Now we are in a position to study some analogues of pseudoprimes in which $a^{n-1} - 1$ is replaced by a Lucas sequence. Recall that odd composite numbers n for which

$$a^{n-1} \equiv 1 \pmod{n}$$

are called Fermat pseudoprimes to base a.

Theorem 2.5.1 (Lucas theorem). Let a and b be integers and put $D = a^2 - 4b \neq 0$. Define the Lucas sequence $\{U_k\}$ with the parameters D, a, b by

$$U_k = \frac{\alpha^k - \beta^k}{\alpha - \beta}, \qquad k \geq 0 \tag{2.61}$$

where α and β are the two roots of $x^2 - ax + b = 0$. If p is an odd prime, $p \nmid b$ and $\left(\dfrac{D}{p}\right) = -1$, where $\left(\dfrac{D}{p}\right)$ is a Jacobi symbol, then $p \mid U_{p+1}$.

The above theorem can be used directly to construct a primality test, often called the *Lucas test*:

Corollary 2.5.1 (Converse of the Lucas theorem – Lucas test). Let n be an odd positive integer. If $n \nmid U_{n+1}$, then n is composite.

Just as there are Fermat probable primes and Fermat pseudoprimes, we also have the concepts of Lucas probable primes and Lucas pseudoprimes.

Definition 2.5.1. An odd positive integer n is called a *Lucas probable prime* with D, a and b, if $n \nmid b$, $\left(\dfrac{D}{n}\right) = -1$ and $n \mid U_{n+1}$. A Lucas probable prime n is called a *Lucas pseudoprime* if n is composite.

Another different but equivalent presentation of Theorem 2.5.1 is as follows:

Theorem 2.5.2. Let n be an odd positive integer, $\varepsilon(n)$ the Jacobi symbol $\left(\dfrac{D}{n}\right)$, and $\delta(n) = n - \varepsilon(n)$. If n is prime and $gcd(n, b) = 1$, then

$$U_{\delta(n)} \equiv 0 \ (\text{mod } n). \tag{2.62}$$

If n is composite, but (2.62) still holds, then n is called a *Lucas pseudoprime with parameters a and b.*

Although Theorem 2.5.2 is true when $\left(\dfrac{D}{n}\right) = 1$, it is best to avoid this situation. A good way to avoid this situation is to select a suitable D such that $\left(\dfrac{D}{n}\right) = -1$. Two methods have been proposed (see Baillie and Wagstaff [15]):

(1) Let D be the first element of the sequence $5, -7, 9, -11, 13, \ldots$ for which $\left(\dfrac{D}{n}\right) = -1$. Let $a = 1$ and $b = (1 - D)/4$.

(2) Let D be the first element of the sequence $5, 9, 13, 17, 21, \ldots$ for which $\left(\dfrac{D}{n}\right) = -1$. Let a be the least odd number exceeding $\sqrt{D}$ and $b = (a^2 - D)/4$.

The first 10 Lucas pseudoprimes found by the first method are

$$323, 377, 1159, 1829, 3827, 5459, 5777, 9071, 9179, 10877,$$

and the first 10 Lucas pseudoprimes found by the second method are:

$$323, 377, 1349, 2033, 2651, 3569, 3599, 3653, 3827, 4991.$$

The most interesting thing about the Lucas test is that if we choose the parameters D, a and b as described in the second method, then the first 50 Carmichael numbers and several other base-2 Fermat pseudoprimes will never be Lucas pseudoprimes (Baillie and Wagstaff [15]). This leads to the general belief that a combination of a strong pseudoprimality test and a Lucas pseudoprimality test (or just a combined test, for short) might be an infallible test for primality. Since to date, no composites have been found to pass such a combined test, it is thus reasonable to conjecture that:

Conjecture 2.5.1. If n is a positive integer greater than 1 which can pass the combination of a strong pseudoprimality test and a Lucas test, then n is prime.

The advantage of the combination of a strong test and a Lucas test seems to be that the two probable prime tests are independent. That is, n being a probable prime of the first type does not affect the probability of n being a probable prime of the second type. In fact, if n is a strong pseudoprime (to a certain base), then n is less likely than a typical composite to be a Lucas pseudoprime (with the parameters a and b), provided a and b are chosen properly, and vice versa. If n passes both a strong test and a Lucas test, we can be more certain that it is prime than if it merely passes several strong tests, or several Lucas tests. Pomerance, Selfridge and Wagstaff [189] issued two challenges for an example of a composite number which passes both a strong pseudoprimality test base 2 and a Lucas test ($620), and/or a proof that no such number exists ($620). At the moment, the prizes are unclaimed; no counter-example has yet been found.

Problems for Section 2.5

Problem 2.5.1. (Rotkiewicz) Define n to be the Fibonacci pseudoprime if

$$F_{n-\epsilon(n)} \equiv 0 \pmod{n}$$

where F_i is the ith Fibonacci number and $\epsilon(n)$ is the Jacobi symbol $\left(\dfrac{a^2 - 4b}{n}\right)$. Prove that if $b = \pm 1$ and a, b are not both 1, then there are infinitely many odd composite Lucas pseudoprimes with parameters a, b.

Problem 2.5.2. (Emma Lehmer) Define n to be the Lucas pseudoprime if

$$L_{n-\epsilon(n)}(a, b) \equiv 0 \pmod{n}$$

where F_i is the ith Fibonacci number, and $\epsilon(n)$ is defined in the previous problem. Prove that there are infinitely many prime p such that $n = F_{2p}$ is a Fibonacci pseudoprime.

Problem 2.5.3. Prove that if n is a prime, F_i is the ith Fibonacci number, and $\left(\frac{n}{5}\right)$ is the Legebdre symbol, then

$$n \mid F_{n-\left(\frac{n}{5}\right)}.$$

Problem 2.5.4. (Pomerance-Selfridge-Wagstaff $620 Prize Problem 1) Prove or disprove the conjecture that no odd composite number is both a strong pseudoprime to base 2 and a Lucas pseudoprime with $a = 1$ and $b = (1-D)/4$, where D is the first element of the sequence $5, -7, 9, -11, 13, \ldots$ for which $\left(\dfrac{D}{n}\right) = -1$. ($500 from Pomerance, $100 from Wagstaff, and $20 from Selfridge.)

Problem 2.5.5. (Pomerance-Selfridge-Wagstaff $620 Prize Problem 2) Find an odd composite number which is both a strong pseudoprime to base 2 and a Lucas pseudoprime with $a = 1$ and $b = (1 - D)/4$, where D is the first element of the sequence $5, -7, 9, -11, 13, \ldots$ for which $\left(\dfrac{D}{n}\right) = -1$. ($500 from Selfridge, $100 from Wagstaff, and $20 from Pomerance.)

Problem 2.5.6. Find an odd composite number which is simultaneously a base 2 strong pseudoprime and a Fibonacci pseudoprime.

Problem 2.5.7. (Atkin \$2500 Prize Problem) Find an odd composite which is simultaneously a base b strong pseudoprime, a Fibonacci pseudoprime, and a Lucas pseudoprime defined in the previous problem.

Problem 2.5.8. Prove that if n is a base b strong pseudoprime, then it is an Euler base b pseudoprime.

Problem 2.5.9. (Miler-Oesterlé-Bach) If n is an odd composite number, then there is a number r, $1 < r < 2(\log n)^2$, such that n is not a base r strong pseudoprime.

2.6 Elliptic Curve Primality Tests

In the past 20 years or so, there have been some surprising applications of elliptic curves to problems in primality testing, integer factorization and public-key cryptography. In 1985, Lenstra [131] announced an elliptic curve factoring method (the formal publication was in 1987), just one year later, Goldwasser and Kilian [82] adapted Lenstra's factoring algorithm to obtain an elliptic curve primality test, and Miller in 1986 [150] and Koblitz in 1987 [120] independently arrived in an idea of elliptic curve cryptography. In this section, we discuss elliptic curve primality testing, and leave elliptic curve factoring and elliptic curve cryptography to Chapters 3 and 4.

First we introduce a test based on Cox [61].

Theorem 2.6.1 (Cox). Let $n \in \mathbb{N}$ with $n > 13$ and $\gcd(n, 6) = 1$, and let $E : y^2 \equiv x^3 + ax + b \pmod{n}$ be an elliptic curve over $\mathbb{Z}/n\mathbb{Z}$. Suppose that

(1) $n + 1 - 2\sqrt{n} \leq |E(\mathbb{Z}_n)| \leq n + 1 + 2\sqrt{n}$.

(2) $|E(\mathbb{Z}_n)| = 2q$, with q an odd prime.

If $P \neq \mathcal{O}_E$ is a point on E and $qP = \mathcal{O}_E$, then n is prime.

Proof. See pp 324–325 of Cox [61].

Example 2.6.1. Suppose we wish to prove that $n = 9343$ is a prime using the above elliptic curve test. First notice that $n > 13$ and $\gcd(n, 6) = 1$. Next, we choose an elliptic curve $y^2 = x^3 + 4x + 4$ over $\mathbb{Z}/n\mathbb{Z}$ with $P = (0, 2)$, and calculate $|E(\mathbb{Z}/n\mathbb{Z})|$, the number of points on the elliptic curve $E(\mathbb{Z}/n\mathbb{Z})$, by e.g., Theorem 1.9.3 in Chapter 1 or by the numerical exhaustive method (by listing all the possible points on the curve). Suppose we use the numerical exhaustive method, we then know that there are 9442 points on this curve:

No.	Points on the curve	No.	Points on the curve
1	$\mathcal{O}_E$	2	$(0, 2)$
3	$(0, 9341)$	4	$(1, 3)$
5	$(1, 9340)$	6	$(3, 1264)$
7	$(3, 8079)$	8	$(7, 3541)$
9	$(7, 5802)$	10	$(10, 196)$
$\vdots$	$\vdots$	$\vdots$	$\vdots$
9439	$(9340, 4588)$	9440	$(9340, 4755)$
9441	$(9341, 3579)$	9442	$(9341, 5764)$

Thus, $|E(\mathbb{Z}_{9343})| = 9442$. Now it is ready to verify the two conditions in Theorem 2.6.1. For the first condition, we have:

$$\lfloor n + 1 - 2\sqrt{n} \rfloor = 9150 < 9442 < 9537 = \lfloor n + 1 + 2\sqrt{n} \rfloor.$$

For the second condition, we have:

$$|E(\mathbb{Z}_{9343})| = 9442 = 2q, \quad q = 4721 \in \text{Primes}.$$

So, both conditions are satisfied. Finally and most importantly, we calculate qP over the elliptic curve $E(\mathbb{Z}_{9343})$ by tabling its values as follows (using Algorithm 2.1.2):

2P	=	(1,9340)	4P	=	(1297,1515)
9P	=	(6583,436)	18P	=	(3816,7562)
36P	=	(2128,1972)	147P	=	(6736,3225)
295P	=	(3799,4250)	590P	=	(7581,7757)
1180P	=	(5279,3262)	2360P	=	(3039,4727)
4721P	=	$\mathcal{O}_E$			

Since $P \in E(\mathbb{Z}_{9343})$ and $P \neq \mathcal{O}_E$, but $qP \in E(\mathbb{Z}_{9343})$ and $qP = \mathcal{O}_E$, we conclude that $n = 9343$ is a prime number!

The main problem with the above test is the calculation of $|E(\mathbb{Z}_n)|$; when n becomes large, finding the value of $|E(\mathbb{Z}_n)|$ is as difficult as proving the primality of n [153]. Fortunately, Goldwasser and Kilian found a way to overcome this difficulty. To introduce the Goldwasser-Kilian method, let us first introduce a useful converse of Fermat's little theorem, which is essentially Pocklington's theorem:

Theorem 2.6.2 (Pocklington's theorem). Let s be a divisor of $n-1$. Let a be an integer prime to n such that

$$\left. \begin{array}{l} a^{n-1} \equiv 1 \pmod{n} \\ \gcd(a^{(n-1)/q}, n) = 1 \end{array} \right\} \tag{2.63}$$

for each prime divisor q of s. Then each prime divisor p of n satisfies

$$p \equiv 1 \pmod{s}. \tag{2.64}$$

Corollary 2.6.1. If $s > \sqrt{N} - 1$, then n is prime.

The Goldwasser-Kilian test can be regarded as an elliptic curve analog of Pocklington's theorem:

Theorem 2.6.3 (Goldwasser-Kilian). Let n be an integer greater than 1 and prime to 6, E an elliptic curve over $\mathbb{Z}/n\mathbb{Z}$, P a point on E, m and s two integers with $s \mid m$. Suppose we have found a point P on E that satisfies $mP = \mathcal{O}_E$, and that for each prime factor q of s, we have verified that $(m/q)P \neq \mathcal{O}_E$. Then if p is a prime divisor of n, $|E(\mathbb{Z}_p)| \equiv 0 \pmod{s}$.

Corollary 2.6.2. If $s > (\sqrt[4]{n} + 1)^2$, then n is prime.

Combining the above theorem with Schoof's algorithm which computes $|E(\mathbb{Z}_p)|$ in time $\mathcal{O}\left((\log p)^{8+\epsilon}\right)$, we obtain the following Goldwasser-Kilian algorithm [82] and [83].

Algorithm 2.6.1 (Goldwasser-Kilian Algorithm). Given a probable prime n, this algorithm will show whether or not n is indeed prime.

[1] Choose a non-singular elliptic curve E over $\mathbb{Z}/n\mathbb{Z}$, for which the number of points m satisfies $m = 2q$, with q a probable prime;

[2] If (E, m) satisfies the conditions of Theorem 2.6.3 with $s = m$, then n is prime, otherwise it is composite;

[3] Perform the same primality proving procedure for q;

[4] Exit.

The running time of the Goldwasser-Kilian algorithm is given in the following two theorems [10]:

Theorem 2.6.4. Suppose that there exist two positive constants c_1 and c_2 such that the number of primes in the interval $[x, x + \sqrt{2x}]$, where $(x \geq 2)$, is greater than $c_1\sqrt{x}(\log x)^{-c_2}$, then the Goldwasser-Kilian algorithm proves the primality of n in expected time $\mathcal{O}\left((\log n)^{10+c_2}\right)$.

Theorem 2.6.5. There exist two positive constants c_3 and c_4 such that, for all $k \geq 2$, the proportion of prime numbers n of k bits for which the expected time of Goldwasser-Kilian is bounded by $c_3((\log n)^{11})$ is at least

$$1 - c_4 2^{-k^{\frac{1}{\log \log k}}}.$$

A serious problem with the Goldwasser-Kilian test is that Schoof's algorithm [217] seems almost impossible to implement. In order to avoid the use of Schoof's algorithm, Atkin and Morain [10] in 1991 developed a new implementation method called ECPP (Elliptic Curve Primality Proving), which uses the properties of elliptic curves over finite fields related to complex multiplication. We summarize the principal properties of ECPP as follows:

Theorem 2.6.6 (Atkin-Morain). Let p be a rational prime number that splits as the product of two principal ideals in a field $\mathcal{K}$: $p = \pi\pi'$ with π, π' integers of $\mathcal{K}$. Then there exists an elliptic curve E defined over $\mathbb{Z}_p$ having complex multiplication by the ring of integers of $\mathcal{K}$, whose cardinality is

$$m = N_{\mathcal{K}}(\pi - 1) = (\pi - 1)(\pi' - 1) = p + 1 - t$$

with $|t| \leq 2\sqrt{p}$ (Hasse's Theorem) and whose invariant is a root of a fixed polynomial $H_D(X)$ (depending only upon D) modulo p.

For more information on the computation of the polynomials H_D, readers are referred to [61] and [156]. Note that there are also some other important improvements on the Goldwasser-Kilian test, notably the Adleman-Huang's primality proving algorithm [4] using hyperelliptic curves.

The Goldwasser-Kilian algorithm begins by searching for a curve and computes its number of points, but the Atkin-Morain ECPP algorithm does exactly the opposite. The following is a brief description of the ECPP algorithm.

Algorithm 2.6.2 (Atkin-Morain ECPP). Given a probable prime n, this algorithm will show whether or not n is indeed prime.

[1] (Initialization) Set $i \leftarrow 0$ and $N_0 \leftarrow n$.

[2] (Building the sequence)

While $N_i > N_{small}$
[2.1] Find a D_i such that $N_i = \pi_i\pi_i'$ in $\mathcal{K} = \mathcal{Q}(\sqrt{-D_i})$;

[2.2] If one of the $w(-D_i)$ numbers $m_1, ..., m_w$ $(m_r = N_K(\omega_r - 1)$ where ω_r is a conjugate of π) is probably factored goto step [2.3] else goto [2.1];

[2.3] Store $\{i, N_i, D_i, \omega_r, m_r, F_i\}$ where $m_r = F_iN_{i+1}$. Here F_i is a completely factored integer and N_{i+1} a probable prime; set $i \leftarrow i + 1$ and goto step [2.1].

[3] (Proving)

For i from k down to 0
[3.1] Compute a root j of $H_{D_i}(X) \equiv 0 \pmod{N_i}$;

[3.2] Compute the equation of the curve E_i of the invariant j and whose cardinality modulo N_i is m_i;

[3.3] Find a point P_i on the curve E_i;

[3.4] Check the conditions of Theorem 2.6.6 with $s = N_{i+1}$ and $m = m_i$.

[4] (Exit) Terminate the execution of the algorithm.

For ECPP, only the following heuristic analysis is known [156].

Theorem 2.6.7. The expected running time of the ECPP algorithm is roughly proportional to $\mathcal{O}\left((\log n)^{6+\epsilon}\right)$ for some $\epsilon > 0$.

Corollary 2.6.3. The ECPP algorithm is in $\mathcal{ZPP}$.

Thus, for all practical purposes, we could just simply use a combined test of a probabilistic test and an elliptic curve test as follows:

Algorithm 2.6.3 (Practical primality testing). Given a random odd positive integer n, this algorithm will make a combined use of probabilistic tests and elliptic curve tests to determine whether or not n is prime:

[1] (Primality Testing – Probabilistic Testing) Use a combination of the strong pseudoprimality test and the Lucas pseudoprimality test to determine if n is a probable prime. (This has been implemented in Maple function *isprime*.) If it is, go to [2], else report that n is composite and go to [3].

[2] (Primality Proving – Elliptic Curve Proving) Use an elliptic curve test (e.g., the ECPP test) to verify whether or not n is indeed a prime. If it is, then report that n is prime, otherwise, report that n is composite.

[3] (Exit) Terminate the algorithm.

Problems for Section 2.6

Problem 2.6.1. Prove Cox's Theorem for elliptic curve primality test (Theorem 2.6.1).

Problem 2.6.2. Prove Pocklington's theorem (Theorem 2.6.2) and its corollary (Corollary 2.6.2).

Problem 2.6.3. Prove the following theorem: Let $n = FR + 1$, where $0 < R < F$. If for some a we have

$$\begin{cases} a^{n-1} \equiv 1 \ (\text{mod } n), \\ \gcd\left(a^{(n-1)/q} - 1, \ n\right) = 1, \ \text{for each distinct prime } q \mid F. \end{cases}$$

Then n is a prime.

Problem 2.6.4. Prove Selfridge's theorem: Let $n > 1$ be an odd integer. If

$$n - 1 = \prod_{i-1}^{k} p_i^{\alpha_i}$$

where p_i, $i = 1, 2, \ldots, k$ are primes and each p_i there exists an a_i such that

$$\begin{cases} a_i^{n-1} \equiv 1 \ (\mathrm{mod}\ n), \\ a_i^{(n-1)/p_i} \not\equiv 1 \ (\mathrm{mod}\ n). \end{cases}$$

Then n is a prime.

Problem 2.6.5. Prove the Goldwasser-Kilian theorem (Theorem 2.6.3) and its corollary (Corollary 2.6.2).

Problem 2.6.6. Show that the Goldwasser-Kilian elliptic curve test algorithm is a $\mathcal{ZPP}$ algorithm.

Problem 2.6.7. Show that the Atkin-Morain ECPP algorithm is a $\mathcal{ZPP}$ algorithm.

Problem 2.6.8. Use Cox's test (Theorem 2.6.1) to prove that 26869 is a prime.

Problem 2.6.9. Use the Goldwasser-Kilian elliptic curve test to show that 907 is a prime.

Problem 2.6.10. Use some variant of ECPP to prove the following four numbers are prime:

(1) $\dfrac{2^{3539}+1}{3}$,

(2) $2177^{580} + 580^{2177}$,

(3) $4405^{2638} + 2638^{4405}$,

(4) $10^{9999} + 33603$.

2.7 Superpolynomial-Time Tests

The tests discussed so far are either too slow or not deterministic. This section discusses some deterministic and nearly polynomial-time tests for primality. First we introduce the APR (Adleman-Pomerance-Rumely) test [3]. This test is strongly influenced by the Solovay-Strassen test. If for some b with $\gcd(a, n) = 1$, we get

$$b^{(n-1)/2} \not\equiv 1 \ (\mathrm{mod}\ n)$$

then n is composite. However, if we carefully choose some b we may get

$$b^{(n-1)/2} \equiv 1 \ (\mathrm{mod}\ n).$$

Define the *mock residue symbol* as follows

$$\left\langle \frac{b}{n} \right\rangle = \begin{cases} \pm 1 & \text{if } (b^{(n-1)/2} \equiv \pm 1 \pmod{n}) \\ 0 & \text{otherwise} \end{cases}$$

then one can prove:

Theorem 2.7.1. If

$$\left\langle \frac{b}{n} \right\rangle = -1$$

and

$$\left\langle \frac{b}{n} \right\rangle^m = \left\langle \frac{c}{n} \right\rangle$$

for certain integer b and c, then

$$\left\langle \frac{b}{r} \right\rangle^m = \left\langle \frac{c}{r} \right\rangle$$

for any prime divisor r of n.

To prove n is prime, one uses a set $\mathcal{I}$ of initial primes p. Define a *Euclidean prime* (the set of such primes are denoted by $\mathcal{Q}$) with respect to $\mathcal{I}$ be a prime q such that $q - 1$ is square-free, every prime factor of $q - 1$ lies in $\mathcal{I}$, and the product of the Euclidean primes exceeds $\sqrt{n}$. Thus, one can has

$$q = 1 + \prod_{p \in S} p > \sqrt{n}, \quad S \subset I$$

and

$$Q = \prod_{q \in \mathcal{Q}} q.$$

By guessing the index r modulo p for each $p \in \mathcal{I}$, a possible divisor r of n can be located modulo q for each $q \in \mathcal{Q}$. By the Chinese Remainder Theorem, r can be found modulo Q. The number of possible divisors to be tested is

$$P = \prod_{p \in \mathcal{I}} p.$$

As the running time of the test is polynomial in the product of the initial primes, it is important for this product to be chosen as small as possible, in fact, it can be chosen such that

$$P = \prod_{p \in \mathcal{I}} p < (\log n)^{c \log \log \log n}$$

for all $n > 100$, where c is a certain computable constant. Suppose r is any prime divisor of n. Let g_j be any fix primitive root of q_j and $I_j = \mathrm{ind}_{g_j}(r)$ be the least integer such that $g_j \equiv r \pmod{q_j}$. The algorithm consists of the following main steps.

[1] Find the positive value of

$$\beta_{ij} \equiv I_j \pmod{q_j}, \quad i = 1, 2, 3, \ldots, m; \ j = 1, 2, 3, \ldots, k \qquad (2.65)$$

where m is the least integer such that if

$$q_j = \prod_{i=1}^{m} p_i^{\alpha_{ij}} + 1, \quad j = 1, 2, 3, \ldots, k$$

are all the primes that can be found with $\alpha_{ij} = 0$ or 1, then

$$\prod_{j=1}^{k} q_j > \sqrt{n}.$$

[2] Use the Chinese Remainder Theorem to find the values of

$$\lambda_j \equiv I_j \pmod{q_j - 1}.$$

[3] Compute

$$r \equiv r_j \equiv q_j^{\lambda_j} \pmod{q_j}, \quad j = 1, 2, 3, \ldots, k$$

[4] Use the Chinese Remainder Theorem to compute s such that

$$r \equiv s \ \left(\bmod \ \prod_{j=1}^{k} q_j \right)$$

where

$$s < \prod_{j=1}^{k} q_j.$$

[5] If s is not a divisor of n, then $r > \sqrt{n}$. If all possible values of p found in this function exceed $\sqrt{n}$, then n is prime.

The most difficult and time-consuming part of the test is the determine of β_{ij} in (2.65), which involves the evaluation of the mock residue symbols in various cyclotomic number fields. However, for $n > 100$, if let $T(n)$ be the bit operations needed to prove the primality, then there exists a positive, absolute and effectively computable constant c such that

$$T(n) < (\log n)^{c \log \log \log n}$$

which is best possible.

Shortly after the publication of APR test, some improvements and modifications have been proposed. The first notable improvement (see [130]) is Lenstra's Gauss sum test, which uses the congruences satisfied by the Gauss sums formed with generators for the character group (see Problems 2.7.4 and 2.7.5) . The test may be described as follows:

[1] Construct the initial prime set $\mathcal{I}$ and Euclidean prime set $\mathcal{Q}$, as in the APR test, with $Q > n$ and $\gcd(n, PQ) = 1$.

[2] For each (p, q) with $p \in I$, $q \in \mathcal{Q}$ and $p \mid (q - 1)$, find a character χ of $(\mathbb{Z}_q)^*$ with order p, and check if the following condition satisfies:

$$\tau(\chi) = \sum_{a=1}^{q-1} \chi(a)\zeta_q^a \in \mathbb{Z}[\zeta_{pq}].$$

If any one of tests fails, then n is not prime. If, during the tests, for any p, no character for which $\chi(n) \neq 1$ has been found, then check

$$\chi(n)^b = \chi(r)$$

for other characters until one is found for which $\chi(n) \neq 1$ or n is showing composite.

[3] For each b, $1 \leq b \leq P$, construct the least positive residue of $x \equiv n^b \pmod{Q}$, check if $x \mid n$. If no divisors are found, then n is prime.

This test is, of course, non-deterministic, since it may need to try to test a large number of characters χ. However, there is a deterministic form of the test which runs in superpolynomial-time. This improved version of APR test is unfortunately not practical on a computer. Lenstra, together with Cohen (see [51]), later gave another version, called the Jacobi sum test, which is computationally practical. Compared with Gauss sum test, the Jacobi sum test offers the following features, among others,

(1) working in a smaller ring $\mathbb{Z}[\zeta_p]$,

(2) replacing the Gauss sums with Jacobi sums,

(3) Permitting the use of character χ with prime power (rather than just prime) order,

(4) Replacing $Q > n$ with $Q > n^{1/2}$ or even $Q > n^{1/3}$,

(5) Using efficient programming techniques.

Both the deterministic and non-deterministic versions of the Jacobi sum test runs in superpolynomial-time.

Problems for Section 2.7

Problem 2.7.1. Prove Theorem 2.7.1: If

$$\left\langle \frac{b}{n} \right\rangle = -1$$

and

$$\left\langle \frac{b}{n} \right\rangle^m = \left\langle \frac{c}{n} \right\rangle$$

for certain integer b and c, then

$$\left\langle \frac{b}{r} \right\rangle^m = \left\langle \frac{c}{r} \right\rangle$$

for any prime divisor r of n.

Problem 2.7.2. (Pomerance-Odlyzko) Let $f(n)$ be the least positive square-free integer such that the product of all primes q, $q - 1 \mid f(n)$ exceeds $\sqrt{n}$. Then there exists positive absolute constants c_1, c_2 such that for all integers $n \geq 2$,

$$(\log n)^{c_1 \log \log \log n} < f(n) < (\log n)^{c_2 \log \log \log n}.$$

Problem 2.7.3. (Adleman-Pomerance-Rumely) Let $T(n)$ be the number of bit operations needed by the APR test. Then there exists positive absolute constants c_3, c_4 such that

$$(\log n)^{c_3 \log \log \log n} < f(n) < (\log n)^{c_4 \log \log \log n}.$$

Problem 2.7.4. Prove the following theorem. Let p and q be primes with $p \mid (q - 1)$, ζ_{pq} the primitive pqth roots of unity in $\mathbb{C}$, g the generator of $(\mathbb{Q}_q)^*$. Let also χ be a character of $(\mathbb{Q}_q)^*$ with order p. Define the Gauss sum as follows

$$\tau(\chi) = \sum_{a=1}^{q-1} \chi(a)\zeta_q^a \in \mathbb{Z}[\zeta_{pq}].$$

If n is a prime, then

$$\tau(\chi)^{n^{p-1}} \equiv \chi(n) \pmod{n\mathbb{Z}[\zeta_{pq}]}.$$

Problem 2.7.5. Prove that if

$$\tau(\chi)^{n^{p-1}} \equiv \chi(n) \pmod{n\mathbb{Z}[\zeta_{pq}]}.$$

is true, then for any prime divisor r of n, there is a number b modulo p, such that

$$\chi(n)^b = \chi(r).$$

2.8 Polynomial-Time Tests

On 6 August 2002, Agrawal, Kayal and Saxena in the Department of Computer Science and Engineering, Indian Institute of Technology, Kanpur, Proposed a deterministic polynomial-time test (AKS test for short) for primality, relying on no unproved assumptions. It was not a great surprise that such a test exists[6], but the relatively easy algorithm and proof is indeed a big surprise.

The key to the AKS test is another simple version of Fermat's Little Theorem:

Figure 2.9. Saxena, Kayal and Agrawal (Photo by courtesy of AKS)

Theorem 2.8.1. Let x be an indeterminate and $\gcd(a, n) = 1$ with $n > 1$. Then

$$n \in \text{Primes} \iff (x - a)^n \equiv (x^n - a) \pmod{n}. \qquad (2.66)$$

Proof. By the binomial theorem, we have

$$(x - a)^n = \sum_{r=0}^{n} \binom{n}{r} x^r (-a)^{n-r}.$$

If n is prime, then $n \mid \binom{n}{r}$ (i.e., $\binom{n}{r} \equiv 0 \pmod{n}$), for $r = 1, 2, \ldots, n - 1$. Thus, $(x - a)^n \equiv (x^n - a^n) \pmod{n}$ and (2.66) follows from Fermat's Little

---

[6] Dixon [73] predicated in 1984 that "the prospect for a polynomial-time algorithm for proving primality seems fairly good, but it may turn out that, on the contrary, factoring is NP-hard".

Theorem. On the other hand, if n is composite, then it has a prime divisor q. Let q^k be the greatest power of q that divides n. Then q^k does not divide $\binom{n}{q}$ and is relatively prime to a^{n-q}, so the coefficient of the term x^q on the left of $(x-a)^n \equiv (x^n - a)$ is not zero, but it is on the right.   □

Remark 2.8.1. In about 1670 Leibniz (1646–1716) used the fact that if n is prime then n divides the binomial coefficient $\binom{n}{r}$, $r = 1, 2, \ldots, n-1$ to show that if n is prime then n divides $(a_1 + a_2 + \cdots + a_m)^n - (a_1^n + a_2^n + \cdots + a_m^n)$. Letting $a_i = 1$, for $i = 1, 2, \ldots, m$, Leibniz proved that n divides $m^n - m$ for any positive integer m.

Example 2.8.1. Let $a = 5$ and $n = 11$. Then we have:

$$(x - 5)^{11} = x^{11} - 55x^{10} + 1375x^9 - 20625x^8 + 206250x^7 - 1443750x^6$$
$$+ 7218750x^5 - 25781250x^4 + 64453125x^3 - 107421875x^2$$
$$+ 107421875x - 48828125$$
$$\equiv x^{11} - 5 \pmod{11}.$$

However, if we let $n = 12$, which is not a prime, then

$$(x - 5)^{12} = x^{12} - 60x^{11} + 1650x^{10} - 27500x^9 + 309375x^8 - 2475000x^7$$
$$+ 14437500x^6 - 61875000x^5 + 193359375x^4 - 429687500x^3$$
$$+ 644531250x^2 - 585937500x + 244140625$$
$$\not\equiv x^{12} - 5 \pmod{12}.$$

Theorem 2.8.1 provides a deterministic test for primality. However, the test cannot be done in polynomial-time because of the intractability of $(x - a)^n$; we need to evaluate n coefficients in the left-hand side of (2.66) in the worst case. A simple way to avoid this computationally intractable problem is to evaluate both sides of (2.66) modulo a polynomial[7] of the form $x^r - 1$: for an *appropriately chosen* small r. Thus, we get a new characterization of primes

$$n \in \text{Primes} \implies (x - a)^n \equiv (x^n - a) \pmod{x^r - 1, \ n}, \qquad (2.67)$$

for all r and n relatively prime to a. A problem with this characterization is that for particular a and r, some composites can satisfy (2.67), too. However, no composite n satisfies (2.67) for all a and r, that is,

$$n \in \text{Composites} \implies \exists \ a, r \text{ such that}$$

$$(x - a)^n \not\equiv (x^n - a) \pmod{x^r - 1, \ n}. \qquad (2.68)$$

---

[7] By analog with congruences in $\mathbb{Z}$, we say that polynomials $f(x)$ and $g(x)$ are congruent modulo $h(x)$ and write $f(x) \equiv g(x) \pmod{h(x)}$, whenever $f(x) - g(x)$ is divisible by $h(x)$. The set (ring) of polynomials modulo $h(x)$ is denoted by $\mathbb{Z}[x]/h(x)$. If all the coefficients in the polynomials are also reduced to n, then we write $f(x) \equiv g(x) \pmod{h(x)}$ as $f(x) \equiv g(x) \pmod{h(x), n}$, and $\mathbb{Z}[x]/h(x)$ as $\mathbb{Z}_n[x]/h(x)$ (or $\mathbb{F}_p[x]/h(x)$ if n is a prime p).

Example 2.8.2. Let $n = 6$ and $a = r = 5$. Then

$$(x - 5)^6 \equiv 3x^4 - 2x^3 + 3x^2 + x + 1 \ (\text{mod } x^5 - 1, 6),$$

$$x^6 - 5 \equiv x + 1 \ (\text{mod } x^5 - 1, 6),$$

$$(x - 5)^6 \not\equiv x^6 - 5 \ (\text{mod } x^5 - 1, 6).$$

The main idea of the AKS test is to restrict the range of a and r enough to keep the complexity of the computation polynomial, while ensuring that no composite n can pass (2.67).

Theorem 2.8.2 (Agrawal-Kayal-Saxena). Let $n \in \mathbb{Z}^+$. Let q and r be prime numbers. Let S be a finite set of integers. Assume that

(1) $q \mid (r - 1)$,

(2) $n^{(r-1)/q} \not\equiv 0, 1 \ (\text{mod } r)$,

(3) $\gcd(a - a', n) = 1$, for all distinct $a, a' \in S$,

(4) $\binom{|S| + q - 1}{|S|} \geq n^{2\lfloor \sqrt{r} \rfloor}$,

(5) $(x - a)^n \equiv (x^n - a) \ (\text{mod } x^r - 1, n)$, for all $a \in S$.

Then n is a prime power.

Proof. We follow the streamlined presentation of [22]. First find a prime factor p of n, with $p^{(r-1)/q} \not\equiv 0, 1 \ (\text{mod } r)$ and $q \mid (r - 1)$. If every prime factor p of n has $p^{(r-1)/q} \equiv 0, 1 \ (\text{mod } r)$, then $n^{(r-1)/q} \equiv 0, 1 \ (\text{mod } r)$. By assumption, we have $(x - a)^n \equiv (x^n - a) \ (\text{mod } x^r - 1, p)$ for all $a \in S$. Substituting x^{n^i} for x, we get $(x^{n^i} - a)^n \equiv (x^{n^{i+1}} - a) \ (\text{mod } x^{n^i r} - 1, p)$, and also $(x^{n^i} - a)^n \equiv (x^{n^{i+1}} - a) \ (\text{mod } x^r - 1, p)$. By induction, $(x - a)^{n^i} \equiv (x^{n^i} - a) \ (\text{mod } x^r - 1, p)$ for all $i \geq 0$. By Fermat's little theorem, $(x - a)^{n^i p^j} \equiv (x^{n^i} - a)^{p^j} \equiv (x^{n^i p^j} - a) \ (\text{mod } x^r - 1, p)$ for all $j \geq 0$. Now consider the products $n^i p^j$, with $0 \leq i, j \leq \lfloor \sqrt{r} \rfloor$ and $1 \leq n^i p^j \leq n^{2\lfloor \sqrt{r} \rfloor}$. There are $(1 + \lfloor \sqrt{r} \rfloor)^2 > r$ such (i, j) pairs, so there are (by the pigeon-hole principle) distinct pairs (i_1, j_1) and (i_2, j_2) for which $t_1 \equiv t_2 \ (\text{mod } r)$ where $t_1 = n^{i_1} p^{j_1}$, $t_2 = n^{i_2} p^{j_2}$. So, $x^{t_1} \equiv x^{t_2} \ (\text{mod } x^r - 1, p)$, therefore, $(x - a)^{t_1} \equiv (x^{t_1} - a) \equiv x^{t_2} - a \equiv (x - a)^{t_2} \ (\text{mod } x^r - 1, p)$ for all $a \in S$. Next find an irreducible polynomial $h(x)$ in $\mathbb{F}_p[x]$ dividing $(x^r - 1)/(x - 1)$ such that $(x - a)^{t_1} \equiv (x - a)^{t_2} \ (\text{mod } h(x), p)$ for all $a \in S$. Define G as a subgroup of $(\mathbb{F}_p[x]/h(x))^*$ generated by $\{x - a : a \in S\}$, then $g^{t_1} = g^{t_2}$ for all $g \in G$. Since G has at least $\binom{|S| + q - 1}{|S|}$ elements (by some combinatorics and the elementary theory of cyclotomic polynomials), we have

$$|t_1 - t_2| < n^{\lfloor \sqrt{r} \rfloor} p^{\lfloor \sqrt{r} \rfloor} \leq n^{2\lfloor \sqrt{r} \rfloor} \leq \binom{|S| + q - 1}{|S|} \leq |G|,$$

so, $t_1 = t_2$, as desired. $\qquad \square$

Remark 2.8.2. There are some new interesting developments and refinements over the above result. The American Institute of Mathematics had a workshop in 24-28 March 2003 on "Future Directions in Algorithmic Number Theory"; the institute has made the lecture notes and a set of problems of the workshop available through `http://www.aimath.org`.

To turn the above theorem into a deterministic polynomial-time test for primality, we first find a small odd prime r such that $n^{(r-1)/q} \equiv 0, 1 \pmod{r}$ and $\binom{q+s-1}{s} > n^{2\lfloor\sqrt{r}\rfloor}$; here q is the largest prime factor of $r - 1$, and s is any moderately large integer. A theorem of Fouvry [77] from analytic number theory implies that a suitable r exists on the order $\mathcal{O}((\log n)^6)$ and s on the order $\mathcal{O}((\log n)^4)$. Given such a triple (q, r, s), we can easily test that n have no prime factors $< s$ and that $(x - a)^n = x^n - b \pmod{x^r - 1, n}$ for all $a \in \{0, 1, 2, \ldots, s - 1\}$. Any failure of the first test reveals a prime factor of n and any failure of the second test proves that n is composite. If n passes both tests then n is a prime power. Here is the algorithm.

Algorithm 2.8.1. (The AKS Algorithm) Give a positive integer $n > 1$, this algorithm will decide whether or not n is prime in deterministic polynomial time.

[1] If $n = a^b$ with $a \in \mathbb{N}$ and $b > 1$, then output COMPOSITE.

[2] Find the smallest r such that $\text{ord}_r(n) > 4(\log n)^2$.

[3] If $1 < \gcd(a, n) < n$ for some $a \le r$, then output COMPOSITE.

[4] If $n \le r$, then output PRIME.

[5] For $a = 1$ to $\lfloor 2\sqrt{\phi(r)} \log n \rfloor$ do

if $(x - a)^n \not\equiv (x^n - a) \pmod{x^r - 1, n}$,

then output COMPOSITE.

[6] Output PRIME.

The algorithm is indeed very simple and short (with only 6 statements), possibly the shortest algorithm for a (big) unsolved problem ever!

Theorem 2.8.3 (Correctness). The above algorithm returns PRIME if and only if n is prime.

Proof. If n is prime, then the steps 1 and 3 will never return COMPOSITE. By Theorem 2.8.1, the `for` loop in step 5 will also never return COMPOSITE. Thus, n can only be identified to be PRIME in either step 4 or step 6. The only if part of the theorem left as an exercise. □

Theorem 2.8.4. The AKS algorithm runs in time $\mathcal{O}((\log n)^{12+\epsilon})$.

Proof. [Sketch] The algorithm has two main steps, 2 and 5. Step 2 finds a suitable r (such an r exists by Fouvry [77] and Baker and Harman [14]), and can be carried out in time $\mathcal{O}((\log n)^{9+\epsilon})$. Step 5 verifies (2.67), and can be performed in time $\mathcal{O}((\log n)^{12+\epsilon})$. So, the overall runtime of the algorithm is $\mathcal{O}((\log n)^{12+\epsilon})$.   □

Remark 2.8.3. Under the assumption of a conjecture on the density of the Sophie Germain primes, the AKS algorithm runs in time $\mathcal{O}((\log n)^{6+\epsilon})$. If a conjecture of Bhatacharjee and Pandey [8] is true, then this can be further reduced to $\mathcal{O}((\log n)^{3+\epsilon})$. Of course, we do not know if these conjectures are true.

Remark 2.8.4. The AKS algorithm is a major breakthrough in computational number theory. However, it can only be of theoretical interest, since its currently runtime is in $\mathcal{O}((\log n)^{12+\epsilon})$, which is much higher than $\mathcal{O}((\log n)^{6+\epsilon})$ for ECPP and $\mathcal{O}((\log n)^3)$ for Miller-Rabin's test. For all practical purposes, we would still prefer to use Miller-Rabin's probabilistic test [191] in the first instance and the zero-error probabilistic test ECPP in the last step of a primality testing process.

Remark 2.8.5. The efficiency of the AKS algorithm for test primality does not have (at least at present) any obvious connections to that of integer factorization, although the two problems are related to each other. The fastest factoring algorithm, namely the Number Field Sieve [132], has expected running time $\mathcal{O}(\exp(c\sqrt[3]{\log n}\sqrt[3]{(\log\log n)^2}))$. We do not know if the simple mathematics used in the AKS algorithm for primality testing can be used to other important mathematical problems, such as the integer factorization problem.

Remark 2.8.6. The efficiency of the AKS algorithm has not yet placed a threat on the security of the factoring base (such as the RSA) cryptographic systems, since the security of RSA depends on the computationally intractability of the integer factorization problem.

Problems for Section 2.8

Problem 2.8.1. (Binomial Theorem) Prove that

$$(x+y)^n = \sum_{i=0}^{n} \binom{n}{r} x^i y^{n-i}$$

where

$$\binom{n}{r} = \frac{n!}{i!(n-i)!} \quad \text{and} \quad m! = m(m-1)(m-2)\cdots 3\cdot 2\cdot 1.$$

---

[8] Bhatacharjee and Pandey conjectured in [23] that if $r \in$ Primes and $\gcd(r,n) = 1$, and $(x-1)^n \equiv x^n - 1 \pmod{x^r - 1, n}$, then either n is prime or $n^2 \equiv 1 \pmod{r}$.

Problem 2.8.2. Prove that

$$(x + y)^n \equiv x^n + y^n \ (\mathrm{mod} \ n)$$

for all variables x and y and primes n.

Problem 2.8.3. Prove that n is prime if and only if

$$(x + 1)^n \equiv x^n + 1 \ (\mathrm{mod} \ n)$$

in $\mathbb{Z}[x]$.

Problem 2.8.4. Show that if a square root x of 1 modulo n which is neither 1 nor -1, then n is composite.

Problem 2.8.5. It is easy to find a square modulo an odd prime p, however, it is hard to find a non-square modulo an odd prime p. Show that, by the Extended Riemann Hypothesis, there is a non-square less than or equal to $2(\log p)^2$.

Problem 2.8.6. (AKS) Let $n \geq 2$ be an integer, r a positive integer $< p$, for which n has order greater than $(\log n)^2$ modulo r. Then n is prime if and only if

(1) n is not a perfect power,

(2) n does not have any prime divisor less than or equal to r,

(3) $(x + a)^n = x^n + a \ (\mathrm{mod} \ n, x^r - 1)$ for each integer a, $1 \leq a \leq \sqrt{r} \log n$.

Problem 2.8.7. (Bernstein) Let $f(x)$ be a monic polynomial $f(x) \in \mathbb{Z}[x]$ of degree $d \geq 1$ and n positive integer. $\mathbb{Z}[x]/(n, f(x))$ is called an almostfield with parameters $(e, v(x))$ if

(1) $e \mid n^d - 1$, with e positive integer,

(2) $v(x)^{n^d-1} \equiv 1 \ (\mathrm{mod} \, n, f(x))$,

(3) $v(x)^{(n^d-1)/q} - 1$ is a unit in $\mathbb{Z}[x]/(n, f(x))$ for all prime $q \mid e$.

If n is a prime and $f(x)$ modulo n is irreducible, then $\mathbb{Z}[x]/(n, f(x))$ is a field. Moreover, any generator $v(x)$ of the multiplicative group of elements of this field satisfies (2) and (3) for each e satisfying (1).

Prove that for $n > 2$ and let $\mathbb{Z}[x]/(n, f(x))$ be an almostfield with parameters $(e, v(x))$ where $e > (2d \log n)^2$, then n is prime if and only if

(a) n is not a perfect power,

(b) $(t - 1)^{n^d} \equiv t^{b^d} \ (\mathrm{mod} \ n, f(x), t^e - v(x))$ in $\mathbb{Z}[x, t]$.

Problem 2.8.8. (Lenstra, 1985) Add one more condition (4) in Problem 2.8.7:

$$(4)\ g(T) := \prod_{i=0}^{d-1} (T - v(x)^{n^i}) \in (\mathbb{Z}[x]/(n, f(x)))[T].$$

Show that $p \equiv n^j \pmod{e}$ for some j, $0 \le j \le d - 1$, and for each prime $p \mid n$.

Problem 2.8.9. (Lenstra-Pomerance) Let $f(x)$ be a monic polynomial $f(x) \in \mathbb{Z}[x]$ of degree $d \ge 1$ and n positive integer. $\mathbb{Z}[x]/(n, f(x))$ is called a pseudofield if

(1) $f(x^n) \equiv 0 \pmod{n, f(x)}$,

(2) $x^{n^d} - x \equiv 0 \pmod{n, f(x)}$,

(3) $x^{n^{d/q}} - x$ is a unit in $\mathbb{Z}[x]/(n, f(x))$ for all prime $q \mid d$.

Prove that for $n \ge 2$ and let $\mathbb{Z}[x]/(n, f(x))$ be a pseudofield with $f(x)$ a monic polynomial of degree d in $((\log n)^2, n)$, then n is prime if and only if

(a) n is not a perfect power,

(b) n does not have any prime divisor less than or equal to d,

(c) $(x + a)^n \equiv x^n + q \pmod{n, f(x)}$, for each integer a, $1 \le a \le \sqrt{d} \log n$.

2.9 Comparison of General Purpose Primality Tests

In this section, we give a comparison of some general purpose primality tests in terms of computational complexity (running time).

Recall that multiplying two $\log n$ bit integers has a running time of

$$\mathcal{O}((\log n)^2)$$

and the fastest known algorithm, the Schönhage-Strassen algorithm, has running time of

$$\begin{aligned} \mathcal{O}(\log n \log \log n \log \log \log n) &= \mathcal{O}((\log n)^{1+\epsilon}) \\ &= \tilde{\mathcal{O}}(\log n). \end{aligned}$$

Thus, if we let $\mathcal{O}((\log n)^\mu)$ be the running time of integer multiplication, then

$$\mathcal{O}((\log n)^\mu) \quad \overset{u=2}{\Longrightarrow} \quad \mathcal{O}((\log n)^2) \qquad \text{use practical multiplication algorithm}$$

$$\overset{u=1+\epsilon}{\Longrightarrow} \quad \mathcal{O}((\log n)^{1+\epsilon}) \quad \text{use fast multiplication algorithm.}$$

The Miller-Rabin test runs in time

$$\mathcal{O}((\log n)^{1+\mu}) \quad \Rightarrow \quad \mathcal{O}((\log n)^3)$$
$$\Rightarrow \quad \mathcal{O}((\log n)^{2+\epsilon})$$
$$= \quad \tilde{\mathcal{O}}((\log n)^2)$$

so it is a very fast (polynomial-time) primality test, as its degree of the complexity is juts one more than integer multiplications. A draw-back of the Miller-Rabin test is that it is probabilistic, not deterministic, that is, there will be a small error of probability when it declares an integer to be prime. However, if we assume the Generalized Riemann Hypothesis (GRH), then the Miller-Rabin test can be made deterministic with running time in

$$\mathcal{O}((\log n)^{3+\mu}) \quad \Rightarrow \quad \mathcal{O}((\log n)^5)$$
$$\Rightarrow \quad \mathcal{O}((\log n)^{4+\epsilon})$$
$$= \quad \tilde{\mathcal{O}}((\log n)^4)$$

It is still polynomial-time complexity, just two degrees higher than its probabilistic version.

The AKS (Agrawal, Kayal and Saxena) test takes time

$$\mathcal{O}((\log n)^{6(1+\mu)}) \quad \Rightarrow \quad \mathcal{O}((\log n)^{18})$$
$$\Rightarrow \quad \mathcal{O}((\log n)^{12+\epsilon})$$
$$= \quad \tilde{\mathcal{O}}((\log n)^{12}).$$

That is, by practical multiplication algorithm, AKS runs in $\mathcal{O}((\log n)^{18})$ whereas by Schönhage-Strassen algorithm in $\tilde{\mathcal{O}}((\log n)^{12})$. It can be show that [218] that the AKS algorithm cannot be expected to be proved to run faster than

$$\mathcal{O}((\log n)^{6(1+\mu)}).$$

However, in practice, it is easy to find a suitable prime of the smallest possible size $\mathcal{O}((\log n)^2)$, thus, the practical running time of the AKS algorithm is

$$\mathcal{O}((\log n)^{3(1+\mu)}).$$

It also can be shown that one cannot find a deterministic test that runs faster than $\mathcal{O}((\log n)^{3(1+\mu)})$. That is, $\mathcal{O}((\log n)^{3(1+\mu)})$ is the fastest possible running time for a deterministic primality test. Recently, H. Lenstra and Pomerance showed that a test having running time in

$$\mathcal{O}((\log n)^{(3+\epsilon)(1+\mu)})$$

is possible, but which is essentially the same as the practical running time $\mathcal{O}((\log n)^{3(1+\mu)})$. Of course, if one is willing to accept a small error of probability, a randomized version of AKS is possible and can be faster, but this is not our point, as if one is willing to use a probabilistic test, one would prefer to use the Miller-Rabin test.

The APR (Adleman, Pomerance and Rumely) cyclotomic (or Jacobi sum) test is a deterministic and nearly polynomial-time test, and it runs in time

$$\mathcal{O}((\log n)^{c \log \log \log n}), \quad c > 0.$$

In fact, Odlyzko and Pomerance have shown that for all large n, the running time is in the interval

$$\left[\mathcal{O}((\log n)^{c_1 \log \log \log n}), \mathcal{O}((\log n)^{c_2 \log \log \log n}) \right],$$

where $c_1, c_2 > 0$. This test was further improved by Lenstra and Cohen, and hence the name APRCL test. It can be used to test numbers of several thousand digits in less than, say, ten minutes.

The elliptic curve test ECPP of Atkin and Morain, based on earlier work Goldwasser and Kilian, runs in time

$$\begin{aligned}
\mathcal{O}((\log n)^{2+2\mu}) \quad &\Rightarrow \quad \mathcal{O}((\log n)^6) \\
&\Rightarrow \quad \mathcal{O}((\log n)^{4+\epsilon}) \\
&= \quad \tilde{\mathcal{O}}((\log n)^4).
\end{aligned}$$

That is, it runs in $\mathcal{O}((\log n)^6)$ if a practical multiplication algorithm is used and $\tilde{\mathcal{O}}((\log n)^4)$ if a fast multiplication algorithm is employed. ECPP is a probabilistic algorithm but with zero error; other names for this type of probabilistic algorithms are ZPP algorithm or Las Vegas algorithm.

In what follows, we present in Table 2.3, summarizing the running times for all the different tests mentioned in this section.

Test	Practical Running Time	Fast Running Time
Miller-Rabin	$\mathcal{O}((\log n)^3)$	$\tilde{\mathcal{O}}((\log n)^2)$
AKS	$\mathcal{O}((\log n)^{18})$	$\tilde{\mathcal{O}}((\log n)^{12})$
APRCL	$\mathcal{O}((\log n)^{c \log \log \log n})$	
ECPP	$\mathcal{O}((\log n)^6)$	$\tilde{\mathcal{O}}((\log n)^4)$

Table 2.3. Running time comparison of some primality tests

Brent [36] at Oxford University (now at the Australian National University) did some numerical experiments for the comparison of the Miller-Rabin, ECPP and AKS tests on a 1 Ghz machine for the number $10^{100} + 267$. Table 2.4 gives times for Magama (a computer algebra system) implementation of of the Miller-Rabin, ECPP and AKS tests, plus our experiment on a Fujitsu P7230 Laptop computer, all for the number $10^{100} + 267$.

Test	Trials	Time
Miller-Rabin	1	0.003 second
Miller-Rabin	10	0.03 second
Miller-Rabin	100	0.3 second
ECPP		2.0 seconds
Maple Test (Miller-Rabin + Lucas)		2.751 seconds
AKS		37 Weeks (Estimated)

Table 2.4. Times for various tests for $10^{100} + 267$

Problems for Section 2.9

Problem 2.9.1. (Problems in computational complexity of primality testing)

(1) (Trial divisions) Show that Primes $\in$ co$\mathcal{NP}$.

(2) (Pratt) Show that Primes $\in \mathcal{NP}$.

(3) (Miller, Solovay-Strassen) Show that, if the Extended Riemann Hypothesis is true, then Primes $\in \mathcal{P}$.

(4) (Rabin, Solovay-Strassen) Show that Primes $\in$ co$\mathcal{RP}$.

(5) (Adleman-Pomerance-Rumely) Show that Primes $\in$ super$P$.

(6) (ECPP, Adleman-Huang) Show that Primes $\in \mathcal{ZPP}$.

(7) (AKS) Show that Primes $\in \mathcal{P}$.

Problem 2.9.2. (Stop Press) Show that one cannot find a deterministic primality test that runs faster than $\mathcal{O}((\log n)^{3(1+\mu)})$.

Problem 2.9.3. Show that if the Generalized Riemann Hypothesis is true, then both the Miller-Rabin test and the Solovay-Strassen test can be performed in deterministic polynomial-time.

2.10 Primality Tests for Special Numbers

The tests discussed so far are for general numbers. If n is a number with a special form such as the Fermat numbers $F_k = 2^{2^k} + 1$ and Mersenne numbers $2^p - 1$, then it is usually much easier to test. In this section, some methods for testing Mersenne primes are discussed.

The Lucas-Lehmer Test. Define recursively the Lucas-Lehmer sequence (of integers) L_k as follows [268]:

$$\begin{cases} L_0 = 4 \\ L_{k+1} = L_k^2 - 2, \quad k \geq 0. \end{cases} \tag{2.69}$$

Then the Lucas-Lehmer sequence begins with integers:

$4, 14, 194, 37634, 1416317954, 2005956546822746114,$
$40238616677410360228256356561102100994,$
$16191462721115671781777559070120513664958590125499158514329308740975788034,$

$$\vdots$$

The famous Lucas-Lehmer test for Mersenne primes is stated as follows:

Theorem 2.10.1. Let $M_p = 2^p - 1$ be a Mersenne number. Then M_p is prime if and only if $M_p \mid L_{p-2}$. That is, M_p is prime if and only if

$$L_{p-2} \equiv 0 \pmod{M_p}. \tag{2.70}$$

This theorem was proposed by Lucas in 1876 [135] and proved by Lehmer in 1935 [129]. Since then, there have been many interesting proofs for this theorem, some of which, say, [205], being based on some deep ideas from algebraic number theory. Clearly, if M_p is composite, then $M_p \nmid L_k$ for $0 \leq k < p - 2$. The following efficient algorithm can be derived directly from this theorem.

Algorithm 2.10.1 (Lucas-Lehmer Test). Given a Mersenne number M_p, the following algorithm will determine whether or not M_p is prime.

> Initialize the value for p
>
> $L \leftarrow 4$
>
> for i from 1 to $p - 2$ do
>
> $\quad L \leftarrow L^2 - 2 \pmod{2^p - 1}$
>
> If $L = 0$ then $2^p - 1$ is prime
>
> $\quad$ else $2^p - 1$ is composite

Algorithm 2.10.1 can be efficiently performed in $\mathcal{O}(p^3)$ bit operations, since the test only requires $\mathcal{O}(p)$ squaring modulo M_p. However, when p becomes very large, even with the help of some special computing techniques, such as FFT (Fast Fourier Transform), it is still very hard to find a new Mersenne prime. This is the reason that there are only 43 Mersenne primes have been found so far. This is also one of the main motivations to find a more powerful test for Mersenne primes.

Example 2.10.1. For $M_{31} = 2^{31} - 1$ we have the following Lucas-Lehmer sequence:

14, 194, 37634, 1416317954, 669670838, 1937259419, 425413602, 842014276,

12692426, 2044502122, 1119438707, 1190075270, 1450757861, 877666528,

630853853, 940321271, 512995887, 692931217, 1883625615, 1992425718,

721929267, 27220594, 1570086542, 1676390412, 1159251674, 211987665,

1181536708, 65536, 0.

Thus, by Theorem 2.10.1, $M_{31} = 2^{31} - 1$ is prime since the last number in the sequence is zero. On the other hand, for $M_{37} = 2^{37} - 1$, we have the following sequence:

14, 194, 37634, 1416317954, 111419319480, 75212031451, 42117743384,

134212256520, 54923239684, 61369726979, 100682126153, 46790825955,

120336432403, 15532303443, 43487582705, 63215664337, 24881968247,

36378170995, 23347868395, 34319987212, 27325339261, 67024860468,

67821607698, 45433743622, 32514699513, 51489094388, 44855569738,

31479590378, 32455804440, 54840899833, 71222372297, 35230286592,

24416019713, 80429963578, 117093979072.

Thus, by Theorem 2.10.1, $M_{37} = 2^{37} - 1$ is not a prime since the last term in the sequence is not zero.

An equivalent form of Theorem 2.10.1 is as follows:

Theorem 2.10.2. Let $M_p = 2^p - 1$ be a Mersenne number. Then M_p is prime if and only if $\gcd(L_k, M_p) = 1$ for $0 \le k \le p - 3$ and $\gcd(L_{p-2}, M_p) > 1$.

Example 2.10.2. Let $M_7 = 2^7 - 1$ be a Mersenne number. Then we have

$$\gcd(\{L_0, L_1, L_2, L_3, L_4\}, M_7)$$
$$= \gcd(\{4, 14, 194, 37634, 1416317954\}, 2^7 - 1)$$
$$= 1,$$

but

$$\gcd(L_5, M_7) \;=\; \gcd(2005956546822746114, 2^7 - 1)$$

$$=\; 127$$

$$>\; 1,$$

thus, by Theorem 2.10.2, M_7 is prime. On the other hand, for $M_{11} = 2^{11} - 1$, we have

$$\gcd(\{L_0, L_1, L_2, L_3, L_4, L_5, L_6, L_7, L_8, L_9\}, M_{11})$$

$$=\; \gcd(\{4, 14, 194, 788, 701, 119, 1877, 240, 282, 1736\}, 2^{11} - 1)$$

$$=\; 1,$$

thus, by Theorem 2.10.2, M_{11} is not prime.

From an algebraic point of view, the Lucas-Lehmer test is based on the successive squaring of a point on the one dimensional algebraic torus over $\mathbb{Q}$, associate to the real quadratic field $k = \mathbb{Q}(\sqrt{3})$. This suggests, as Gross noted[94], some other tests for Mersenne numbers, using different algebraic groups, e.g., elliptic curve groups, based on successive squaring of a point on an elliptic curve. This is the idea underpinning the elliptic curve test for Mersenne numbers considered in this paper.

The Elliptic Curve Test. Define a sequence of rational numbers G_k (see [94] and [227]) as follows:

$$\begin{cases} G_0 = -2 \\ G_{k+1} = \dfrac{(G_k^2 + 12)^2}{4 \cdot G_k \cdot (G_k^2 - 12)}, & k \geq 0. \end{cases} \tag{2.71}$$

Then, the sequence begins with rational numbers:

$$-2, 4, \frac{49}{4}, \frac{6723649}{1731856}, \frac{6593335793533896979873913089}{42923806093478210389068 2944}, \dots$$

Remarkably enough, this sequence, as in the case of the Lucas-Lehmer sequence, is intimately connected to the primality testing of Mersenne numbers. The Mersenne number $M_p = 2^p - 1$ is prime if and only if $\gcd(G_k(G_k^2 - 12), M_p) = 1$ for $0 \leq k \leq p - 2$, and $\gcd(G_{p-1}, M_p) > 1$. This test, in fact, involves the successive squaring of a point on an elliptic curve E over $\mathbb{Q}$:

$$E/\mathbb{Q}: \; y^2 = x^3 - 12x, \tag{2.72}$$

with discriminant $\Delta = 2^{12} \cdot 3^3$ and conductor $N = 2^5 \cdot 3^2 = 288$ [94].

Theorem 2.10.3. Let $M_p = 2^p - 1$ be a Mersenne number. Then M_p is prime if and only if $\gcd(G_k(G_k^2 - 12), M_p) = 1$ for $0 \leq k \leq p - 2$, and $\gcd(G_{p-1}, M_p) > 1$.

Remark 2.10.1. Theorem 2.10.3 is just an elliptic curve analog of Theorem 2.10.1.

We note that Theorem 2.10.2 and Algorithm 4.4.1 can be modified and simplified to produce a practical primality test for Mersenne primes, that is as *simple*, *elegant* and *efficient* as the Lucas-Lehmer test.

Theorem 2.10.4. Let $M_p = 2^p - 1$ be a Mersenne number. Then M_p is prime if and only if $M_p \mid G_{p-1}$, or equivalently,

$$G_{p-1} \equiv 0 \pmod{M_p}. \tag{2.73}$$

Remark 2.10.2. Theorem 2.10.4 is an elliptic curve analog of Theorem 2.10.2. The test based on Theorem 2.10.4 is competitive to the test based on Theorem 2.10.2. More importantly and more interestingly, in the case that M_p is a composite, the test is much better and quicker than the test based on Theorem 2.10.2.

Algorithms and Complexities. The test based on (and derived from) Theorem 4.4.1 for Mersenne primes can be easily converted to an algorithm (and implemented by a computer algebra system such as Maple) as follows:

Algorithm 2.10.2. This algorithm provides a test for primality of Mersenne numbers M_p based on elliptic curves.

> Initialize the value for p
>
> $G \leftarrow -2$
>
> for i from 1 to $p - 2$ do
>
> $\quad G \leftarrow (G^2 + 12)^2/(4 \cdot G \cdot (G^2 - 12)) \pmod{(M_p)}$
>
> $\quad \gcd \leftarrow \gcd(G(G^2 - 12), M_p)$
>
> $\quad G \leftarrow (G^2 + 12)^2/(4 \cdot G \cdot (G^2 - 12)) \pmod{(M_p)}$
>
> $\quad \gcd \leftarrow \gcd(G, M_p)$
>
> If all $\gcd = 1$ (except the last one) and the last $\gcd > 1$
>
> $\quad$ then M_p is prime
>
> $\quad$ else M_p is not prime

The complexity of the elliptic curve is almost the same as that of Lucas-Lehmer test, since it also requires $\mathcal{O}(p)$ successive squaring of a point on an elliptic curve modulo M_p and the gcd computation.

Example 2.10.3. For $M_{13} = 2^{13} - 1$ we have:

$G_1 = 4$ and $\gcd(G_1(G_1^2 - 12), 2^p - 1) = \gcd(16, 8191) = 1$

$G_2 = 2060$ and $\gcd(G_2(G_2^2 - 12), 2^p - 1) = \gcd(8741791280, 8191) = 1$

$G_3 = 4647$ and $\gcd(G_3(G_3^2 - 12), 2^p - 1) = \gcd(100350092259, 8191) = 1$

$$G_4 = 6472 \text{ and } \gcd(G_4(G_4^2 - 12), 2^p - 1) = \gcd(271091188384, 8191) = 1$$

$$\vdots$$

$$G_{10} = 3036 \text{ and } \gcd(G_{10}(G_{10}^2 - 12), 2^p - 1) = \gcd(27983674224, 8191) = 1$$

$$G_{11} = 362 \text{ and } \gcd(G_{11}(G_{11}^2 - 12), 2^p - 1) = \gcd(47433584, 8191) = 1$$

$$G_{12} = 0 \text{ and } \gcd(G_{12}, 2^p - 1) = \gcd(0, 8191) = 8191 > 1$$

Thus, by Theorem 2.10.2, M_{13} is prime.

Theorem 2.10.4, however, can be converted to a very efficient algorithm as follows:

Algorithm 2.10.3 (Elliptic Curve Test). This algorithm provides a test for primality of Mersenne numbers M_p based on elliptic curves. The procedure is similar to that of Algorithm 2.10.1 for the Lucas-Lehmer test.

> Initialize the value for p
>
> $G \leftarrow -2$
>
> for i from 1 to $p - 1$ do
>
> $G \leftarrow (G^2 + 12)^2 / (4 \cdot G \cdot (G^2 - 12)) \pmod{M_p}$
>
>   If G does not exist, then stop the Algorithm, M_p is not prime
>
> end_do
>
> If $G = 0$ then M_p is prime
>
>   else M_p is not prime

The most important feature of Algorithm 2.10.3 is that it can stop anytime whenever G does not exist (the multiplication inverse modulo M_p does not exist which implies that M_p is not prime). Unlike Algorithm 2.10.1 and even Algorithm 2.10.2, Algorithm 2.10.3 does not need to perform the loops from 1 up to $p-1$. Thus, if the complexity, measured by bit operations, for Algorithm 2.10.1 and even Algorithm 2.10.2 is $\mathcal{O}(p^3)$, then the complexity for Algorithm 2.10.3 is $\mathcal{O}(\lambda^3)$, with $\lambda \ll p$. It is this feature that makes it very suitable to test the primality for Mersenne primes, since most of the Mersenne numbers are not prime, and, in fact, very few of the Mersenne numbers are prime. Algorithm 2.10.3 can stop the useless loops at a very earlier stage of the test if the Mersenne number is not prime. On the other hand, no matter M_p is prime or not, the Lucas-Lehmer test will need to perform the test for $p - 1$ time. Thus, for the Lucas-Lehmer test, most of the computation times are spent on the useless loops on the uninteresting Mersenne composite numbers. Therefore, the elliptic curve test of Algorithm 2.10.3 is very practical and efficient for finding new Mersenne primes than any other existing method; these can be seen from the following two examples.

Recursion k	LL Test $L_k \bmod M_{31}$	EC Test $G_k \bmod M_{31}$
0	4	-2
1	14	4
2	194	536870924
3	37634	242940031
4	1416317954	1997781005
5	669670838	1070166402
6	1937259419	1316556811
7	425413602	1539940455
8	842014276	2000813575
9	12692426	361728374
10	2044502122	602038520
11	1119438707	1031405401
12	1190075270	553331261
13	1450757861	2131672851
14	877666528	1777654456
15	630853853	321910276
16	940321271	1025466243
17	512995887	102277241
18	692931217	1545817800
19	1883625615	1322945035
20	1992425718	1314104653
21	721929267	427090555
22	27220594	326614488
23	1570086542	2116120380
24	1676390412	1193217808
25	1159251674	47494329
26	211987665	73295707
27	1181536708	1114213282
28	65536	40029428
$29 = p - 2$	0	388539999
$30 = p - 1$	M_{31} is prime STOP	0
		M_{31} is prime STOP

Table 2.5. Comparisons of Lucas-Lehmer and Elliptic Curve tests for M_{31}

Example 2.10.4. Let $M_{61} = 2^{61} - 1$. Then by Algorithm 4.4.2, we have the following sequence of G:

$-2, 4, 576460752303423500, 2273229660002968910, 61942924192900875,$

$227188641361726013, 584955155938028078, 1625992443540546788,$

$2281402858238123895, 96833079332604917, 190667854843477861, \ldots,$

$754634924515350806, 580226091912597714, 485337391446314291,$

$523773010142196837, 2066643542538005360, 0.$

Hence, by Theorem 4.4.1, $M_{61} = 2^{61} - 1$ is prime. On the other hand, for $M_{59} = 2^{59} - 1$ with 59 prime, then by Algorithm 4.4.2, we have

$-2, 4, 144115188075855884, 525264983758271552, 359803122487490816,$

$307956120154006017, 125273927552899636, 455734762531475987,$

$443763109181002475, 516877268683260273, 459543861857489578, \ldots,$

$118819180406808830, 343866746792758738, 379530138081340611,$

$158582427978263517, 89062835136920118.$

Hence, by Theorem 4.4.1, $M_{59} = 2^{59} - 1$ is not prime since the last term in the squence is not zero.

Example 2.10.5. For comparison purpose, we tabulate the results for primality testing of M_{31} and M_{23}, by using the two methods, in Tables 2.5 and 2.6. As can be seen from Table 2.5 the computational costs of the two tests for M_{31} are the same. However for results of M_{23} in Table 2.6 a big saving has been made using the elliptic curve test over the Lucas-Lehmer test, since, for a composite Mersenne number, the Lucas-Lehmer test will still need to go through all the loops from 1 to $p - 2$, whereas the elliptic curve test will stop whenever G_k does not exist. This is because the computation of G_k is based on fractions modulo M_p which may not exist (because M_p is composite). Whenever a modular inverse does not exist the computation stops and hence a saving is made in computation time; the earlier the computation stops the bigger the saving can be made. This is a significant advantage of adopting the elliptic curve test over the Lucas-Lehmer test. As the Mersenne primes become parser and sparser, the elliptic curve test can skip those composite M_p and concentrate on the prime M_p. For example, for M_{23}, the Lucas-Lehmer test will need 21 tests whereas the Elliptic curve test will only need 3 tests to reach an answer that M_{23} is not a prime.

Remark 2.10.3. As can be seen from Tables 2.5 and 2.6 that if $M_p = 2^p - 1$ is prime, we do need to perform $p - 2$ recursions/loops in order to decide whether or not M_p is prime. However, if M_p is composite, then we do not need to perform all the $p-2$ loops; we can conclude that M_p is not a prime at a very earlier stage of the loops, hence a big saving in time can be made. For

Recursion k	LL Test $L_k \bmod M_{23}$	EC Test $G_k \bmod M_{23}$
0	4	-2
1	14	4
2	194	2097164
3	37634	$G_3 = \perp$
4	7031978	M_{23} is not prime STOP - saving
5	7033660	- saving
6	1176429	- saving
7	7643358	- saving
8	3179743	- saving
9	2694768	- saving
10	763525	- saving
11	4182158	- saving
12	7004001	- saving
13	1531454	- saving
14	5888805	- saving
15	1140622	- saving
16	4321431	- saving
17	7041324	- saving
18	2756392	- saving
19	1280050	- saving
20	6563009	- saving
$21 = p - 2$	$6107895 \neq 0$	- saving
	M_{23} is not prime STOP	-

Table 2.6. Comparisons of Lucas-Lehmer and Elliptic Curve tests for M_{23}

example, for $p = 23$, only three loops (not $p-2 = 21$ loops) are needed in order to decide that $M_{23} = 2^{23} - 1$ is not prime. As most of the Mersenne numbers are not prime but composite, and in fact for $p = 2, 3, 5, 7, \ldots, 25964951$, there are 1622441 primes (i.e., $\pi(25964951) = 1622441$), but only 43 such p (primes) that can make $M_p = 2^p - 1$ to be prime. This is why we say that the complexity of the elliptic curve test is $\mathcal{O}(\lambda^3)$ with $\lambda \ll p$, because it can quickly skip the Mersenne composite numbers and only concentrate on the verification of the Mersenne primes. Although at the beginning of the test we do not know which M_p is indeed prime, we can quickly know it is not prime by the elliptic curve test; whenever the current M_p is not prime, we just simply throw it away and choose a next M_p to test. Our complexity analysis and computing experiments show that the elliptic curve test is more efficient than the Lucas-Lehmer test and is a good candidate as a replacement to the

Lucas-Lehmer test for Mersenne primes. Of course, much needs to be done in order to make the elliptic curve test more practical and more useful.

Methods for test primality for other kinds of special numbers are given in the following problems.

Problems for Section 2.10

Problem 2.10.1. (Pepin test) Let $F_k = 2^{2^k} + 1$ be Fermat numbers with $k \geq 1$.

(1) Prove that F_k is prime if and only if

$$5^{(F_k - 1)/2} \equiv -1 \pmod{F_k}.$$

(2) Using the test to show that $F_4 = 2^{2^4} + 1 = 65537$ is prime.

(3) Prove that the test can be performed in $\mathcal{O}((\log n)^3)$ bit operations.

Problem 2.10.2. Show that for all $k \geq 0$, $F_{k+1} = F_0 F_1 \cdot F_k + 2$, and hence, $\gcd(F_j, f_j) = 1$, for $i \neq j$.

Problem 2.10.3. (Proth Test) Let $n = f \cdot 2^k + 1$ with f odd and $1 \leq f < 2^k$.

(1) Show if there is an integer a such that

$$a^{(n-1)/2} \equiv -1 \pmod{n}$$

then n is prime.

(2) Use the test to show that $180(2^{127} - 1)^2 + 1$ is prime.

(3) Show that the Proth test uses the same amount of computer time as the Lucas-Lehmer test (Algorithm 4.34) for Mersennne primes, except for an additional $\mathcal{O}(n \log n)$ bit operations.

Problem 2.10.4. Let $3 \nmid k$, $2^n + 1 \geq k$ and $2^n_1 > 3$. Then $k2^{n+1}$ is prime if and only if $3^{k2^{n-1}} \equiv -1 \pmod{k2^n + 1}$.

Problem 2.10.5. Find a formula for U_n

(1) if

$$\begin{cases} U_0 = 0, \\ U_1 = 1, \\ U_n = 2U_{n-1} - U_{n-2}. \end{cases}$$

(2) if

$$\begin{cases} U_0 = 0, \\ U_1 = 1, \\ U_n = 2U_{n-1} - U_{n-2}. \end{cases}$$

Problem 2.10.6. Let the sequence U_n be determined by

$$\begin{cases} U_1 = 0, \\ U_2 = 2, \\ U_3 = 3, \\ U_{n+1} = U_{n-1} - U_{n-2}, \quad n \geq 3. \end{cases}$$

Show that if p is prime, then $p \mid u_p$.

2.11 Prime Number Generation

Large Random Prime Generation. It is clear that one of the most important tasks in the construction of RSA cryptosystems is to find two large primes, say each with at least 100 digits. An algorithm for finding two 100 digit primes can be described as follows:

Algorithm 2.11.1 (Large prime generation). This algorithm generates prime numbers with 100 digits; it can be modified to generate any length of the required prime numbers:

[1] (Initialization) Randomly generate an odd integer n with say, for example, 100 digits;

[2] (Primality Testing – Probabilistic Method) Use a combination of the Miller–Rabin test and a Lucas test to determine if n is a probable prime. If it is, goto Step [3], else goto Step [1] to get another 100-digit odd integer.

[3] (Primality Proving – Elliptic Curve Method) Use the elliptic curve method to verify whether or not n is indeed a prime. If it is, then report that n is prime, and save it for later use; or otherwise, goto Step [1] to get another 100-digit odd integer.

[4] (done?) If you need more primes, goto Step [1], else terminate the algorithm.

How many primes with 100 digits do we have? By Chebyshev's inequality (2.18), if N is large, then

$$0.92129 \frac{N}{\ln N} < \pi(N) < 1.1056 \frac{N}{\ln N}. \tag{2.74}$$

Hence

$$0.92129\frac{10^{99}}{\ln 10^{99}} < \pi(10^{99}) < 1.1056\frac{10^{99}}{\ln 10^{99}},$$

$$0.92129\frac{10^{100}}{\ln 10^{100}} < \pi(10^{100}) < 1.1056\frac{10^{100}}{\ln 10^{100}}.$$

The difference $\pi(10^{100}) - \pi(10^{99})$ will give the number of primes with exactly 100 digits, we have

$$3.596958942 \cdot 10^{97} < \pi(10^{100}) - \pi(10^{99}) < 4.076949099 \cdot 10^{97}.$$

The above algorithm for large prime generation depends on primality testing and proving. However, there are methods which do not rely on primality testing and proving. One such method is based on Pocklington's theorem, that can automatically lead to primes, say with 100 digits (Ribenboim [194]). We re-state the theorem in a slightly different way as follows:

Theorem 2.11.1 (Pocklington-Ribenboim). Let p be an odd prime, k a natural number such that p does not divide k and $1 < k < 2(p+1)$ and let $N = 2kp + 1$. Then the following conditions are equivalent:

(1) N is prime.

(2) There exists a natural number a, $2 \le a < N$, such that

$$a^{kp} \equiv -1 \pmod{N}, \text{ and} \tag{2.75}$$

$$\gcd(a^k + 1, N) = 1. \tag{2.76}$$

Algorithm 2.11.2 (Large prime number generation). This algorithm, based on Theorem 2.11.1, generates large prime numbers without the use of primality testing:

[1] Choose, for example, a prime p_1 with $d_1 = 5$ digits. Find $k_1 < 2(p_1+1)$ such that $p_2 = 2k_1p_1 + 1$ has $d_2 = 2d_1 = 10$ digits or $d_2 = 2d_1 - 1 = 9$ digits and there exists $a_1 < p_2$ satisfying the conditions $a_1^{k_1 p_1} \equiv -1 \pmod{p_2}$ and $\gcd(a_1^{k_1} + 1, p_2) = 1$. By Pocklington's Theorem, p_2 is prime.

[2] Repeat the same procedure starting from p_2 to obtain the primes $p_3, p_4, \ldots$. In order to produce a prime with 100 digits, the process must be iterated five times. In the last step, k_5 should be chosen so that $2k_5p_5 + 1$ has 100 digits.

As pointed out in Ribenboim [194], for all practical purposes, the above algorithm for producing primes of a given size will run in polynomial time, even though this has not yet been supported by a proof.

Example 2.11.1. The following two examples exhibit the generation of 100-digit random primes, by showing the intermediate results as well.

(1) Starting from $p_1 = 97711$, the following have been produced:

$a_1 = 5,$
$k_1 = 22548,$
$p_2 = 4406375257;$

$a_2 = 5,$
$k_2 = 5672269218,$
$p_3 = 49988293466475878053;$

$a_3 = 3,$
$k_3 = 30278440082130267825,$
$p_4 = 30271350970652676311828776867898590 89451;$

$a_4 = 5,$
$k_4 = 18138542572554147127811646614204775 80766,$
$p_5 = 10981563766198237482064153956177259 64757319640785_$
$89556192484826353307841421989 33;$

$a_5 = 3,$
$k_5 = 60397763629481446952,$
$p_6 = 13265237852658384156030973594451425 88419492086693787 5_$
$78139264439708820318686315226 3340789357 41004433.$

(2) Starting from $p_1 = 97711$, the following have been produced:

$a_1 = 5,$
$k_1 = 30036,$
$p_2 = 5869695193;$

$a_2 = 3,$
$k_2 = 2436120302,$
$p_3 = 28598567252438216573;$

$a_3 = 3,$
$k_3 = 23378902714462628788,$
$p_4 = 13372062431355397301247229226000970 07049;$

$a_4 = 5,$
$k_4 = 48481122417757027006271196635773449 17472,$
$p_5 = 12965851914248613763710377243469059 4448791299892 5_$
$4423118286594309361798214520 257;$

$a_5 = 5,$
$k_5 = 42141446863948470528,$
$p_6 = 10927995189802650358117769874941404 62248308834526365 8_$
$56288164109429090190918511730 9963185528 46971393.$

Remark 2.11.1. This method of Pocklington-Ribenboim will lead automatically to a prime, although the proof is hard. There are many other faster methods that lead to probable primes; we shall discuss some of these methods later in this section.

Safe Prime Generation. In some cryptosystems, it is necessary to generate some large safe primes, not just large random primes.

Definition 2.11.1. Two prime numbers p and q are called safe primes if $(p-1)/2$ and $(q-1)/2$ are also prime numbers.

Safe prime is useful in the selection of RSA modulus $n = pq$ (will discuss in Chapter 4) as if p and q are safe primes (i.e., p and q are not close together), then RSA is not vulnerable to the Fermat factoring attacks (will discuss in Chapter 3). Here is an algorithm for generating probable safe primes.

Algorithm 2.11.3. This algorithm generates a pair of two α bit-length probable safe primes p and q.

[1] Choose an α bit odd random positive integer n and pre-determined smooth bound B.

[2] Trial divide n by all primes $p \leq B$. If n is divisible by any such p, go to step [1] to choose another $\alpha - 1$ bit odd random positive integer n.

[3] Test the primality of n using the strong pseudoprimality test (Algorithm 2.4.2); if n is declared to be a probable prime, then let $p = n$ and output p as a probable safe prime, otherwise go to Step [1].

[4] Compute $q = 2n + 1$, and use the strong pseudoprimality test on q. If q is declared to be a probable prime, output q as a probable safe prime, otherwise repeat this step to generate and test a new q.

[5] Exit. Terminate the algorithm with p and q both safe primes.

Strong Prime Generation. It may be useful to define and generate strong primes in some cryptographic applications, although it may not be necessary for RSA cryptosystem [202].

Definition 2.11.2. A prime is a strong prime if the following conditions satisfy:

(1) p is a large prime, of course,

(2) The largest prime factor, denoted by p^-, of $p - 1$ is large. That is,

$$p - 1 = a^- p^-$$

for some integers a^- and p^-.

(3) The largest prime factor, denoted by p^{--}, of $p^- - 1$ is large. That is,

$$p^- - 1 = a^{--} p^{--}$$

for some integers a^{--} and p^{--}.

(4) The largest prime factor, denoted by p^+, of $p + 1$ is large. That is

$$p + 1 = a^+ p^+$$

for some integers a^+ and p^+.

It may also require $p^+ - 1$ to have a large prime factor as follows:

(5) The largest prime factor, denoted by p^{+-}, of $p^+ - 1$ is large. That is

$$p^+ - 1 = a^{+-} p^{+-}$$

for some integers a^+ and p^+.

The following algorithm, due to Gordon (see [87] and [88]), generates a strong prime p that is just slightly harder than finding large random primes of the same size.

Algorithm 2.11.4. This algorithm generates a required bit-length probable strong prime p.

[1] Generate two large probable primes p^{--} and p^+ of roughly equal bit-length (say, each with α bit) using the strong test.

[2] Compute

$$p^- = a^{--} p^{--} + 1$$

for some a^{--}.

[3] Set

$$p_0 = \left((p^+)^{p^- - 1} - (p^-)^{p^+ - 1} \right) \pmod{p^- p^+}.$$

[4] Compute

$$p = p_0 + a p^- p^+$$

for some integer a.

Let $T(\alpha) = \mathcal{O}(\alpha^3)$ be the time to test an α bit number for primality, then Gordon's algorithm takes

$$\alpha T(\alpha) + 3(\alpha/2)(T(\alpha/2) = 1.1875(\alpha T(\alpha))$$

bit operations, which justifies that finding strong primes requires only 19% more time than the ordinary algorithm for finding a random prime [202].

Problems for Section 2.11

Problem 2.11.1. Prove that for each prime factor p of $n - 1$, if there is a number x_p such that $x_p^{(n-1)/p} \not\equiv 1 \pmod{n}$ but $x_p^{n-1} \equiv 1 \pmod{n}$, then n is prime.

Problem 2.11.2. (Legendre) Prove that

$$\pi(x) - \pi(\sqrt{x}) + 1 = S_0 - S_1 + S_2 - \cdots + (-1)^L S_L$$

where

$$S_0 = \lfloor x \rfloor, \quad s_l = \sum_{q_1 < \cdots < q_l < \sqrt{t}} \left\lfloor \frac{x}{q_1 \cdots q_l} \right\rfloor, \quad 1 \leq l \leq L.$$

Problem 2.11.3. Use Legendre's formula in Problem 2.11.2 to compute $\pi(500)$.

Problem 2.11.4. Estimate the number of primes with 1024 and 2048 bits.

Problem 2.11.5. (Wagstaff) Prove that for all $k \geq 1$, there is no prime number of the form

$$2^{2^k} - 1 - (2^i + 2^j)$$

where

$$1 \leq i < j \leq 2^k - 2.$$

Problem 2.11.6. Prove or disprove there are infinitely many safe primes.

Problem 2.11.7. Prove or disprove the following conjecture: there is an algorithm that can determine all primes $\leq n$ in $\mathcal{O}(\log n)$ bit operations.

Problem 2.11.8. (Knuth) Let $0 \leq d \leq 9$. Find the largest 50-digit prime that has the maximum possible number of decimal digits equal to d.

Problem 2.11.9. Design and implement a practical algorithm for generating 200 digit

(1) probable primes

(2) strong probable primes

(3) provable primes

Problem 2.11.10. A prime p is called a successor of a prime q, if $q = 2^k p + 1$ for some $k \geq 0$. For example,

$$2 \to 3 \to 7 \to 29 \to 59 \to 1889 \to 3779 \to 7559 \to 4058207223809 \to \cdots$$

is the chain of the first a few such prime numbers. Generate the first 100 such prime numbers in the chain.

2.12 Chapter Notes and Further Reading

Primality testing and large random prime number generation are the most frequently used operations in public-key cryptography. In this chapter, we have introduced the most widely used and also the most recent algorithms (e.g., the Agrawal-Kayal-Saxena test [5]) for primality testing and prime number generation. As can be seen, the complexity of the Agrawal-Kayal-Saxena test is $\mathcal{O}((\log n)^{12+\epsilon})$; to make the test practical, much work still needs to be done. Primality testing is a very lively subject, there are many books and papers in this field. The books by Bach and Shallit [12], Brassard [29], Cohen [50], Crandall and Pomerance [62], Wagstaff [251], and Yan [268] have a detailed discussion on algorithms for primality testing. The general books on algorithms such as Cormen, Ceiserson and Rivest [59], Knuth [119], and Wilf [258] also have chapters on number theoretic algorithms, including primality testing. Dixon [73] provides a survey of algorithms for primality testing (as well as factoring) up to 1984, whereas Bernstein [22], and Lenstra and Pomerance [134] gives some new developments of primality testing after Agrawal-Kayal-Saxena. Readers may find the books by Kranakis [126] and Riesel [200] helpful although they are little bit out of date.

We have also listed many related references (papers and books) on primality testing in the Bibliography section at the end of this book, see, e.g., Adleman [2], Adleman, Pomerance and Rumely [3], Adleman and Huang [4], Alford, Granville and Pomerance [6], Atkin and Morain [10], Bhatacharjee and Pandey [23], Brent ([33] and [36]), Cohen [50], Cohen and Lenstra [51], Cox [61], Davenport [65], Dixon [73], Goldwasser and Kilian ([82] and [83]), Granville ([89], [90] and [91]), Gross [94], Kilian [118], Lenstra [130], Miller [149], Morain ([157] and [158]), Pinch [173], Pomerance ([180] and [188]), Pomerance, Selfridge and Wagstaff [189], Pratt [190], Rabin [191], Ribenboim [194], Schoof [218], Solovay and Strassen [239], Wagon [249], Wagstaff [251], Williams [264], and Yan ([266] and [268]).

Random, fast and reliable prime number generation is another important topic related to primality testing, which is important in the construction of RSA related cryptography (RSA related cryptography will be discussed in Chapter 4). Readers who are interested in prime number generation are suggested to consult Maure [137], Mollin [154], Papanikolaou and Yan [170], and Ribenboim [194]. The papers by Wagstaff ([250] and [252]) discussed methods for generating prime numbers with a fixed number of one bits or zero bits in their binary representation.

3. Integer Factorization and Discrete Logarithms

Of all the problems in the theory of numbers to which computers have been applied, probably none has been influenced more than of factoring.

HUGH C. WILLIAMS [260]

3.1 Introduction

The *integer factorization problem* (IFP) is to find a nontrivial factor f (not necessarily prime) of a composite integer n. That is,

$$\left. \begin{array}{ll} \text{Input}: & n \in \mathbb{Z}^+_{>1} \\[2mm] \text{Output}: & f \text{ such that } f \mid n. \end{array} \right\} \tag{3.1}$$

Clearly, if there is an algorithm to test whether or not an integer n is a prime, and an algorithm to find a nontrivial factor f of a composite integer n, then by recursively calling the primality testing algorithm and the integer factorization algorithm, it should be easy to find the prime factorization of

$$n = p_1^{\alpha_1} p_2^{\alpha_2} \cdots p_k^{\alpha_k}.$$

In this chapter we shall mainly be concerned with the integer factorization algorithms (the primality testing algorithms have been discussed in the previous chapter). Generally speaking, the most useful factoring algorithms fall into one of the following two main classes (Brent [33]):

(A) General purpose factoring algorithms: the running time depends mainly on the size of N, the number to be factored, and is not strongly dependent on the size of the factor p found. Examples are:

(1) *Lehman's method*, which has a rigorous worst-case running time bound $\mathcal{O}\left(n^{1/3+\epsilon}\right)$.

(2) *Shanks' SQUare FOrm Factorization method* SQUFOF, which has expected running time $\mathcal{O}\left(n^{1/4}\right)$.

(3) *Shanks' class group method*, which has running time $\mathcal{O}\left(n^{1/5+\epsilon}\right)$.

(4) *Continued FRACtion (CFRAC) method*, which under plausible assumptions has expected running time

$$\mathcal{O}\left(\exp\left(c\sqrt{\log n \log\log n}\right)\right) = \mathcal{O}\left(n^{c\sqrt{\log\log n/\log n}}\right),$$

where c is a constant (depending on the details of the algorithm); usually $c = \sqrt{2} \approx 1.414213562$.

(5) *Multiple Polynomial Quadratic Sieve (MPQS)*, which under plausible assumptions has expected running time

$$\mathcal{O}\left(\exp\left(c\sqrt{\log n \log\log n}\right)\right) = \mathcal{O}\left(n^{c\sqrt{\log\log n/\log n}}\right),$$

where c is a constant (depending on the details of the algorithm); usually $c = \dfrac{3}{2\sqrt{2}} \approx 1.060660172$.

(6) *Number Field Sieve (NFS)*, which under plausible assumptions has the expected running time

$$\mathcal{O}\left(\exp\left(c\sqrt[3]{\log n}\sqrt[3]{(\log\log n)^2}\right)\right),$$

where $c = (64/9)^{1/3} \approx 1.922999427$ if GNFS (a general version of NFS) is used to factor an arbitrary integer n, whereas $c = (32/9)^{1/3} \approx 1.526285657$ if SNFS (a special version of NFS) is used to factor a special integer n such as $n = r^e \pm s$, where r and s are small, $r > 1$ and e is large. This is substantially and asymptotically faster than any other currently known factoring method.

(B) Special purpose factoring algorithms: The running time depends mainly on the size of p (the factor found) of n. (We can assume that $p \leq \sqrt{n}$.) Examples are:

(1) *Trial division*, which has running time $\mathcal{O}\left(p(\log n)^2\right)$.

(2) *Pollard's ρ-method* (also known as Pollard's "rho" algorithm), which under plausible assumptions has expected running time $\mathcal{O}\left(p^{1/2}(\log n)^2\right)$.

(3) *Lenstra's Elliptic Curve Method (ECM)*, which under plausible assumptions has expected running time

$$\mathcal{O}\left(\exp\left(c\sqrt{\log p \log\log p}\right) \cdot (\log n)^2\right),$$

where $c \approx 2$ is a constant (depending on the details of the algorithm).

The term $\mathcal{O}\left((\log n)^2\right)$ is for the cost of performing arithmetic operations on numbers which are $\mathcal{O}(\log n)$ or $\mathcal{O}\left((\log n)^2\right)$ bits long; the second can be theoretically replaced by $\mathcal{O}\left((\log n)^{1+\epsilon}\right)$ for any $\epsilon > 0$.

Note that there is a quantum factoring algorithm, original proposed by Shor, which can run in polynomial-time

$$\mathcal{O}((\log n)^{2+\epsilon}).$$

However, this quantum algorithm requires to be run on a quantum computer, which is not available at present.

In practice, algorithms in both categories are important. It is sometimes very difficult to say whether one method is better than another, but it is generally worth attempting to find small factors with algorithms in the second class before using the algorithms in the first class. That is, we could first try the *trial division algorithm*, then use some other method such as NFS. This fact shows that the trial division method is still useful for integer factorization, even though it is simple. In this chapter we shall introduce some most useful and widely used factoring algorithms.

It is interesting to note that integer factorization is related to many other number theoretic problems, that is, if we can find an algorithm for integer factorization, then with some modifications, this algorithm can always be used for some other problems, such as the discrete logarithms, the quadratic residuosity problem and the square root problem, etc. So, in this chapter, we shall also introduce some of the algorithms for these problems.

Problems for Section 3.1

Problem 3.1.1. Explain why general purpose factoring algorithms are slower than special purpose factoring algorithms, or why the special numbers are easy to factor than general numbers.

Problem 3.1.2. Show that addition of two $\log n$ bit integers can be performed in $\mathcal{O}(\log n)$ bit operations.

Problem 3.1.3. Show that multiplication of two $\log n$ bit integers can be performed in $\mathcal{O}((\log n)^2)$ bit operations.

Problem 3.1.4. Show that there is an algorithm which can multiply two $\log n$ bit integers in

$$\mathcal{O}(\log n \log \log n \log \log \log n) = \mathcal{O}((\log n)^{1+\epsilon})$$

bit operations.

Problem 3.1.5. Estimate the bit operations for computing $n!$.

3.2 Simple Factoring Methods

The simplest factoring algorithm is the trial division method, which tries all the possible divisors of n to obtain its complete prime factorization:

$$n = p_1 p_2 \cdots p_t, \qquad p_1 \leq p_2 \leq \cdots \leq p_t. \qquad (3.2)$$

The following is the algorithm:

Algorithm 3.2.1 (Factoring by trial divisions). This algorithm tries to factor an integer $n > 1$ using trial divisions by all the possible divisors of n.

[1] Input n and set $t \leftarrow 0$, $k \leftarrow 2$.

[2] If $n = 1$, then go to Step [5].

[3] $q \leftarrow n/k$ and $r \leftarrow n \pmod{k}$.
 If $r \neq 0$, go to Step [4].
 $t \leftarrow t + 1$, $p_t \leftarrow k$, $n \leftarrow q$, go to Step [2].

[4] If $q > k$, then $k \leftarrow k + 1$, and go to Step [3].
 $t \leftarrow t + 1$; $p_t \leftarrow n$.

[5] Exit: terminate the algorithm.

Exercise 3.2.1. Use Algorithm 3.2.1 to factor $n = 2759$.

An immediate improvement of Algorithm 3.2.1 is to make use of an auxiliary sequence of *trial divisors*:

$$2 = d_0 < d_1 < d_2 < d_3 < \cdots \qquad (3.3)$$

which includes all primes $\leq \sqrt{n}$ (possibly some composites as well if it is convenient to do so) and at least one value $d_k \geq \sqrt{n}$. The algorithm can be described as follows:

Algorithm 3.2.2 (Factoring by Trial Division). This algorithm tries to factor an integer $n > 1$ using trial divisions by an auxiliary sequence of trial divisors.

[1] Input n and set $t \leftarrow 0$, $k \leftarrow 0$.

[2] If $n = 1$, then go to Step [5].

[3] $q \leftarrow n/d_k$ and $r \leftarrow n \pmod{d_k}$.
 If $r \neq 0$, go to Step [4].
 $t \leftarrow t + 1$, $p_t \leftarrow d_k$, $n \leftarrow q$, go to Step [2].

[4] If $q > d_k$, then $k \leftarrow k + 1$, and go to Step [3].
 $t \leftarrow t + 1$; $p_t \leftarrow n$.

[5] Exit: terminate the algorithm.

Exercise 3.2.2. Use Algorithm 3.2.2 to factor $n = 2759$; assume that we have the list L of all primes $\leq \lfloor \sqrt{2759} \rfloor = 52$ and at least one $\geq \sqrt{n}$, that is, $L = \{2, 3, 5, 7, 11, 13, 17, 19, 23, 29, 31, 37, 41, 43, 47, 53\}$.

Theorem 3.2.1. Algorithm 3.2.2 requires a running time in

$$\mathcal{O}\left(\max\left(p_{t-1}, \ \sqrt{p_t}\right)\right).$$

If a primality test between steps [2] and [3] were inserted, the running time would then be in $\mathcal{O}(p_{t-1})$, or $\mathcal{O}\left(\dfrac{p_{t-1}}{\ln p_{t-1}}\right)$ if one does trial division only by primes, where p_{t-1} is the second largest prime factor of n.

The trial division test is very useful for removing small factors, but it should not be used for factoring completely, except when n is very small, say, for example, $n < 10^8$.

Now suppose n is any odd integer (if n were even we could repeatedly divide by 2 until an odd integer is obtained). If $n = pq$, where $p \leq q$ are both odd, then by setting $x = \frac{1}{2}(p + q)$ and $y = \frac{1}{2}(q - p)$ we find that $n = x^2 - y^2 = (x + y)(x - y)$, or $y^2 = x^2 - n$. The following algorithm tries to find $n = pq$ using the above idea.

Algorithm 3.2.3 (Fermat's factoring algorithm). Given an odd integer $n > 1$, then this algorithm determines the largest factor $\leq \sqrt{n}$, of n.

[1] Input n and set $k \leftarrow \lfloor \sqrt{n} \rfloor + 1$, $y \leftarrow k \cdot k - n$, $d \leftarrow 1$

[2] If $\lfloor \sqrt{y} \rfloor = \sqrt{y}$ go to Step [4] else $y \leftarrow y + 2k + d$ and $d \leftarrow d + 2$

[3] If $\lfloor \sqrt{y} \rfloor < n/2$ go to Step [2] else print "No Factor Found" and go to Step [5]

[4] $x \leftarrow \sqrt{n + y}$, $y \leftarrow \sqrt{y}$, print $x - y$ and $x + y$, the nontrivial factors of n

[5] Exit: terminate the algorithm.

Exercise 3.2.3. Use the Fermat method to factor $n = 278153$.

Theorem 3.2.2 (The Complexity of Fermat's Method). *The Fermat method will try as many as* $\dfrac{n+1}{2} - \sqrt{n}$ *arithmetic steps to factor* n, *that is, it is of complexity* $\mathcal{O}\left(\dfrac{n+1}{2} - \sqrt{n}\right)$.

In what follows, we shall introduce some special-purpose factoring methods. By special-purpose we mean that the methods depend, for their success, upon some special properties of the number being factored. The usual property is that the factors of the number are small. However, other properties might include the number or its factors having a special mathematical form.

For example, if p is a prime number and if $2^p - 1$ is a composite number, then all of the factors of $2^p - 1$ must be congruent to 1 modulo $2p$. For example, if $2^{11} - 1 = 23 \cdot 89$, then $23 \equiv 89 \equiv 1 \pmod{22}$. Certain factoring algorithms can take advantage of this special form of the factors. Special-purpose methods do not always succeed, but it is useful to try them first, before using the more powerful, general methods, such as CFRAC, MPQS or NFS.

In 1975 John M. Pollard proposed a very efficient Monte Carlo method, now widely known as Pollard's "rho" or ρ-method, for finding a small non-trivial factor d of a large integer n. Trial division by all integers up to $\sqrt{n}$ is guaranteed to factor completely any number up to N. For the same amount of work, Pollard's "rho" method will factor any number up to n^2 (unless we are unlucky). The method uses an iteration of the form

$$\left.\begin{aligned}
x_0 &= \mathrm{random}(0,\ n-1), \\
x_i &\equiv f(x_{i-1}) \pmod{n}, \qquad i = 1, 2, 3, \ldots
\end{aligned}\right\} \tag{3.4}$$

where x_0 is a random starting value, n is the number to be factored, and $f \in \mathbb{Z}[x]$ is a polynomial with integer coefficients; usually, we just simply choose $f(x) = x^2 \pm a$ with $a \neq -2, 0$. Then starting with some initial value x_0, a "random" sequence $x_1, x_2, x_3, \ldots$ is computed modulo n in the following way:

$$\left.\begin{aligned}
x_1 &= f(x_0), \\
x_2 &= f(f(x_0)) = f(x_1), \\
x_3 &= f(f(f(x_0))) = f(f(x_1)) = f(x_2), \\
&\ \ \vdots \\
x_i &= f(x_{i-1}).
\end{aligned}\right\} \tag{3.5}$$

Let d be a nontrivial divisor of n, where d is small compared with n. Since there are relatively few congruence classes modulo d (namely, d of them), there will probably exist integers x_i and x_j which lie in the same congruence class modulo d, but belong to different classes modulo N; in short, we will have

$$\left.\begin{aligned}
x_i &\equiv x_j \pmod{d}, \\
x_i &\not\equiv x_j \pmod{n}.
\end{aligned}\right\} \tag{3.6}$$

Since $d \mid (x_i - x_j)$ and $n \nmid (x_i - x_j)$, it follows that $\gcd(x_i - x_j,\ n)$ is a nontrivial factor of n. In practice, a divisor d of n is not known in advance, but it can most likely be detected by keeping track of the integers x_i, which we do know; we simply compare x_i with the earlier x_j, calculating $\gcd(x_i - x_j,\ n)$ until a nontrivial gcd occurs. The divisor obtained in this way is not necessarily the smallest factor of n and indeed it may not be prime. The possibility exists that when a gcd greater that 1 is found, it may also turn out to be equal to n itself, though this happens very rarely.

Example 3.2.1. For example, let $n = 1387 = 19 \cdot 73$, $f(x) = x^2 - 1$ and $x_1 = 2$. Then the "random" sequence $x_1, x_2, x_3, \ldots$ is as follows:

$$2, 3, 8, 63, 1194, \overline{1186, 177, 814, 996, 310, 396, 84, 120, 529, 1053, 595, 339}$$

where the repeated values are overlined. Now we find that

$$x_3 \equiv 6 \pmod{19},$$
$$x_3 \equiv 63 \pmod{1387},$$

$$x_4 \equiv 16 \pmod{19},$$
$$x_4 \equiv 1194 \pmod{1387},$$

$$x_5 \equiv 8 \pmod{19},$$
$$x_5 \equiv 1186 \pmod{1387},$$

$$\vdots$$

So we have

$$\gcd(63 - 6, 1387) = \gcd(1194 - 16, 1387) = \gcd(1186 - 8, 1387) = \cdots = 19.$$

Of course, as mentioned earlier, d is not known in advance, but we can keep track of the integers x_i which we do know, and simply compare x_i with all the previous x_j with $j < i$, calculating $\gcd(x_i - x_j, n)$ until a nontrivial gcd occurs:

$$\gcd(x_1 - x_0, n) = \gcd(3 - 2, 1387) = 1,$$
$$\gcd(x_2 - x_1, n) = \gcd(8 - 3, 1387) = 1,$$
$$\gcd(x_2 - x_0, n) = \gcd(8 - 2, 1387) = 1,$$
$$\gcd(x_3 - x_2, n) = \gcd(63 - 8, 1387) = 1,$$
$$\gcd(x_3 - x_1, n) = \gcd(63 - 3, 1387) = 1,$$
$$\gcd(x_3 - x_0, n) = \gcd(63 - 2, 1387) = 1,$$
$$\gcd(x_4 - x_3, n) = \gcd(1194 - 63, 1387) = 1,$$
$$\gcd(x_4 - x_2, n) = \gcd(1194 - 8, 1387) = 1,$$
$$\gcd(x_4 - x_1, n) = \gcd(1194 - 3, 1387) = 1,$$
$$\gcd(x_4 - x_0, n) = \gcd(1194 - 2, 1387) = 1,$$
$$\gcd(x_5 - x_4, n) = \gcd(1186 - 1194, 1387) = 1,$$
$$\gcd(x_5 - x_3, n) = \gcd(1186 - 63, 1387) = 1,$$
$$\gcd(x_5 - x_2, n) = \gcd(1186 - 8, 1387) = 19.$$

So after 13 comparisons and calculations, we eventually find the divisor 19.

As k increases, the task of computing $\gcd(x_i - x_j, n)$ for all $j < i$ becomes very time-consuming; for $n = 10^{50}$, the computation of $\gcd(x_i - x_j, n)$ would require about $1.5 \cdot 10^6$ bit operations, as the complexity for computing one gcd is $\mathcal{O}((\log n)^3)$. Pollard actually used Floyd's method to detect a cycle in a long sequence $\langle x_i \rangle$, which just looks at cases in which $x_i = x_{2i}$. To see how it works, suppose that $x_i \equiv x_j \pmod{n}$, then

$$\left.\begin{aligned}
x_{i+1} &\equiv f(x_i) \equiv f(x_j) \equiv x_{j+1} \pmod{d}, \\
x_{i+2} &\equiv f(x_{i+1}) \equiv f(x_{j+1}) \equiv x_{j+2} \pmod{d}, \\
&\vdots \\
x_{i+k} &\equiv f(x_{i+k-1}) \equiv f(x_{j+k-1}) \equiv x_{j+k} \pmod{d}.
\end{aligned}\right\} \tag{3.7}$$

If $k = j - i$, then $x_{2i} \equiv x_i \pmod{d}$. Hence, we only need look at $x_{2i} - x_i$ (or $x_i - x_{2i}$) for $i = 1, 2, \ldots$. That is, we only need to check one gcd for each i. Note that the sequence $x_0, x_1, x_2, \ldots$ modulo a prime number p, say, looks like a circle with a tail; it is from this behavior that the method gets its name (see Figure 3.1 for a graphical sketch; it looks like the Greek letter ρ).

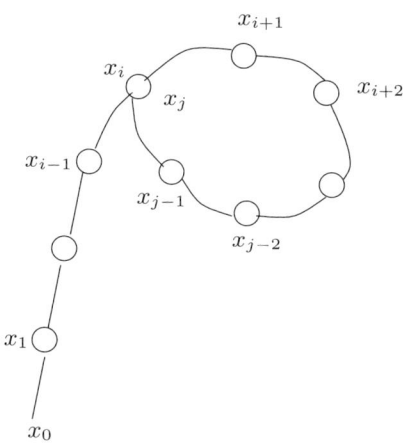

Figure 3.1. Illustration of the ρ-method

Example 3.2.2. Again, let $n = 1387 = 19 \cdot 73$, $f(x) = x^2 - 1$ and $x_1 = 2$. By comparing pairs x_i and x_{2i}, for $i = 1, 2, \ldots$, we have:

$$\gcd(x_1 - x_2,\ n) = \gcd(3 - 8,\ 1387) = 1,$$
$$\gcd(x_2 - x_4,\ n) = \gcd(8 - 1194,\ 1387) = 1,$$
$$\gcd(x_3 - x_6,\ n) = \gcd(63 - 177,\ 1387) = 19.$$

So only after 3 comparisons and gcd calculations, the divisor 19 of 1387 is found.

In what follows, we shall show that to compute $y_i = x_{2i}$, we do not need to compute $x_{i+1}, x_{i+2}, \ldots, x_{2i-1}$ until we get x_{2i}. Observe that

$$
\left.\begin{aligned}
y_1 &= x_2 = f(x_1) = f(f(x_0)) = f(f(y_0)), \\
y_2 &= x_4 = f(x_3) = f(f(x_2)) = f(f(y_1)), \\
y_3 &= x_6 = f(x_5) = f(f(x_4)) = f(f(y_2)), \\
&\vdots \\
y_i &= x_{2i} = f(f(y_{i-1})).
\end{aligned}\right\} \tag{3.8}
$$

So at each step, we compute

$$
\left.\begin{aligned}
x_i &= f(x_{i-1}) \quad (\bmod\ n), \\
y_i &= f(f(y_{i-1})) \quad (\bmod\ n).
\end{aligned}\right\} \tag{3.9}
$$

Therefore, only three evaluations of f will be required.

Example 3.2.3. Let once again $n = 1387 = 19 \cdot 73$, $f(x) = x^2 - 1$ and $x_0 = y_0 = 2$. By comparing pairs x_i and x_{2i}, for $i = 1, 2, \ldots$, we get:

$f(y_0) = 2^2 - 1 = 3,$

$f(f(y_0)) = 3^2 - 1 = 8 = y_1$

$\qquad \Longrightarrow\ \gcd(y_1 - x_1,\ N) = \gcd(3 - 8,\ 1387) = 1$

$f(y_1) = 8^2 - 1 = 63,$

$f(f(y_1)) = 63^2 - 1 = 1194 = y_2$

$\qquad \Longrightarrow\ \gcd(y_2 - x_2,\ N) = \gcd(8 - 1194,\ 1387) = 1$

$f(y_2) = 1194^2 - 1 \bmod 1387 = 1186,$

$f(f(y_2)) = 1186^2 - 1 \bmod 1387 = 177 = y_3$

$\qquad \Longrightarrow\ \gcd(y_3 - x_3,\ N) = \gcd(63 - 177,\ 1387) = 19.$

The divisor 19 of 1387 is then found.

Remark 3.2.1. There is an even more efficient algorithm, due to Richard Brent, that looks only at the following differences and the corresponding gcd results:

$$
\begin{aligned}
x_1 - x_3 &\Longrightarrow \gcd(x_1 - x_3,\ n), \\
x_3 - x_6 &\Longrightarrow \gcd(x_3 - x_6,\ n), \\
x_3 - x_7 &\Longrightarrow \gcd(x_3 - x_7,\ n), \\
x_7 - x_{12} &\Longrightarrow \gcd(x_7 - x_{12},\ n), \\
x_7 - x_{13} &\Longrightarrow \gcd(x_7 - x_{13},\ n), \\
x_7 - x_{14} &\Longrightarrow \gcd(x_7 - x_{14},\ n),
\end{aligned}
$$

$$x_7 - x_{15} \implies \gcd(x_7 - x_{15}, \ n),$$

$$\vdots$$

and in general:

$$x_{2^n - 1} - x_j, \quad 2^{n+1} - 2^{n-1} \leq j \leq 2^{n+1} - 1. \qquad (3.10)$$

Brent's algorithm is about 24 percent faster than Pollard's original version.

Now we are in a position to present an algorithm for the ρ-method.

Algorithm 3.2.4 (Brent-Pollard's ρ-method). Let n be a composite integer greater than 1. This algorithm tries to find a nontrivial factor d of n, which is small compared with $\sqrt{n}$. Suppose the polynomial to use is $f(x) = x^2 + 1$.

[1] (Initialization) Choose a seed, say $x_0 = 2$, a generating function, say $f(x) = x^2 + 1 \pmod{n}$. Choose also a value for t not much bigger than $\sqrt{d}$, perhaps $t < 100\sqrt{d}$.

[2] (Iteration and Computation) Compute x_i and y_i in the following way:

$$
\begin{aligned}
x_1 &= f(x_0), \\
x_2 &= f(f(x_0)) = f(x_1), \\
x_3 &= f(f(f(x_0))) = f(f(x_1)) = f(x_2), \\
&\vdots \\
x_i &= f(x_{i-1}).
\end{aligned}
$$

$$
\begin{aligned}
y_1 &= x_2 = f(x_1) = f(f(x_0)) = f(f(y_0)), \\
y_2 &= x_4 = f(x_3) = f(f(x_2)) = f(f(y_1)), \\
y_3 &= x_6 = f(x_5) = f(f(x_4)) = f(f(y_2)), \\
&\vdots \\
y_i &= x_{2i} = f(f(y_{i-1})).
\end{aligned}
$$

and simultaneously compare x_i and y_i by computing $d = \gcd(x_i - y_i, \ n)$.

[3] (Factor Found?) If $1 < d < n$, then d is a nontrivial factor of n, print d, and go to Step [5].

[4] (Another Search?) If $x_i = y_i \pmod{n}$ for some i or $i \geq \sqrt{t}$, then go to Step [1] to choose a new seed and a new generator and repeat.

[5] (Exit) Terminate the algorithm.

Example 3.2.4. The 8th Fermat number $F_8 = 2^{2^8} + 1$ was factored by Brent and Pollard in 1980 by using Brent-Pollard's "rho" method:

$$2^{2^8} + 1 = 2^{256} + 1 = 1238926361552897 \cdot p_{63}.$$

Now let us move on to the complexity of the ρ-method. Let p be the smallest prime factor of N, and j the smallest positive index such that $x_{2j} \equiv x_j \pmod{p}$. Making some plausible assumptions, it is easy to show that the expected value of j is $\mathcal{O}(\sqrt{p})$. The argument is related to the well-known "birthday" paradox: suppose that $1 \leq k \leq n$ and that the numbers $x_1, x_2, \ldots, x_k$ are independently chosen from the set $\{1, 2, \ldots, n\}$. Then the probability that the numbers x_k are distinct is

$$\left(1 - \frac{1}{n}\right) \cdot \left(1 - \frac{2}{n}\right) \cdots \left(1 - \frac{k-1}{n}\right) \quad \sim \quad \exp\left(\frac{-k^2}{2n}\right). \qquad (3.11)$$

Note that the x_i's are likely to be distinct if k is small compared with $\sqrt{n}$, but unlikely to be distinct if k is large compared with $\sqrt{n}$. Of course, we cannot work out $x_i \bmod p$, since we do not know p in advance, but we can detect x_j by taking greatest common divisors. We simply compute $d = \gcd(x_{2i} - x_i, n)$ for $i = 1, 2, \ldots$ and stop when a $d > 1$ is found.

Conjecture 3.2.1 (Complexity of the ρ-method). Let p be a prime dividing n and $p = \mathcal{O}(\sqrt{p})$, then the ρ-algorithm has has expected running time

$$\mathcal{O}(\sqrt{p}) = \mathcal{O}(\sqrt{p}(\log n)^2) = \mathcal{O}(n^{1/4}(\log n)^2) \qquad (3.12)$$

to find the prime factor p of n.

Remark 3.2.2. The ρ-method is an improvement over trial division, because in trial division, $\mathcal{O}(p) = \mathcal{O}(n^{1/4})$ divisions is needed to find a small factor p of n. But of course, one disadvantage of the ρ-algorithm is that its running time is only a conjectured expected value, not a rigorous bound.

Pollard in 1974 invented also another simple but effective factoring algorithm, now widely known as Pollard's "$p-1$" method, which can be described as follows:

Algorithm 3.2.5. [Pollard's "$p-1$" Method] Let $n > 1$ be a composite number. This algorithm attempts to find a nontrivial factor of n.

[1] (Initialization) Pick out $a \in \mathbb{Z}/n\mathbb{Z}$ at random. Select a positive integer k that is divisible by many prime powers, for example, $k = \text{lcm}(1, 2, \ldots, B)$ for a suitable bound B (the larger B is the more likely the method will be to succeed in producing a factor, but the longer the method will take to work).

[2] (Exponentiation) Compute $a_k = a^k \bmod n$.

[3] (Compute GCD) Computing $d = \gcd(a_k - 1, \ n)$.

[4] (Factor Found?) If $1 < d < N$, then d is a nontrivial factor of n, output d and go to Step [6].

[5] (Start Over?) If d is not a nontrivial factor of n and if you still want to try more experiments, then go to Step [1] to start all over again with a new choice of a and/or a new choice of k, else go to Step [6].

[6] (Exit) Terminate the algorithm.

The "$p - 1$" algorithm is usually successful in the fortunate case where n has a prime divisor p for which $p - 1$ has no large prime factors. Suppose that $(p - 1) \mid k$ and that $p \nmid a$. Since $|(\mathbb{Z}/p\mathbb{Z})^*| = p - 1$, we have $a^k \equiv 1 \pmod{p}$, thus $p \mid \gcd(a_k - 1, \ n)$. In many cases, we have $p = \gcd(a_k - 1, \ n)$, so the method finds a nontrivial factor of n.

Example 3.2.5. Use the "$p - 1$" method to factor the number $n = 540143$. Choose $B = 8$ and hence $k = 840$. Choose also $a = 2$. Then we have

$$\gcd(2^{840} - 1 \bmod 540143, \ 540143) = \gcd(53046, \ 540143) = 421.$$

Thus 421 is a (prime) factor of 540143. In fact, $421 \cdot 1283$ is the complete prime factorization of 540143. It is interesting to note that by using the "$p - 1$" method Baillie in 1980 found the prime factor

$$p_{25} = 1155685395246619182673033$$

of the Mersenne number $M_{257} = 2^{257} - 1$. In this case

$$p_{25} - 1 = 2^3 \cdot 3^2 \cdot 19^2 \cdot 47 \cdot 67 \cdot 257 \cdot 439 \cdot 119173 \cdot 1050151.$$

In the worst case, where $(p - 1)/2$ is prime, the "$p - 1$" algorithm is no better than trial division. Since the group has fixed order $p - 1$ there is nothing to be done except try a different algorithm. Note that there is a similar method to "$p - 1$", called "$p + 1$", proposed by H. C. Williams in 1982. It is suitable for the case where N has a prime factor p for which $p + 1$ has no large prime factors.

Problems for Section 3.2

Problem 3.2.1. (Sierpinski) Prove that there are infinitely many integers k such that

$$n = k \cdot 2^n + 1, \quad n = 1, 2, 3, \ldots$$

are all composite. Find three such composites.

Problem 3.2.2. Show that the sequence $2^{2^n} + 3$, $n = 1, 2, 3, \ldots$ contains infinitely many composites.

Problem 3.2.3. Use Fermat factoring method to factor the number 1254713.

Problem 3.2.4. Suppose $n = pq$ with p, q prime and $p < q < 2p$. Let δ be the number defined by $q/p = 1 + \delta$ such that $0 < \delta < 1$. Show that the number of steps used in the Fermat factoring method is approximately $p\delta^2/8$.

Problem 3.2.5. Let $n = pq = 18886013$ such that

$$|\frac{p}{q} - 3| < \frac{1}{100}.$$

Find the factorization of n.

Problem 3.2.6. Use $p - 1$ method to find the smallest prime factor of the 9th Fermat number $F_9 = 2^{2^9} + 1$.

Problem 3.2.7. Use $p - 1$ method to find three prime factors of $2^{71} - 1$.

Problem 3.2.8. Use the ρ factoring method to factor 4087 using $x_0 = 2$ and $f(x) = x^2 - 1$.

Problem 3.2.9. (Corman, et al) Let $x_i = f(x_{i-1})$, $i = 1, 2, 3, \ldots$. Let also $t, u > 0$ be the smallest numbers in the sequence $x_{t+i} = x_{t+u+i}$, $i = 0, 1, 2, \ldots$, where t and u are called the lengths of the ρ tail and cycle, respectively. Give an efficient algorithm to determine t and u exactly, and analyze the running time of your algorithm.

3.3 Elliptic Curve Method (ECM)

In Section 2.6, we discussed the application of elliptic curves to primality testing. In this section, we shall introduce a factoring method depending on the use of elliptic curves. The method is actually obtained from Pollard's "$p - 1$" algorithm: if we can choose a *random* group G with order g close to p, we may be able to perform a computation similar to that involved in Pollard's "$p - 1$" algorithm, working in G rather than in F_p. If all prime factors of g are less than the bound B then we find a factor of n. Otherwise, we repeat this procedure with a different group G (and hence, usually, a different g) until a factor is found. This is the motivation of the ECM method, invented by H. W. Lenstra in 1987.

Algorithm 3.3.1 (Lenstra's Elliptic Curve Method). Let $n > 1$ be a composite number, with $\gcd(n, 6) = 1$. This algorithm attempts to find a non-trivial factor of n. The method depends on the use of elliptic curves and is the analogue to Pollard's "$p - 1$" method.

[1] (Choose an Elliptic Curve) Choose a random pair (E, P), where E is an elliptic curve $y^2 = x^3 + ax + b$ over $\mathbb{Z}/n\mathbb{Z}$, and $P(x, y) \in E(\mathbb{Z}/n\mathbb{Z})$ is a point on E. That is, choose $a, x, y \in \mathbb{Z}/n\mathbb{Z}$ at random, and set $b \leftarrow y^2 - x^3 - ax$. If $\gcd(4a^3 + 27b^2, n) \neq 1$, then E is not an elliptic curve, start all over and choose another pair (E, P).

[2] (Choose an Integer k) Just as in the "$p-1$" method, select a positive integer k that is divisible by many prime powers, for example, $k = \operatorname{lcm}(1, 2, \ldots, B)$ or $k = B!$ for a suitable bound B; the larger B is the more likely the method will succeed in producing a factor, but the longer the method will take to work.

[3] (Calculate kP) Calculate the point $kP \in E(\mathbb{Z}/n\mathbb{Z})$. We use the following formula to compute $P_3(x_3, y_3) = P_1(x_1, y_1) + P_2(x_2, y_2)$ modulo n:

$$(x_3, y_3) = (\lambda^2 - x_1 - x_2 \mod n, \quad \lambda(x_1 - x_3) - y_1 \mod n)$$

where

$$\lambda = \begin{cases} \dfrac{m_1}{m_2} \equiv \dfrac{3x_1^2 + a}{2y_1}(\mod \ n), & \text{if } P_1 = P_2 \\[3mm] \dfrac{m_1}{m_2} \equiv \dfrac{y_1 - y_2}{x_1 - x_2} (\mod \ n), & \text{otherwise.} \end{cases}$$

The computation of $kP \mod n$ can be done in $\mathcal{O}(\log k)$ doublings and additions.

[4] (Computing GCD) If $kP \equiv \mathcal{O}_E \pmod{n}$, then compute $d = \gcd(m_2, \ n)$, else go to Step [1] to make a new choice for "a" or even for a new pair (E, P).

[5] (Factor Found?) If $1 < d < n$, then d is a nontrivial factor of n, output d and go to Step [7].

[6] (Start Over?) If d is not a nontrivial factor of n and if you still wish to try more elliptic curves, then go to Step [1] to start all over again, else go to Step [7].

[7] (Exit) Terminate the algorithm.

As for the "$p-1$" method, one can show that a given pair (E, P) is likely to be successful in the above algorithm if n has a prime factor p for which $\mathbb{Z}/p\mathbb{Z}$ is composed of small primes only. The probability for this to happen increases with the number of pairs (E, P) that one tries.

Example 3.3.1. Use the ECM method to factor the number $n = 187$.

[1] Choose $B = 3$, and hence $k = \text{lcm}(1, 2, 3) = 6$. Let $P = (0, 5)$ be a point on the elliptic curve $E : y^2 = x^3 + x + 25$ which satisfies $\gcd(N, 4a^3 + 27b^2) = \gcd(187, 16879) = 1$ (note that here $a = 1$ and $b = 25$).

[2] Since $k = 6 = 110_2$, we compute $6P = 2(P + 2P)$ in the following way:

[a] Compute $2P = P + P = (0, 5) + (0, 5)$:

$$
\begin{cases}
\lambda = \dfrac{m_1}{m_2} = \dfrac{1}{10} \equiv 131 \ (\text{mod } 187) \\
x_3 = 144 \ (\text{mod } 187) \\
y_3 = 18 \ (\text{mod } 187)
\end{cases}
$$

So $2P = (144, 18)$ with $m_2 = 10$ and $\lambda = 131$.

[b] Compute $3P = P + 2P = (0, 5) + (144, 18)$:

$$
\begin{cases}
\lambda = \dfrac{m_1}{m_2} = \dfrac{13}{144} \equiv 178 \ (\text{mod } 187) \\
x_3 = 124 \ (\text{mod } 187) \\
y_3 = 176 \ (\text{mod } 187)
\end{cases}
$$

So $3P = (124, 176)$ with $m_2 = 144$ and $\lambda = 178$.

[c] Compute $6P = 2(3P) = 3P + 3P = (124, 176) + (124, 176)$:

$$
\lambda = \frac{m_1}{m_2} = \frac{46129}{352} \equiv \frac{127}{165} \equiv \mathcal{O}_E \ (\text{mod } 187).
$$

This time $m_1 = 127$ and $m_2 = 165$, so the modular inverse for $127/165$ modulo 187 does not exist; but this is exactly what we want! – this type of failure is called a "pretended failure".

[3] Compute $d = \gcd(n, m_2) = \gcd(187, 165) = 11$. Since $1 < 11 < 187$, 11 is a (prime) factor of 187. In fact, $187 = 11 \cdot 17$.

Example 3.3.2. To factor $n = 7560636089$, we calculate $kP = k(1, 3)$ with $k = \text{lcm}(1, 2, 3, \dots, 19)$ on $y^2 \equiv x^3 - x + 7 \ (\text{mod } n)$:

$3P = (1329185554, 395213649)$	$6P = (646076693, 5714212282)$
$12P = (5471830359, 5103472059)$	$24P = (04270711, 3729197625)$
$49P = (326178740, 3033431040)$	$99P = (5140727517, 2482333384)$
$199P = (1075608203, 3158750830)$	$398P = (4900089049, 2668152272)$
$797P = (243200145, 2284975169)$	$1595P = (3858922333, 4843162438)$
$3191P = (7550557590, 1472275078)$	$6382P = (4680335599, 1331171175)$
$12765P = (6687327444, 7233749859)$	$25530P = (6652513841, 6306817073)$
$51061P = (6578825631, 5517394034)$	$102123P = (1383310127, 2036899446)$

$204247P = (3138092894, 2918615751)$ $408495P = (6052513220, 1280964400)$

$816990P = (2660742654, 3418862519)$ $1633980P = (7023086430, 1556397347)$

$3267961P = (5398595429, 795490222)$ $6535923P = (4999132, 4591063762)$

$13071847P = (3972919246, 7322445069)$ $26143695P = (3597132904, 3966259569)$

$52287391P = (2477960886, 862860073)$ $104574782P = (658268732, 3654016834)$

$209149565P = (6484065460, 287965264)$ $418299131P = (1622459893, 4833264668)$

$836598262P = (7162984288, 487850179)$ $1673196525P = \mathcal{O}_E$.

Now, $1673196525P$ is the point at infinity since $5398907681/1016070716$ mod n is impossible. Hence, $\gcd(1016070716, 7560636089) = 15121$ gives a factor of n.

Example 3.3.3. The following are some ECM factoring examples. In 1995 Richard Brent at the Australian National University completed the factorization of the 10th Fermat number using ECM:

$$2^{2^{10}} + 1 = 2^{1024} + 1 = 45592577 \cdot 6487031809 \cdot p_{40} \cdot p_{252}$$

where the 40-digit prime p_{40} was found using ECM, and p_{252} was proved to be a 252-digit prime. Brent also completed the factorization of the 11th Fermat number (with 617-digit) $F_{11} = 2^{2^{11}} + 1$ using ECM:

$$F_{11} = 319489 \cdot 974849 \cdot 167988556341760475137 \cdot 3560841906445833920513 \cdot p_{564}$$

where the 21-digit and 22-digit prime factors were found using ECM, and p_{564} is a 564-digit prime. Other recent ECM-records include a 38-digit prime factor (found by A. K. Lenstra and M. S. Manasse) in the 112-digit composite $(11^{118} + 1)/(2 \cdot 61 \cdot 193121673)$, a 40-digit prime factor of $26^{126} + 1$, a 43-digit prime factor of the partition number $p(19997)$ and a 44-digit prime factor of the partition number $p(19069)$ in the RSA Factoring Challenge List, and a 47-digit prime in c_{135} of $5^{2^8} + 1 = 2 \cdot 1655809 \cdot p_{38} \cdot c_{135}$.

Both Lenstra's ECM algorithm and Pollard's "$p - 1$" algorithm can be speeded up by the addition of a second phase. The idea of the second phase in ECM is to find a factor in the case that the first phase terminates with a group element $P \neq I$, such that $|\langle P \rangle|$ is reasonably small (say $\mathcal{O}(m^2)$). (Here $\langle P \rangle$ is the cyclic group generated by P.) There are several possible implementations of the second phase. One of the simplest uses a pseudorandom walk in $\langle P \rangle$. By the birthday paradox argument, there is a good chance that two points in the random walk will coincide after $\mathcal{O}(\sqrt{|\langle P \rangle|})$ steps, and when this occurs a nontrivial factor of N can usually be found (see [32] and [155] for more detailed information on the implementation issues of ECM.

Conjecture 3.3.1 (Complexity of the ECM method). Let p be the smallest prime dividing n. Then the ECM method will find p of n, under some plausible assumptions, in expected running time

$$\mathcal{O}\left(\exp\left(\sqrt{(2+o(1))\log p \log\log p}\right)\cdot(\log n)^2\right) \tag{3.13}$$

In the worst case, when n is the product of two prime factors of the same order of magnitude, we have

$$\mathcal{O}\left(\exp\left(\sqrt{(2+o(1))\log n \log\log n}\right)\right)$$

$$= \mathcal{O}\left(n^{\sqrt{(2+o(1))\log\log n/\log n}}\right). \tag{3.14}$$

Remark 3.3.1. The most interesting feature of ECM is that its running time depends very much on p (the factor found) of n, rather than N itself. So one advantage of the ECM is that one may use it, in a manner similar to trial divisions, to locate the smaller prime factors p of a number n which is much too large to factor completely.

Problems for Section 3.3

Problem 3.3.1. Show that $\#(a,b) = p^2 - p$, where $\#(a,b)$ denotes the number of integer pairs (a,b) such that $0 \le a,b < p$, for which $4a^3 \not\equiv 27b^2 \pmod{p}$ is exactly $p^2 - p$.

Problem 3.3.2. Show that if $p > 3$, $4a^3 + 27b^2 \equiv 0 \pmod{p}$, $p \nmid a$, then the root r of the congruence $-2ar \equiv 3b \pmod{p}$ is a repeated rood modulo p of the polynomial $x^3 - ax - b$.

Problem 3.3.3. Let $p > 2$ be prime. Suppose that x and y are integers such that

$$x^2 + y^2 \equiv 1 \pmod{p}, \quad x \not\equiv 1 \pmod{p}.$$

Let u be determined by the congruence

$$(1+x)u \equiv y \pmod{p}.$$

Show that

$$u^2 + 1 \equiv 0 \pmod{p}.$$

Problem 3.3.4. Canfield-Erdös-Pomerance theorem: Let n be a positive integer which is not a prime power and not divisible by 2 or 3. if α is a real number, then the probability that a random positive integer $s \le x$ has all its prime factors $\le L(x)^\alpha$ is $L(x)^{-1/(2\alpha)+o(1)}$ for $x \to \infty$, where $L(x) = e^{\sqrt{\log x \log\log x}}$ with x a real number $> e$.

We also need the following conjecture: Let $x = p$, the probability that a random integer has all its prime factors $\le L(x)^\alpha$ in the small interval $(x+1-\sqrt{x}, x+1+\sqrt{x})$ is $L(p)^{-1/(2\alpha)+o(1)}$ for $p \to \infty$.

By the Canfield-Erdös-Pomerance theorem and the conjecture, show that

(1) the probabilistic time estimate for ECM to find the smallest prime factor p of n is

$$e^{\sqrt{2+o(1)\log p \log\log p}}$$

(2) the probabilistic time estimate for ECM to find n is

$$e^{\sqrt{1+o(1)\log n \log\log n}}.$$

Problem 3.3.5. Use Algorithm 3.3.1 to factor the three integers 17531, 218548425731 and 190387615311371.

Problem 3.3.6. Modify and Improve Algorithm 3.3.1 to a practical factoring algorithm for large integer n.

Problem 3.3.7. Modify and Improve Algorithm 3.3.1 to a parallel practical factoring algorithm for large integer n.

3.4 General Factoring Congruence

In the next three sections, we shall introduce three widely used *general* purpose integer factorization methods, namely, the continued fraction method (abbreviated CFRAC), the quadratic sieve (abbreviated QS) and the number field sieve (abbreviated NFS). By a general purpose factoring method, we mean one that will factor *any* integer of a given size in about the same time as any other of that size. The method will take as long, for example, to split a 100-digit number into the product of a 1-digit and a 99-digit prime, as it will to split a different number into the product of two 50-digit primes. These methods do not depend upon any special properties of the number or its factors.

The CFRAC method, as well as other powerful general purpose factoring methods such as the Quadratic Sieve (QS) and the Number Field Sieve (NFS), makes use of the simple but important observation that if we have two integers x and y such that

$$x^2 \equiv y^2 \pmod{n}, \ 0 < x < y < n, \ x \neq y, \ x + y \neq n, \tag{3.15}$$

then $\gcd(x - y, \ n)$ and $\gcd(x + y, \ n)$ are possibly nontrivial factors of n, because $n \mid (x + y)(x - y)$, but $n \nmid (x + y)$ and $n \nmid (x - y)$. The congruence (3.15) is often called Legendre's congruence. So, to use Legendre's congruence for factorization, we simply perform the following two steps:

[1] Find a nontrivial solution to the congruence $x^2 \equiv y^2 \pmod{n}$.

[2] Compute the factors d_1 and d_2 of N by using Euclid's algorithm:

$$(d_1, d_2) = (\gcd(x + y, \ n), \ \gcd(x - y, \ n)).$$

Example 3.4.1. Let $n = 119$. Since, $12^2 \bmod 119 = 5^2 \bmod 119$, we have

$$(d_1, d_2) = (\gcd(12 + 5, 119), \ \gcd(12 - 5, 119)) = (17, 7).$$

In fact, $119 = 7 \cdot 17$.

The best method for constructing congruences of the form (3.15) starts by accumulating several congruences of the form

$$\left(A_i = \prod p_k^{e_k}\right) \equiv \left(B_i = \prod p_j^{e_j}\right) \pmod{n}. \tag{3.16}$$

Some of these congruences are then multiplied in order to generate squares on both sides (Montgomery [159]). We illustrate this idea in the following example.

Example 3.4.2. Let $n = 77$. Then, on the left hand side of the following table, we collect eight congruences of the form (3.16) over the prime factor base FB $= \{-1, 2, 3, 5\}$ (note that we include -1 as a "prime" factor); the right hand side of the table contains the exponent vector information of $v(A_i)$ and $v(B_i)$ modulo 2.

$45 = 3^2 \cdot 5$	$\equiv$	$-32 = -2^5$	$(0\ 0\ 0\ 1) \equiv (1\ 1\ 0\ 0)$
$50 = 2 \cdot 5^2$	$\equiv$	$-27 = -3^3$	$(0\ 1\ 0\ 0) \equiv (1\ 0\ 1\ 0)$
$72 = 2^3 \cdot 3^2$	$\equiv$	-5	$(0\ 1\ 0\ 0) \equiv (1\ 0\ 0\ 1)$
$75 = 3 \cdot 5^2$	$\equiv$	-2	$(0\ 0\ 1\ 0) \equiv (1\ 1\ 0\ 0)$
$80 = 2^4 \cdot 5$	$\equiv$	3	$(0\ 0\ 0\ 1) \equiv (0\ 0\ 1\ 0)$
$125 = 5^3$	$\equiv$	$48 = 2^4 \cdot 3$	$(0\ 0\ 0\ 1) \equiv (0\ 0\ 1\ 0)$
$320 = 2^6 \cdot 5$	$\equiv$	$243 = 3^5$	$(0\ 0\ 0\ 1) \equiv (0\ 0\ 1\ 0)$
$384 = 2^7 \cdot 3$	$\equiv$	-1	$(0\ 1\ 1\ 0) \equiv (1\ 0\ 0\ 0)$

Now we multiply some of these congruences in order to generate squares on both sides; both sides will be squares precisely when the sum of the exponent vectors is the zero vector modulo 2. We first multiply the sixth and seventh congruences and get:

$125 = 5^3$	$\equiv$	$48 = 2^4 \cdot 3$	$(0\ 0\ 0\ 1)$	$\equiv$	$(0\ 0\ 1\ 0)$
$320 = 2^6 \cdot 5$	$\equiv$	$243 = 3^5$	$(0\ 0\ 0\ 1)$	$\equiv$	$(0\ 0\ 1\ 0)$
			$\downarrow\downarrow\downarrow\downarrow$		$\downarrow\downarrow\downarrow\downarrow$
			$(0\ 0\ 0\ 0)$		$(0\ 0\ 0\ 0)$

Since the sum of the exponent vectors is the zero vector modulo 2, we find squares on both sides:

$$5^3 \cdot 2^6 \cdot 5 \equiv 2^4 \cdot 3 \cdot 3^5 \iff (5^2 \cdot 2^3)^2 \equiv (2^2 \cdot 3^3)^2$$

and hence we have $\gcd(5^2 \cdot 2^3 \pm 2^2 \cdot 3^3, \ 77) = (77, 1)$, but this does not split 77, so we try to multiply some other congruences, for example, the fifth and the seventh, and get:

$80 = 2^4 \cdot 5$	$\equiv$	3	$(0\ 0\ 0\ 1)$	$\equiv$	$(0\ 0\ 1\ 0)$
$320 = 2^6 \cdot 5$	$\equiv$	$243 = 3^5$	$(0\ 0\ 0\ 1)$	$\equiv$	$(0\ 0\ 1\ 0)$
			$\downarrow\downarrow\downarrow\downarrow$ $(0\ 0\ 0\ 0)$		$\downarrow\downarrow\downarrow\downarrow$ $(0\ 0\ 0\ 0)$

The sum of the exponent vectors is the zero vector modulo 2, so we find

$$2^4 \cdot 5 \cdot 2^6 \cdot 5 \equiv 3 \cdot 3^5 \iff (2^5 \cdot 5)^2 \equiv (3^3)^2$$

and compute $\gcd(2^5 \cdot 5 \pm 3^3,\ 77) = (11, 7)$. This time, it splits 77. Once we split N, we stop the process. Just for the purpose of illustration, we try one more example, which will also split N.

$45 = 3^2 \cdot 5$	$\equiv$	$-32 = -2^5$	$(0\ 0\ 0\ 1)$	$\equiv$	$(1\ 1\ 0\ 0)$
$50 = 2 \cdot 5^2$	$\equiv$	$-27 = -3^3$	$(0\ 1\ 0\ 0)$	$\equiv$	$(1\ 0\ 1\ 0)$
$75 = 3 \cdot 5^2$	$\equiv$	-2	$(0\ 0\ 1\ 0)$	$\equiv$	$(1\ 1\ 0\ 0)$
$320 = 2^6 \cdot 5$	$\equiv$	$243 = 3^5$	$(0\ 0\ 0\ 1)$	$\equiv$	$(0\ 0\ 1\ 0)$
$384 = 2^7 \cdot 3$	$\equiv$	-1	$(0\ 1\ 1\ 0)$	$\equiv$	$(1\ 0\ 0\ 0)$
			$\downarrow\downarrow\downarrow\downarrow$ $(0\ 0\ 0\ 0)$		$\downarrow\downarrow\downarrow\downarrow$ $(0\ 0\ 0\ 0)$

So we have

$$3^2 \cdot 5 \cdot 2 \cdot 5^2 \cdot 3 \cdot 5^2 \cdot 2^6 \cdot 5 \cdot 2^7 \cdot 3 \equiv -2^5 \cdot -3^3 \cdot -2 \cdot 3^5 \cdot -1 \iff (2^7 \cdot 3^2 \cdot 5^3)^2 \equiv (2^3 \cdot 3^4)^2,$$

thus $\gcd(2^7 \cdot 3^2 \cdot 5^3 \pm 2^3 \cdot 3^4,\ 77) = (7, 11)$.

Based on the above idea, the trick, common to the CFRAC, QS and NFS, is to find a congruence (also called a *relation*) of the form

$$x_k^2 \equiv (-1)^{e_{0k}} p_1^{e_{1k}} p_2^{e_{2k}} \cdots p_m^{e_{mk}} \pmod{n}, \tag{3.17}$$

where each p_i is a "small" prime number (the set of all such p_i, for $1 \leq i \leq m$, forms a *factor base*, denoted by FB). If we find sufficiently many such congruences, by Gaussian elimination over $\mathbb{Z}/2\mathbb{Z}$ we may hope to find a relation of the form

$$\sum_{1 \leq k \leq n} \epsilon_k (e_{0k}, e_{1k}, e_{2k}, \ldots, e_{mk}) \equiv (0, 0, 0, \ldots, 0) \pmod{2}, \tag{3.18}$$

where ϵ is either 1 or 0, and then

$$x = \prod_{1 \leq k \leq n} x_k^{\epsilon_k}, \tag{3.19}$$

$$y = (-1)^{v_0} p_1^{v_1} p_2^{v_2} \cdots p_m^{v_m}, \tag{3.20}$$

where

$$\sum_k \epsilon_k(e_{0k}, e_{1k}, e_{2k}, \ldots, e_{mk}) = 2(v_0, v_1, v_2, \ldots, v_m). \qquad (3.21)$$

It is clear that we now have $x^2 \equiv y^2 \pmod{n}$. This splits N if, in addition, $x \not\equiv \pm y \pmod{n}$.

Problems for Section 3.4

Problem 3.4.1. Show that

$$n^{c(\log\log n/\log n)^{1/2}} = \exp(c(\log n \log\log n)^{1/2}).$$

Problem 3.4.2. Show that

$$(\log n)^{c \log\log\log n} = \exp(c \log\log n \log\log\log n).$$

Problem 3.4.3. Show that if $n = p^k$ with p prime and $k \geq 1$, then $(p-1) \mid (n-1)$.

Problem 3.4.4. Let $n = p_1 p_2 \cdots p_k$, where $p_1, p_2, \ldots, p_k$ are distinct odd primes. Let also $y \in (\mathbb{Z}/n\mathbb{Z})^*$. Then the congruence

$$x^2 \equiv y^2 \pmod{n}$$

has exact 2^k solutions modulo n, two of them are

$$x = \begin{cases} y, \\ -y. \end{cases}$$

Problem 3.4.5. Let x and y be randomly chosen so that

$$x^2 \equiv y^2 \pmod{n}.$$

Show that the chance of

$$x \not\equiv \pm y \pmod{n}$$

is greater than $1/2$. That is, the chance for

$$1 < \gcd(x - y, n) < n, \quad \text{and} \quad \gcd(x - y, n) \mid n$$

is greater than $1/2$.

Problem 3.4.6. Find a suitable pair of integers (x, y) such that

$$x^2 \equiv y^2 \pmod{139511931371319137}$$

and then factor 139511931371319137.

3.5 Continued FRACtion Method (CFRAC)

The continued fraction method is perhaps the first *modern, general* purpose integer factorization method, although its original idea may go back to M. Kraitchik in the 1920s or even earlier to A. M. Legendre. It was used by D. H. Lehmer and R. E. Powers to devise a new technique in the 1930s, however the method was not very useful and applicable at the time because it was unsuitable for desk calculators. About 40 years later, it was first implemented on a computer by M. A. Morrison and J. Brillhart, who used it to successfully factor the seventh Fermat number

$$F_7 = 2^{2^7} + 1 = 59649589127497217 \cdot 5704689200685129054721$$

on the morning of 13 September 1970.

The Continued FRACtion (CFRAC) method looks for small values of $|W|$ such that $x^2 \equiv W \pmod{n}$ has a solution. Since W is small (specifically $W = \mathcal{O}(\sqrt{n})$), it has a reasonably good chance of being a product of primes in our factor base FB. Now if W is small and $x^2 \equiv W \pmod{n}$, then we can write $x^2 = W + knd^2$ for some k and d, hence $(x/d)^2 - kn = W/d^2$ will be small. In other words, the rational number x/d is an approximation of $\sqrt{kn}$. This suggests looking at the continued fraction expansion of $\sqrt{kn}$, since continued fraction expansions of real numbers give good rational approximations. This is exactly the idea behind the CFRAC method! We first obtain a sequence of approximations (i.e., convergents) P_i/Q_i to $\sqrt{kn}$ for a number of values of k, such that

$$\left| \sqrt{kn} - \frac{P_i}{Q_i} \right| \leq \frac{1}{Q_i^2}. \tag{3.22}$$

Putting $W_i = P_i^2 - Q_i^2 kn$, then we have

$$W_i = (P_i + Q_i\sqrt{kn})(P_i - Q_i\sqrt{kn}) \sim 2Q_i\sqrt{kn}\frac{1}{Q_i} \sim 2\sqrt{kn}. \tag{3.23}$$

Hence, the $P_i^2 \bmod n$ are small and more likely to be smooth, as desired. Then, we try to factor the corresponding integers $W_i = P_i^2 - Q_i^2 kn$ over our factor base FB; at each success, we obtain a new congruence of the form

$$P_i^2 \equiv W_i \Longleftrightarrow x^2 \equiv (-1)^{e_0} p_1^{e_1} p_2^{e_2} \cdots p_m^{e_m} \pmod{n}. \tag{3.24}$$

Once we have obtained at least $m + 2$ such congruences, by Gaussian elimination over $\mathbb{Z}/2\mathbb{Z}$ we have obtained a congruence $x^2 \equiv y^2 \pmod{n}$. That is, if $(x_1, e_{01}, e_{11}, \ldots, e_{m1}), \ldots, (x_r, e_{0r}, e_{1r}, \ldots, e_{mr})$ are solutions of (3.24) such that the vector sum

$$(e_{01}, e_{11}, \ldots, e_{m1}) + \cdots + (e_{0r}, e_{1r}, \ldots, e_{mr}) = (2e_0', 2e_1', \ldots, 2e_m') \tag{3.25}$$

is even in each component, then

$$\begin{cases} x \equiv x_1 x_2 \cdots x_r \pmod{n} \\ y \equiv (-1)^{e'_0} p_1^{e'_1} \cdots p_m^{e'_m} \pmod{n} \end{cases} \tag{3.26}$$

is a solution to (3.15), except for the possibility that $x \equiv \pm y \pmod{n}$, and hence (usually) a nontrivial splitting of n.

Example 3.5.1. We now illustrate, by an example, the idea of CFRAC factoring. Let $n = 1037$. Then $\sqrt{1037} = [32, \overline{4, 1, 15, 3, 3, 15, 1, 4, 64}]$. The first ten continued fraction approximations of $\sqrt{1037}$ are:

Convergent P/Q	$P^2 - n \cdot Q^2 := W$
$32/1$	$-13 = -13$
$129/4$	$49 = 7^2$
$161/5$	$-4 = -2^2$
$2544/79 \equiv 470/79$	$19 = 19,$
$7793/242 \equiv 534/242$	$-19 = -19$
$25923/805 \equiv 1035/805$	$4 = 2^2$
$396638/12317 \equiv 504/910$	$-49 = -7^2$
$422561/13122 \equiv 502/678$	$13 = 13$
$2086882/64805 \equiv 438/511$	$-1 = -1$
$133983009/4160642 \equiv 535/198$	$13 = 13$

Now we search for squares on both sides either just by a single congruence or by a combination (i.e., multiplying together) of several congruences, and find that

$$129^2 \equiv 7^2 \iff \gcd(1037,\ 129 \pm 7) = (17, 61),$$

$$1035^2 \equiv 2^2 \iff \gcd(1037,\ 1035 \pm 2) = (1037, 1),$$

$$129^2 \cdot 1035^2 \equiv 7^2 \cdot 2^2 \iff \gcd(1037,\ 129 \cdot 1035 \pm 7 \cdot 2) = (61, 17),$$

$$161^2 \cdot 504^2 \equiv (-1)^2 \cdot 2^2 \cdot 7^2 \iff \gcd(1037,\ 161 \cdot 504 \pm 2 \cdot 7) = (17, 61),$$

$$502^2 \cdot 535^2 \equiv 13^2 \iff \gcd(1037\ 502 \cdot 535 \pm 13) = (1037, 1).$$

Three of them yield a factorization of $1037 = 17 \cdot 61$.

Exercise 3.5.1. Use the continued fraction expansion

$$\sqrt{1711} = [41, \overline{2, 1, 2, 1, 13, 16, 2, 8, 1, 2, 2, 2, 2, 2, 1, 8, 2, 16, 13, 1, 2, 1, 2, 82}]$$

and the factor base FB $= \{-1, 2, 3, 5\}$ to factor the integer 1711.

It is clear that the CFRAC factoring algorithm is essentially just a continued fraction algorithm for finding the continued fraction expansion

$[q_0, q_1, \ldots, q_k, \cdots]$ of $\sqrt{kn}$, or the P_k and Q_k of such an expansion. In what follows, we shall briefly summarize the CFRAC method just discussed above in the following algorithmic form:

Algorithm 3.5.1 (CFRAC factoring). Given a positive integer n and a positive integer k such that kn is not a perfect square, this algorithm tries to find a factor of n by computing the continued fraction expansion of $\sqrt{kn}$.

[1] Let n be the integer to be factored and k any small integer (usually 1), and let the factor base, FB, be a set of small primes $\{p_1, p_2, \ldots, p_r\}$ chosen such that it is possible to find some integer x_i such that $x_i^2 \equiv kn \pmod{p_i}$. Usually, FB contains all such primes less than or equal to some limit. Note that the multiplier $k > 1$ is needed only when the period is short. For example, Morrison and Brillhart used $k = 257$ in factoring F_7.

[2] Compute the continued fraction expansion $[q_0, \overline{q_1, q_2, \ldots, q_r}]$ of $\sqrt{kn}$ for a number of values of k. This gives us good rational approximations P/Q. The recursion formulas to use for computing P/Q are as follows:

$$\frac{P_0}{Q_0} = \frac{q_0}{1},$$

$$\frac{P_1}{Q_1} = \frac{q_0 q_1 + 1}{q_1},$$

$$\vdots$$

$$\frac{P_i}{Q_i} = \frac{q_i P_{i-1} + P_{i-2}}{q_i Q_{i-1} + Q_{i-2}}, \quad i \geq 2.$$

This can be done by a continued fraction algorithm such as Algorithm 1.4.7 introduced in Chapter 1.

[3] Try to factor the corresponding integer $W = P^2 - Q^2 kn$ in factor base FB. Since $W < 2\sqrt{kn}$, each of these W only about half the length of kn. If we succeed, we get a new congruence. For each success, we obtain a congruence

$$x^2 \equiv (-1)^{e_0} p_1^{e_1} p_2^{e_2} \cdots p_m^{e_m} \pmod{n},$$

since, if P_i/Q_i is the ith continued fraction convergent to $\sqrt{kn}$ and $W_i = P_i^2 - N \cdot Q_i^2$, then

$$P_i^2 \equiv W_i \pmod{n}. \tag{3.27}$$

[4] Once we have obtained at least $m + 2$ such congruences, then by Gaussian elimination over $\mathbb{Z}/2\mathbb{Z}$ we obtain a congruence $x^2 \equiv y^2 \pmod{n}$. That is, if $(x_1, e_{01}, e_{11}, \ldots, e_{m1}), \ldots, (x_r, e_{0r}, e_{1r}, \ldots, e_{mr})$ are solutions of (3.24) such that the vector sum defined in (3.25) is even in each component, then

$$\begin{cases} x \equiv x_1 x_2 \cdots x_r \pmod{n} \\ y \equiv (-1)^{e'_0} p_1^{e'_1} \cdots p_m^{e'_m} \pmod{n} \end{cases}$$

is a solution to $x^2 \equiv y^2 \pmod{n}$, except for the possibility that $x \equiv \pm y \pmod{n}$, and hence we have

$$(d_1, d_2) = (\gcd(x + y, \ n), \ \gcd(x - y, \ n))$$

which are then possibly nontrivial factors of n.

Example 3.5.2. The 7th Fermat number $F_7 = 2^{2^7} + 1$ was factored by Morrison and Brillhart in 1970 using the CFRAC method:

$$F_7 = 2^{128} + 1 = 59649589127497217 \cdot 5704689200685129054721.$$

The idea of CFRAC may go back to Kraitchik in the 1929's or even earlier, and in the 1930's Lehmer, Power, and others made it suitable for a computer to find x and y. However, it was Morrison and Brilhart who used it to successfully factor F_7. It was the most powerful factoring method, in the 1970's. However, it was soon replaced by another powerful factoring method, the Quadratic Sieve, in the 1980, which we shall discuss in the next section.

Problems for Section 3.5

Problem 3.5.1. Let

$$x = 1 + \cfrac{1}{4 + \cfrac{1}{1 + \cfrac{1}{1 + \cfrac{1}{8 + \cfrac{1}{1 + \cfrac{1}{3 + \cfrac{1}{2 + \cfrac{1}{1 + \cfrac{1}{14 + \cfrac{1}{1 + \cfrac{1}{2 + \cfrac{1}{22}}}}}}}}}}}}$$

be the continued fraction expansion of x. Find the successive convergents P_i/Q_i of this continued fraction.

Problem 3.5.2. Let the successive convergents P_i/Q_i of the continued fraction of x be as follows:

$$\left[2, 3, \frac{14}{5}, \frac{17}{6}, \frac{65}{23}, \frac{82}{29}, \frac{967}{342}, \frac{1049}{371}, \frac{7261}{2568}, \frac{15571}{5507}, \frac{925950}{327481}, \frac{3719371}{1315431} \right].$$

Find the continued fraction expansion of x.

Problem 3.5.3. Let n be a positive integer that is not a perfect square. Let P_k/Q_k the kth convergent of the simple continued fraction expansion of $\sqrt{n}$. Then

$$P_k^2 - nQ_k^2 = (-1)^k W_{k+1}, \quad 0 < W_i < 2\sqrt{n}.$$

Problem 3.5.4. Let

$$x^2 \equiv (-1)^{e_0} p_1^{e_1} p_2^{e_2} \cdots p_k^{e_k}. \tag{3.28}$$

If

$$(x_1, e_{01}, e_{11}, \ldots, e_{m1}), \ldots, (x_r, e_{0r}, e_{1r}, \ldots, e_{mr})$$

are solutions to (3.28) such that

$$(e_{01}, e_{11}, \ldots, e_{m1}) + \cdots + (e_{0r}, e_{1r}, \ldots, e_{mr}) = 2(e_1', e_2', \ldots, e_m'),$$

then

$$\begin{cases} x \equiv (x_1, x_2, \ldots, x_r) \ (\text{mod } n) \\ y \equiv (-1)^{e_1'} p_1^{e_2'} \cdots p_m^{e_m'} \ (\text{mod } n) \end{cases}$$

is a solution to

$$x^2 \equiv y^2 \ (\text{mod } n), \quad x \not\equiv \pm y \ (\text{mod } n).$$

Problem 3.5.5. Give a heuristic analysis of the running time of the CFRAC method

$$\mathcal{O}(\exp(\sqrt{2 \log n \log \log n})).$$

Problem 3.5.6. Implement the CFRAC algorithm on a computer.

Problem 3.5.7. Use your CFRAC program above to factor the integers 193541963 and 19354196373153173137.

3.6 Quadratic Sieve (QS)

The idea of the quadratic sieve (QS) was first introduced by Carl Pomerance in 1982. QS is somewhat similar to CFRAC except that instead of using continued fractions to produce the values for $W_k = P_k^2 - n \cdot Q_k^2$, it uses expressions of the form

$$W_k = (k + \lfloor \sqrt{n} \rfloor)^2 - n \equiv (k + \lfloor \sqrt{n} \rfloor)^2 \ (\text{mod } n). \tag{3.29}$$

Here, if $0 < k < L$, then

$$0 < W_k < (2L + 1)\sqrt{n} + L^2. \tag{3.30}$$

If we get

$$\prod_{i=1}^{t} W_{n_i} = y^2, \tag{3.31}$$

then we have $x^2 \equiv y^2 \pmod{n}$ with

$$x \equiv \prod_{i=1}^{t} (\lfloor \sqrt{n} \rfloor + n_i) \pmod{n}. \tag{3.32}$$

Once such x and y are found, there is a good chance that $\gcd(x - y, n)$ is a nontrivial factor of n.

Example 3.6.1. Use the quadratic sieve method (QS) to factor $n = 2041$. Let $W(x) = x^2 - n$, with $x = 43, 44, 45, 46$. Then we have:

$W(43)$	$=$	$-2^6 \cdot 3$
$W(44)$	$=$	$-3 \cdot 5 \cdot 7$
$W(45)$	$=$	-2^4
$W(46)$	$=$	$3 \cdot 5^2$

p	$W(43)$	$W(44)$	$W(45)$	$W(46)$
-1	1		1	0
2	0		0	0
3	1		0	1
5	0		0	0

which leads to the following congruence:

$$(43 \cdot 45 \cdot 46)^2 \equiv (-1)^2 \cdot 2^{10} \cdot 3^2 \cdot 5^2 = (2^5 \cdot 3 \cdot 5)^2.$$

This congruence gives the factorization of $2041 = 13 \cdot 157$, since

$$\gcd(2041, \ 43 \cdot 45 \cdot 46 + 2^5 \cdot 3 \cdot 5) = 157, \quad \gcd(2041, \ 43 \cdot 45 \cdot 46 - 2^5 \cdot 3 \cdot 5) = 13.$$

For the purpose of implementation, we can use the same set FB as that used in CFRAC and the same idea as that described above to arrange that (3.31) holds.

The most widely used variation of the quadratic sieve is perhaps the Multiple Polynomial Quadratic Sieve (MPQS), first proposed by Peter Montgomery [234] in 1986. The idea of the MPQS is as follows: To find the (x, y) pair in

$$x^2 \equiv y^2 \pmod{n} \tag{3.33}$$

we try to find triples (U_i, V_i, W_i), for $i = 1, 2, \ldots$, such that

$$U_i^2 \equiv V_i^2 W_i \pmod{n} \tag{3.34}$$

where W is easy to factor (at least easier than N). If sufficiently many congruences (3.34) are found, they can be combined, by multiplying together a subset of them, in order to get a relation of the form (3.33). The version of the MPQS algorithm described here is based on [199].

Algorithm 3.6.1 (Multiple Polynomial Quadratic Sieve). Given a positive integer $n > 1$, this algorithm will try to find a factor N using the multiple polynomial quadratic sieve.

[1] Choose B and M, and compute the factor base FB.
Note: M is some fixed integer so that we can define: $U(x) = a^2x + b$, $V = a$ and $W(x) = a^2x^2 + 2bx + c$, $x \in [-M, M)$, such that a, b, c satisfy the following relations:

$$a^2 \approx \sqrt{2n/M}, \quad b^2 - n = a^2c, \quad |b| < (a^2)/2. \qquad (3.35)$$

Note: Since the potential prime divisors p of a given quadratic polynomial $W(x)$ may be characterized as: if $p \mid W(x)$, then

$$a^2W(x) = (a^2x + b)^2 - n \equiv 0 \,(\text{mod } p). \qquad (3.36)$$

That is, the congruence $t^2 - n \equiv 0 \,(\text{mod } p)$ should be solvable. So, the factor base FB (consisting of all primes less than a bound B) should be chosen in such a way that $t^2 \equiv n \,(\text{mod } p)$ is solvable. There are L primes p_j, $j = 1, 2, \ldots, L$ in FB; this set of primes is fixed in the whole factoring process.

[2] Generate a new quadratic polynomial $W(x)$.
Note: The quadratic polynomial $W(x)$ in

$$(U(x))^2 \equiv V(x)W(x) \,(\text{mod } n) \qquad (3.37)$$

assumes extreme values in $x = 0, \pm M$ such that $|W(0)| \approx |W(\pm M)| \approx M\sqrt{n/2}$. If $M \ll n$, then $W(x) \ll n$, thus $W(x)$ is easier to factor than n.

[3] Solve $W(x) \equiv 0 \,(\text{mod } q)$ for all $q = p^e < B$, for all primes $p \in$ FB, and save the solutions for each q.

[4] Initialize the sieving array $\text{SI}[-M, M)$ to zero.

[5] Add $\log p$ to all elements $\text{SI}(j)$, $j \in [-M, M]$, for which $W(j) \equiv 0 \,(\text{mod } q)$, for all $q = p^e < B$, and for all primes $p \in$ PFB.
Note: Now we can collect those $x \in [-M, M)$ for which $W(x)$ is only composed of prime factors $< B$.

[6] Selecting those $j \in [-M, M)$ for which $\text{SI}(j)$ is closed to $\log(n/2\sqrt{n/2})$.

[7] If the number of $W(x)$-values collected in Step 6 is $< L + 2$, then go to Step 2 to construct a new polynomial $W(x)$.
Note: If at least $L + 2$ completely factorized W-values have been collected, then the (x, y)-pairs satisfying (3.33) may be found as follows: For x_i, $i = 1, 2, \ldots, L + 2$,

$$W(x_i) = (-1)^{\alpha_{i0}} \prod_{j=1}^{L} p_j^{\alpha_{ij}}, \quad i = 1, 2, \ldots, L + 2. \qquad (3.38)$$

[8] Perform Gaussian elimination on the matrix of exponents (mod 2) of $W(x)$.
   Note: Associated with each $W(x_i)$, we define the vector α_i as follows

$$\alpha_i^T = (\alpha_{i0}, \alpha_{i1}, \ldots, \alpha_{iL}) \pmod{2}. \tag{3.39}$$

Since we have more vectors α_i (at least $L + 2$) than components ($L + 1$), there exists at least one subset S of the set $\{1, 2, \ldots, L + 2\}$ such that

$$\sum_{i \in S} \alpha_i \equiv 0 \pmod{2},$$

so that

$$\prod_{i \in S} W(x) = Z^2.$$

Hence, from (3.37) it follows that

$$\left[\prod_{i \in S} (a^2 x_i + b) \right] \equiv Z^2 \prod_{i \in S} a^2 \pmod{n}$$

which is of the required form $x^2 \equiv y^2 \pmod{n}$.

[9] Compute gcd.
   Note: Now we can calculate $\gcd(x \pm y, n)$ to find the prime factors of n.

Example 3.6.2. MPQS has been used to obtain many spectacular factorizations. One such factorization is the 103-digit composite number

$$\frac{2^{361} + 1}{3 \cdot 174763} = 6874301617534827509350575768454356245025403 \cdot p_{61}.$$

The other record of the MPQS is the factorization of the RSA-129 in April 1994, a 129 digit composite number:

RSA-129 = 114381625757888867669235779976146612010218296721242362562561842935706935245733897830597123563958705058989075147599290026879543541

= $p_{64} \cdot q_{65}$

= 3490529510847650949147849619903898133417764638493387843990820577 · 32769132993266709549961988190834461413177642967992942539798288533.

It was estimated in Gardner [78] in 1977 that the running time required to factor numbers with about the same size as RSA-129 would be about 40 quadrillion years using the best algorithm and fastest computer at that time. It was factorized by Derek Atkins, Michael Graff, Arjen Lenstra, Paul Leyland, and more than 600 volunteers from more than 20 countries, on all continents except Antarctica. To factor this number, they used the double large prime variation of the multiple polynomial quadratic sieve factoring method. The sieving step took approximately 5000 mips years.

Conjecture 3.6.1 (The complexity of the QS/MPQS Method). *If n is the integer to be factored, then under certain reasonable heuristic assumptions, the QS/MPQS method will factor n in time*

$$\mathcal{O}\left(\exp\left((1+o(1))\sqrt{\log n \log\log n}\ \right)\right)$$

$$= \mathcal{O}\left(n^{(1+o(1))\sqrt{\log\log n/\log n}}\ \right). \tag{3.40}$$

The MPQS is however not the fastest factoring algorithm at present; the fastest factoring algorithm in use today is the Number Field Sieve, which is the subject matter of our next section.

Problems for Section 3.6

Problem 3.6.1. A number is smooth if all of its prime factors are small; a number is B-smooth if all of its prime factors are $\leq B$. Let $\pi(B)$ the numbers of primes in the interval $[1, B]$ and $u_1, u_2, \ldots, u_k$ be positive B-smooth integers with $k > \pi(B)$. Then some non-empty subset of $\{u_i\}$ has product which is a square.

Problem 3.6.2. Let ϵ an arbitrary small positive integer, and

$$L(x) = \exp(\sqrt{\log x \log\log x}).$$

If $L(x)^{\sqrt{2}+\epsilon}$ is chosen from $[1, x]$ independently and uniformly, then the probability that some nonempty subset product is a square tends to 1 as $x \to \infty$, whereas the probability that some nonempty subset product is a square tends to 0 as $x \to \infty$ if $L(x)^{\sqrt{2}-\epsilon}$ is chosen.

Problem 3.6.3. Let $\psi(x, y)$ be y-smooth numbers up to x. Then the expected number of choices of random integers in $[1, x]$ to find one y-smooth number is

$$\frac{x}{\psi(x, y)}$$

and to find $\pi(y) + 1$ such y-smooth numbers is

$$\frac{x(\pi(y) + 1)}{\psi(x, y)}.$$

Problem 3.6.4. Let $u_1, u_2, \ldots$ are y-smooth number, each be factored as follows

$$u_i = 2^{\alpha_{i,1}} 3^{\alpha_{i,2}} \cdots p_k^{\alpha_{i,k}}.$$

Then

$$\prod_{i \in I} u_i = \beta^2, \quad \text{for some positive integer } \beta$$

if and only if

$$\sum_{i \in I} (\alpha_{i,1}, \alpha_{i,2}, \ldots, \alpha_{i,k}) = 0$$

as a vector in $(\mathbb{Z}/2\mathbb{Z})^k$. Moreover, such a nontrivial subset is guaranteed amongst $u_1, u_2, \ldots, u_{k+1}$.

Problem 3.6.5. Let ϵ be any fixed, positive real number. Then

$$\psi(x, L(x)^\epsilon) = x L(x)^{-1/(2\epsilon) + o(1)}, \quad \text{as } x \to \infty.$$

Problem 3.6.6. Deduce that the Quadratic Sieve is a deterministic factoring algorithm with the following conjectured complexity

$$\exp((1 + o(1))(\log n \log \log n)^{1/2}).$$

3.7 Number Field Sieve (NFS)

Before introducing the number field sieve (NFS), it will be interesting to briefly review some important milestones in the development of the factoring methods. In 1970, it was barely possible to factor "hard" 20-digit numbers. In 1980, by using the CFRAC method, factoring of 50-digit numbers was becoming commonplace. In 1990, the QS method had doubled the length of the numbers that could be factored by CFRAC, with a record having 116 digits. In the spring of 1996, the NFS method had successfully split a 130-digit RSA challenge number in about 15% of the time the QS would have taken. At present, the number field sieve (NFS) is the champion of all known factoring methods. NFS was first proposed by John Pollard in a letter to A. M. Odlyzko, dated 31 August 1988, with copies to R. P. Brent, J. Brillhart, H. W. Lenstra, C. P. Schnorr and H. Suyama, outlining an idea of factoring certain big numbers via *algebraic number fields*. His original idea was not for any large composite, but for certain "pretty" composites that had the property that they were close to powers. He illustrated the idea with a factorization of the seventh Fermat number $F_7 = 2^{2^7} + 1$ which was first factored by CFRAC in 1970. He also speculated in the letter that "if F_9 is still unfactored, then it might be a candidate for this kind of method eventually?" The answer now is of course "yes", since F_9 was factored by using NFS in 1990. It is worthwhile pointing out that NFS is not only a method suitable for factoring numbers in a special form like F_9, but also a general purpose factoring method for any integer of a given size. There are, in fact, two forms of NFS: the *special* NFS (SNFS), tailored specifically for integers of the form $N = c_1 r^t + c_2 s^u$, and the *general* NFS (GNFS), applicable to any arbitrary numbers. Since NFS uses some ideas from algebraic number theory, a brief introduction to some basic concepts of algebraic number theory is in order.

Definition 3.7.1. A complex number α is an *algebraic number* if it is a root of a polynomial

$$f(x) = a_0 x^k + a_1 x^{k-1} + a_2 x^{k-2} \cdots + a_k = 0 \qquad (3.41)$$

where $a_0, a_1, a_2, \ldots, a_k \in \mathbb{Q}$ and $a_0 \neq 0$. If $f(x)$ is irreducible over $\mathbb{Q}$ and $a_0 \neq 0$, then k is the degree of x.

Example 3.7.1. Two examples of algebraic numbers are as follows:

(1) rational numbers, which are the algebraic numbers of degree 1.

(2) $\sqrt{2}$, which is of degree 2 because we can take $f(x) = x^2 - 2 = 0$ ($\sqrt{2}$ is irrational).

Any complex number that is not algebraic is said to be *transcendental* such as π and e.

Definition 3.7.2. A complex number β is an *algebraic integer* if it is a root of a monic polynomial

$$x^k + b_1 x^{k-1} + b_2 x^{k-2} \cdots + b_k = 0 \qquad (3.42)$$

where $b_0, b_1, b_2, \ldots, b_k \in \mathbb{Z}$.

Remark 3.7.1. A quadratic integer is an algebraic integer satisfying a monic quadratic equation with integer coefficients. A cubic integer is an algebraic integer satisfying a monic cubic equation with integer coefficients.

Example 3.7.2. Some examples of algebraic integers are as follows:

(1) ordinary (rational) integers, which are the algebraic integers of degree 1. i.e., they satisfy the monic equations $x - a = 0$ for $a \in \mathbb{Z}$.

(2) $\sqrt[3]{2}$ and $\sqrt[5]{3}$, because they satisfy the monic equations $x^3 - 2 = 0$ and $x^3 - 5 = 0$, respectively.

(3) $(-1 + \sqrt{-3})/2$, because it satisfies $x^2 + x + 1 = 0$.

(4) Gaussian integer $a + b\sqrt{-1}$, with $a, b \in \mathbb{Z}$.

Clearly, every algebraic integer is an algebraic number, but the converse is not true.

Proposition 3.7.1. A rational number $r \in \mathbb{Q}$ is an algebraic integer if and only if $r \in \mathbb{Z}$.

Proof. If $r \in \mathbb{Z}$, then r is a root of $x - r = 0$. Thus r is an algebraic integer Now suppose that $r \in \mathbb{Q}$ and r is an algebraic integer (i.e., $r = c/d$ is a root of (3.42), where $c, d \in \mathbb{Z}$; we may assume $\gcd(c, d) = 1$). Substituting c/d into (3.42) and multiplying both sides by d^n, we get

$$c^k + b_1 c^{k-1} d + b_2 c^{k-2} d^2 \cdots + b_k d^k = 0.$$

It follows that $d \mid c^k$ and $d \mid c$ (since $\gcd(c, d) = 1$). Again since $\gcd(c, d) = 1$, it follows that $d = \pm 1$. Hence $r = c/d \in \mathbb{Z}$. It follows, for example, that $2/5$ is an algebraic number but not an algebraic integer. $\square$

Remark 3.7.2. The elements of $\mathbb{Z}$ are the only rational numbers that are algebraic integers, We shall refer to the elements of $\mathbb{Z}$ as *rational integers* when we need to distinguish them from other algebraic integers that are not rational. For example, $\sqrt{2}$ is an algebraic integer but not a rational integer.

The most interesting results concerned with the algebraic numbers and algebraic integers are the following theorem.

Theorem 3.7.1. The set of algebraic numbers forms a field, and the set of algebraic integers forms a ring.

Proof. See pp 67–68 of Ireland and Rosen [109]. $\square$

Lemma 3.7.1. Let $f(x)$ is an irreducible monic polynomial of degree d over integers and m an integer such that $f(m) \equiv 0 \pmod{n}$. Let α be a complex root of $f(x)$ and $\mathbb{Z}[\alpha]$ the set of all polynomials in α with integer coefficients. Then there exists a unique mapping $\Phi : \mathbb{Z}[\alpha] \mapsto \mathbb{Z}_n$ satisfying

(1) $\Phi(ab) = \Phi(a)\Phi(b)$, $\forall a, b \in \mathbb{Z}[\alpha]$;
(2) $\Phi(a + b) = \Phi(a) + \Phi(b)$, $\forall a, b \in \mathbb{Z}[\alpha]$;
(3) $\Phi(za) = z\Phi(a)$, $\forall a \in \mathbb{Z}[\alpha], z \in \mathbb{Z}$;
(4) $\Phi(1) = 1$;
(5) $\Phi(\alpha) = m \pmod{n}$.

Now we are in a position to introduce the number field sieve (NFS). Note that there are two main types of NFS: NFS (general NFS) for general numbers and SNFS (special NFS) for numbers with special forms. The idea, however, behind the GNFS and SNFS are the same:

[1] Find a monic irreducible polynomial $f(x)$ of degree d in $\mathbb{Z}[x]$, and an integer m such that $f(m) \equiv 0 \pmod{n}$.

[2] Let $\alpha \in \mathbb{C}$ be an algebraic number that is the root of $f(x)$, and denote the set of polynomials in α with integer coefficients as $\mathbb{Z}[\alpha]$.

[3] Define the mapping (ring homomorphism): $\Phi : \mathbb{Z}[\alpha] \mapsto \mathbb{Z}_n$ via $\Phi(\alpha) = m$ which ensures that for any $f(x) \in \mathbb{Z}[x]$, we have $\Phi(f(\alpha)) \equiv f(m) \pmod{n}$.

[4] Find a finite set U of coprime integers (a, b) such that

$$\prod_{(a,b)\in U} (a - b\alpha) = \beta^2, \quad \prod_{(a,b)\in U} (a - bm) = y^2$$

for $\beta \in \mathbb{Z}[\alpha]$ and $y \in \mathbb{Z}$. Let $x = \Phi(\beta)$. Then

$$
\begin{aligned}
x^2 &\equiv \Phi(\beta)\Phi(\beta) \\
&\equiv \Phi(\beta^2) \\
&\equiv \Phi\left(\prod_{(a,b)\in U} (a - b\alpha)\right) \\
&\equiv \prod_{(a,b)\in U} \Phi(a - b\alpha) \\
&\equiv \prod_{(a,b)\in U} (a - bm) \\
&\equiv y^2 \qquad (\mathrm{mod}\ n)
\end{aligned}
$$

which is of the required form of the factoring congruence, and hopefully, a factor of n can be found by calculating $\gcd(x \pm y, n)$.

There are many ways to implement the above idea, all of which follow the same pattern as we discussed previously in CFRAC and QS/MPQS: by a sieving process one first tries to find congruences modulo n by working over a factor base, and then do a Gaussian elimination over $\mathbb{Z}/2\mathbb{Z}$ to obtain a congruence of squares $x^2 \equiv y^2 \pmod{n}$. We give in the following a brief description of the NFS algorithm [159].

Algorithm 3.7.1. Given an odd positive integer n, the NFS algorithm has the following four main steps in factoring n:

[1] (Polynomials Selection) Select two irreducible polynomials $f(x)$ and $g(x)$ with small integer coefficients for which there exists an integer m such that

$$f(m) \equiv g(m) \equiv 0 \pmod{n} \tag{3.43}$$

The polynomials should not have a common factor over $\mathbb{Q}$.

[2] (Sieving) Let α be a complex root of f and β a complex root of g. Find pairs (a, b) with $\gcd(a, b) = 1$ such that the integral norms of $a - b\alpha$ and $a - b\beta$:

$$N(a - b\alpha) = b^{\deg(f)} f(a/b), \qquad N(a - b\beta) = b^{\deg(g)} g(a/b) \tag{3.44}$$

are smooth with respect to a chosen factor base. (The principal ideals $a - b\alpha$ and $a - b\beta$ factor into products of prime ideals in the number field $\mathbb{Q}(\alpha)$ and $\mathbb{Q}(\beta)$, respectively.)

[3] (Linear Algebra) Use techniques of linear algebra to find a set $U = \{a_i, b_i\}$ of indices such that the two products

$$\prod_U (a_i - b_i\alpha), \qquad\qquad \prod_U (a_i - b_i\beta) \qquad (3.45)$$

are both squares of products of prime ideals.

[4] (Square root) Use the set S in (3.45) to find an algebraic numbers $\alpha' \in \mathbb{Q}(\alpha)$ and $\beta' \in \mathbb{Q}(\beta)$ such that

$$(\alpha')^2 = \prod_U (a_i - b_i\alpha), \qquad (\beta')^2 = \prod_U (a_i - b_i\beta) \qquad (3.46)$$

Define $\Phi_\alpha : \mathbb{Q}(\alpha) \to \mathbb{Z}_n$ and $\Phi_\beta : \mathbb{Q}(\beta) \to \mathbb{Z}_n$ via $\Phi_\alpha(\alpha) = \Phi_\beta(\beta) = m$, where m is the common root of both f and g. Then

$$
\begin{aligned}
x^2 &\equiv \Phi_\alpha(\alpha')\Phi_\alpha(\alpha') \\
&\equiv \Phi_\alpha((\alpha')^2) \\
&\equiv \Phi_\alpha\left(\prod_{i \in U}(a_i - b_i\alpha)\right) \\
&\equiv \prod_U \Phi_\alpha(a_i - b_i\alpha) \\
&\equiv \prod_U (a_i - b_i m) \\
&\equiv \Phi_\beta(\beta')^2 \\
&\equiv y^2 \qquad (\bmod\ n)
\end{aligned}
$$

which is of the required form of the factoring congruence, and hopefully, a factor of N can be found by calculating $\gcd(x \pm y, n)$.

Example 3.7.3. We first give a rather simple NFS factoring example. Let $n = 14885 = 5 \cdot 13 \cdot 229 = 122^2 + 1$. So we put $f(x) = x^2 + 1$ and $m = 122$, such that

$$f(x) \equiv f(m) \equiv 0 \ (\bmod\ n).$$

If we choose $|a|, |b| \leq 50$, then we can easily find (by sieving) that

(a, b)	Norm$(a + bi)$	$a + bm$
$\vdots$	$\vdots$	$\vdots$
$(-49, 49)$	$4802 = 2 \cdot 7^4$	$5929 = 7^2 \cdot 11^2$
$\vdots$	$\vdots$	$\vdots$
$(-41, 1)$	$1682 = 2 \cdot 29^2$	$81 = 3^4$
$\vdots$	$\vdots$	$\vdots$

(Readers should be able to find many such pairs of (a_i, b_i) in the interval, that are smooth up to e.g. 29.) So, we have

$$
\begin{aligned}
(49 + 49i)(-41 + i) &= (49 - 21i)^2, \\
f(49 - 21i) &= 49 - 21m \\
&= 49 - 21 \cdot 122 \\
&= -2513, \\
&\updownarrow \\
&\alpha \\
5929 \cdot 81 &= (2^2 \cdot 7 \cdot 11)^2 \\
&= 693^2 \\
&= 693. \\
&\updownarrow \\
&\beta
\end{aligned}
$$

Thus,

$$
\begin{aligned}
\gcd(\alpha \pm \beta, n) &= \gcd(-2513 \pm 693, 14885) \\
&= (65, 229).
\end{aligned}
$$

In the same way, if we wish to fact $n = 84101 = 290^2 + 1$, then we let $m = 290$, and $f(x) = x^2 + 1$ so that

$$
f(x) \equiv f(m) \equiv 0 \pmod{n}.
$$

We tabulate the sieving process as follows:

(a, b)	Norm$(a + bi)$	$a + bm$
$\vdots$	$\vdots$	$\vdots$
$-50, 1$	$2501 = 41 \cdot 61$	$240 = 2^4 \cdot 3 \cdot 5$
$\vdots$	$\vdots$	$\vdots$
$-50, 3$	$2509 = 13 \cdot 193$	$820 = 2^2 \cdot 5 \cdot 41$
$\vdots$	$\vdots$	$\vdots$
$-49, 43$	$4250 = 2 \cdot 5^3 \cdot 17$	$12421 = 12421$
$\vdots$	$\vdots$	$\vdots$
$-38, 1$	$1445 = 5 \cdot 17^2$	$252 = 2^2 \cdot 3^2 \cdot 7$
$\vdots$	$\vdots$	$\vdots$
$-22, 19$	$845 = 5 \cdot 13^2$	$5488 = 2^4 \cdot 7^3$
$\vdots$	$\vdots$	$\vdots$
$-118, 11$	$14045 = 5 \cdot 53^2$	$3072 = 2^{10} \cdot 3$
$\vdots$	$\vdots$	$\vdots$
$218, 59$	$51005 = 5 \cdot 101^2$	$17328 = 2^4 \cdot 3 \cdot 19^2$
$\vdots$	$\vdots$	$\vdots$

Clearly, $-38 + i$ and $-22 + 19i$ can produce a product square, since

$$
\begin{aligned}
(-38 + i)(-22 + 19i) &= (31 - 12i)^2, \\
f(31 - 12i) &= 31 - 12m \\
&= -3449, \\
&\updownarrow \\
&x \\
252 \cdot 5488 &= (2^3 \cdot 3 \cdot 7^2)^2 \\
&= 1176^2, \\
&\updownarrow \\
&y \\
\gcd(x \pm y, n) &= \gcd(-3449 \pm 1176, 84101) \\
&= (2273, 37).
\end{aligned}
$$

In fact, $84101 = 2273 \times 37$. Note that $-118 + 11i$ and $218 + 59i$ can also produce a product square, since

$$(-118 + 11i)(218 + 59i) \;=\; (14 - 163i)^2,$$

$$f(14 - 163i) \;=\; 14 - 163m$$

$$=\; -47256,$$

$$\updownarrow$$

$$x$$

$$3071 \cdot 173288 \;=\; (2^7 \cdot 3 \cdot 19)^2$$

$$=\; 7296^2,$$

$$\updownarrow$$

$$y$$

$$\gcd(x \pm y, n) \;=\; \gcd(-47256 \pm 7296, 84101)$$

$$=\; (37, 2273).$$

Example 3.7.4. Next we present a little bit more complicated example. Use NFS to factor $n = 1098413$. First notice that $n = 1098413 = 12 \cdot 45^3 + 17^3$, which is in a special form and can be factored by using SNFS.

[1] (Polynomials Selection) Select the two irreducible polynomials $f(x)$ and $g(x)$ and the integer m as follows:

$$m = \frac{17}{45},$$

$$f(x) = x^3 + 12 \quad\Longrightarrow\quad f(m) = \left(\frac{17}{45}\right)^3 + 12 \equiv 0 \;(\text{mod } n),$$

$$g(x) = 45x - 17 \quad\Longrightarrow\quad g(m) = 45\left(\frac{17}{45}\right) - 17 \equiv 0 \;(\text{mod } n).$$

[2] (Sieving) Suppose after sieving, we get $U = \{a_i, b_i\}$ as follows:

$$U = \{(6, -1), (3, 2), (-7, 3), (1, 3), (-2, 5), (-3, 8), (9, 10)\}.$$

That is, the chosen polynomial that produces a product square can be constructed as follows (as an exercise. readers may wish to choose some other polynomial which can also produce a product square):

$$\prod_U (a_i + b_i x) = (6-x)(3+2x)(-7+3x)(1+3x)(-2+5x)(-3+8x)(9+10x).$$

Let $\alpha = \sqrt[3]{-12}$ and $\beta = \frac{17}{45}$. Then

$$\prod_U (a - b\alpha) = 7400772 + 1138236\alpha - 10549\alpha^2$$

$$= (2694 + 213\alpha - 28\alpha^2)^2$$

$$= \left(\frac{5610203}{2025} \right)$$

$$= 270729^2,$$

$$\prod_U (a - b\beta) = \frac{2^8 \cdot 11^2 \cdot 13^2 \cdot 23^2}{3^{12} \cdot 5^4}$$

$$= \left(\frac{52624}{18225} \right)^2$$

$$= 875539^2.$$

So, we get the required square of congruence:

$$270729^2 \equiv 875539^2 \ (\text{mod } 1098413).$$

Thus,

$$\gcd(270729 \pm 875539, 1098413) = (563, 1951).$$

That is,

$$1098413 = 563 \cdot 1951.$$

Example 3.7.5. Finally, we give some large factoring examples using NFS.

(1) SNFS examples: One of the largest numbers factored by SNFS is

$$n = (12^{167} + 1)/13 = p_{75} \times p_{105}$$

It was announced by P. Montgomery, S. Cavallar and H. te Riele at CWI in Amsterdam on 3 September 1997. They used the polynomials $f(x) = x^5 - 144$ and $g(x) = 12^{33}x + 1$ with common root $m \equiv 12^{134} \ (\text{mod } n)$. The factor base bound was 4.8 million for f and 12 million for g. Both large prime bounds were 150 million, with two large primes allowed on each side. They sieved over $|a| \leq 8.4$ million and $0 < b \leq 2.5$ million. The sieving lasted 10.3 calendar days; 85 SGI machines at CWI contributed a combined 13027719 relations in 560 machine-days. It took 1.6 more calendar days to process the data. This processing included 16 CPU-hours on a Cray C90 at SARA in Amsterdam to process a 1969262×1986500 matrix with 57942503 nonzero entries. The other large number factorized by using SNFS is the 9th Fermat number:

$$F_9 = 2^{2^9} + 1 = 2^{512} + 1 = 2424833 \cdot p_{49} \cdot p_{99},$$

a number with 155 digits; it was completely factored in April 1990. The most wanted factoring number of special form at present is the 12th Fermat number $F_{12} = 2^{2^{12}} + 1$; we only know its partial prime factorization:

$$F_{12} = 114689 \cdot 26017793 \cdot 63766529 \cdot 190274191361 \cdot 1256132134125569 \cdot c_{1187}$$

and we want to find the prime factors of the remaining 1187-digit composite.

(2) GNFS examples: Three large general numbers RSA-130 (in April 1996), RSA-140 (in February 1999), RSA-155 (August 1999) and RSA-174 (December 2003) were factorized using GNFS:

(a) RSA-130 = $p_{65} \cdot q_{65}$:
39685999459597454290161126162883786067576449112810064832555_
157243,
45534498646735972188403686897274408864356301263205069600999_
044599,

(b) RSA-140 = $p_{70} \cdot q_{70}$:
33987174230284385545301236276138758356339864959695974234909_
29302771479,
62642001874012850961516549482644422193020371786235090191116_
60653946049,

(c) RSA-155 = $p_{78} \cdot q_{79}$:
10263959282974110577205419657399167590071656780803806680334_
1933521790711307779,
10660348838016845482092722036001287867920795857598929152227_
0608237193062808643,

(d) RSA-174 = $p_{87} \cdot q_{87}$:
39807508642406493739712550055038649119906436234252670840638_
5189575946388957261768583317,
47277214610743530253622307197304822463291469530209711645985_
2171130520711256363590397527.

Remark 3.7.3. Prior to the NFS, all modern factoring methods had an expected running time of at best

$$\mathcal{O}\left(\exp\left((c + o(1)) \sqrt{\log n \log \log n} \right) \right).$$

For example, Dixon's random square method has the expected running time

$$\mathcal{O}\left(\exp\left((\sqrt{2} + o(1)) \sqrt{\log n \log \log n} \right) \right).$$

whereas the multiple polynomial quadratic sieve (MPQS) takes time

$$\mathcal{O}\left(\exp\left((1 + o(1)) \sqrt{\log \log n / \log n} \right) \right).$$

Because of the Canfield-Erdős-Pomerance theorem, some people even believed that this could not be improved, except maybe for the term $(c + o(1))$, but the invention of the NFS has changed this belief.

Conjecture 3.7.1 (Complexity of NFS). Under some reasonable heuristic assumptions, the NFS method can factor an integer N in time

$$\mathcal{O}\left(\exp\left((c + o(1))\sqrt[3]{\log n}\sqrt[3]{(\log\log n)^2}\right)\right) \tag{3.47}$$

where $c = (64/9)^{1/3} \approx 1.922999427$ if GNFS is used to factor an arbitrary integer N, whereas $c = (32/9)^{1/3} \approx 1.526285657$ if SNFS is used to factor a special integer N.

Problems for Section 3.7

Problem 3.7.1. Use NFS to factor the following two numbers:

(1) $n = 5667228962 = 75281^2 + 1$

(2) $n = 71511302477 = 12 \cdot 1813^3 + 17^3$

Problem 3.7.2. Give a heuristic analysis that the Number Field Sieve has the following conjectured complexity

$$\exp(((64/9)^{1/3} + o(1))(\log n)^{1/3}(\log\log n)^{1/3}).$$

Problem 3.7.3. Show also that the above constant $(64/9)^{1/3}$ can be slightly improved to $(32/9)^{1/3}$ for special numbers.

Problem 3.7.4. Use NFS to factor the special number $111557 = 334^2 + 1$.

Problem 3.7.5. (General Factoring Challenge Problems) The RSA Security Inc offers prizes for the first individual or organization who can factor the following challenge numbers (Note that in the following RSA numbers, the value for x in RSA-x represents the number of bits, not the number of digits; this is just the RSA convention):

(1) RSA-704 (212 digits, 704 bits), $30,000
7403756347956171282804679609742957314259318888923128908493623263_
8972765034028266276891996419625117843995894330502127585370118968_
0982867331732731089309005525051168770632990723963807867100860969_
62537934650563796359,

(2) RSA-768 (232 digits, 768 bits), $50,000
1230186684530117755130494958384962720772853569595334792197322452_
1517264005072636575187452021997864693899564749427740638459251925_
5732630345373154826850791702612214291346167042921431160222124047_

9274737794080665351419597459856902143413,

(3) RSA-896 (270 digits, 896 bits), \$75,000
41202343698665954385553136533257594817981169984432798284545562644
33876445565248426198098870423161841879261420247188869492560931774
63750334211309823974851509449091069102698610318627041148808669704
56490290365365886743373172081310410519086425479328260139125762404
33946373269391,

(4) RSA-1024 (309 digits, 1024 bits), \$100,000
13506641086599522334960321627880596993888147560566702752448514384
51526510604859533833940287150571909441798207282164471551373680414
97039641917430464965892742562393410208643832021103729587257623584
50964311056407350150818751067659462920556368552947521350085287944
16377328533906109750544334999811150056977236890927563,

(5) RSA-1536 (463 digits, 1536 bits), \$150,000
18476997032117414743068356202001644030185493386634101714717857744
91065169671116124985933768430543574458561606154457179405222971774
32524660960646946071249623720442022269756756687378427562389508764
46784409332851574965788434150884755282981867264513398633649319084
08467199043187438128336350279547028265329780293491615581188104984
44908319545000984839377522725705257859194499387007369575568843693_
38127796130892303925696952532616208236764903160365513714479139324
347169566988069,

(6) RSA-2048 (617 digits, 2048 bits), \$200,000
25195908475657893494027183240048398571429282126204032027777137834
60436620207075955562640185258807844069182906412495150821892985594
14917618450280848912007284499268739280728777673597141834727026184
96375014971824691165077613379859095700097330459748808428401797424
91006424586918171951187461215151726546322822168699875491824224334
63725908514186546204357679842338718477444792073993423658482382424
81198163815010674810451660377306056201619676256133844143603833904
44149526344321901146575444541178424020924616515723350778707749817_
12577246796292638635637328991215483143816789988504044536402352738
81951378636564391212010397122822120720357.

Problem 3.7.6. (Knuth Factoring Problem) In [119], Knuth proposed the
following 211 digits factoring challenge number, marked with difficulty degree
50, the hardest problem in his book:

77903022885101595423624756547055783624857676209739839410844022224
13572872511709998585048387648131944340510932265136815168574119934
47755868542740942256445000879127232585749337061853958340278434054
8208881085485078737.

Problem 3.7.7. In this problem, we list the smallest unfactored (not completely factored) Fermat numbers for you to try to find the complete factorization for each of these numbers:

$$F_{12} = 114689 \cdot 26017793 \cdot 63766529 \cdot 190274191361 \cdot$$
$$1256132134125569 \cdot c_{1187},$$

$$F_{13} = 2710954639361 \cdot 2663848877152141313 \cdot 36031098445229199 \cdot$$
$$319546020820551643220672513 \cdot c_{2391},$$

$$F_{14} = c_{4933},$$

$$F_{15} = 1214251009 \cdot 2327042503868417 \cdot$$
$$168768817029516972383024127016961 \cdot c_{9808},$$

$$F_{16} = 825753601 \cdot 188981757975021318420037633 \cdot c_{19694},$$

$$F_{17} = 31065037602817 \cdot c_{39444},$$

$$F_{18} = 13631489 \cdot 81274690703860512587777 \cdot c_{78884},$$

$$F_{19} = 70525124609 \cdot 646730219521 \cdot c_{157804},$$

$$F_{20} = c_{315653},$$

$$F_{21} = 4485296422913 \cdot c_{631294},$$

$$F_{22} = c_{1262612},$$

$$F_{23} = 167772161 \cdot c_{2525215},$$

$$F_{24} = c_{5050446}.$$

3.8 Quantum Factoring Algorithm

The idea of quantum algorithms for IFP is in fact very simple: one can factor n if one can find the order of an element x modulo n (or more precisely, the order of an element x in the multiplicative group $\mathcal{G} = (\mathbb{Z}/n\mathbb{Z})^*$), denoted by $\mathrm{ord}_N(x)$. Recall that the order r of x in the multiplicative group $\mathcal{G}$ is the smallest positive integer r such that $x^r \equiv 1 \pmod{n}$. Finding the order of an element x in $\mathcal{G}$ is, in theory, not a problem: just keep multiplying until you get to "1", the identity element of the multiplicative group $\mathcal{G}$. For example, let $n = 179359$, $x = 3 \in \mathcal{G}$, and $\mathcal{G} = (\mathbb{Z}/179359\mathbb{Z})^*$, such that $\gcd(3, 179359) = 1$.

To find the order $r = \mathrm{ord}_{179359}(3)$, we just keep multiplying until we get to "1":

$$3^1 \bmod 179359 \;=\; 3$$
$$3^2 \bmod 179359 \;=\; 9$$
$$3^3 \bmod 179359 \;=\; 27$$
$$\vdots$$
$$3^{1000} \bmod 179359 \;=\; 31981$$
$$3^{1001} \bmod 179359 \;=\; 95943$$
$$3^{1002} \bmod 179359 \;=\; 108470$$
$$\vdots$$
$$3^{14716} \bmod 179359 \;=\; 99644$$
$$3^{14717} \bmod 179359 \;=\; 119573$$
$$3^{14718} \bmod 179359 \;=\; 1.$$

Thus, the order r of 3 in the multiplicative group $\mathcal{G} = (\mathbb{Z}/179359\mathbb{Z})^*$ is 14718, that is, $\mathrm{ord}_{179359}(3) = 14718$. Once the order $\mathrm{ord}_N(x)$ is found, it is then trivial to factor n by just calculating

$$\gcd(x^{r/2} \pm 1, N),$$

For instance, for $x = 3$, $r = 14718$ and $n = 179359$, we have

$$\gcd(3^{14718/2} \pm 1, 179359) = (67, 2677),$$

and hence the factorization of

$$n = 179359 = 67 \cdot 2677.$$

However, in practice, the above computation for finding the order of $x \in (\mathbb{Z}/n\mathbb{Z})^*$ may not work, since for an element x in a large group $\mathcal{G}$ with n having more than 200 digits, the computation of r may require more than 10^{150} multiplications. Even if these multiplications could be carried out at the rate of 1000 billion per second, it would take approximately $3 \cdot 10^{80}$ years to arrive at the answer. There is however a "quick" way to find the order of an element x in the multiplicative group $\mathcal{G}$ modulo n if the order $|\mathcal{G}|$ (where $|\mathcal{G}| = |(\mathbb{Z}/n\mathbb{Z})^*| = \phi(n)$) of $\mathcal{G}$ as well as the prime factorization of $|\mathcal{G}|$ are known, since, by Lagrange's theorem, $r = \mathrm{ord}_N(x)$ is a divisor of $|\mathcal{G}|$. Of course, as we know, the number $\lambda(n)$ is the largest possible order of an element x in the group $\mathcal{G}$. So, once we have the value of $\lambda(n)$, it is relatively easy to find $\mathrm{ord}_n(x)$, the order of the element $x \in \mathcal{G}$. For example, let $n = 179359$, then $\lambda(179359) = 29436$. Therefore, $\text{ord}_{179359}(3) \leq 29436$. In fact, $\text{ord}_{179359}(3) = 14718$, which of course is a divisor of 29436. However, there are no efficient algorithms at present for calculating either $\phi(n)$ or $\lambda(n)$. Therefore, the "quick" ways for computing $\text{ord}_n(x)$ by either $\phi(n)$ or $\lambda(n)$ are essentially useless in practice. This partly explains why integer factorization is difficult. Fortunately, Shor discovered in 1994 an efficient quantum algorithm to find the order of an element $x \in (\mathbb{Z}/n\mathbb{Z})^*$ and hence *possibly* the factorization of n. The main idea of Shor's method is as follows:First of all, we create two quantum registers for our machine: Register-1 and Register-2. Of course, we can create just one single quantum memory register partitioned into two parts. Secondly, we create in Register-1, a superposition of the integers $a = 0, 1, 2, 3, \ldots$ which will be the arguments of $f(a) = x^a \pmod{n}$, and load Register-2 with all zeros. Thirdly, we compute in Register-2, $f(a) = x^a \pmod{n}$ for each input a. (Since the values of a are kept in Register-1, this can be done reversibly). Fourthly, we perform the discrete Fourier transform on Register-1. Finally we observe both registers of the machine and find the order r that satisfies $x^r \equiv 1 \pmod{n}$. The following is a brief description of the quantum algorithm for IFP ([224] and [265]):

Algorithm 3.8.1 (Quantum algorithm for integer factorization).
Given integers x and n, the algorithm will find the order of x, i.e., the smallest positive integer r such that $x^r \equiv 1 \pmod{n}$. Assume our machine has two quantum registers: Register-1 and Register-2, which hold integers in binary form.

[1] [Initialize] Find a number q, a power of 2, with $n^2 < q < 2n^2$.

[2] [Prepare information for quantum registers] Put in Register-1 the uniform superposition of states representing numbers $a \pmod{q}$, and load Register-2 with all zeros. This leaves the machine in the state $|\Psi_1\rangle$:

$$|\Psi_1\rangle = \frac{1}{\sqrt{q}} \sum_{a=0}^{q-1} |a\rangle |0\rangle. \tag{3.48}$$

(Note that the joint state of both registers are represented by $|\text{Register-1}\rangle$ and $|\text{Register-2}\rangle$). What this step does is put each bit in Register-1 into the superposition

$$\frac{1}{\sqrt{2}} (|0\rangle + |1\rangle).$$

[3] [Create quantum-parallelly all powers] Compute $x^a \pmod{n}$ in Register-2. This leaves the machine in state $|\Psi_2\rangle$:

$$|\Psi_2\rangle = \frac{1}{\sqrt{q}} \sum_{a=0}^{q-1} |a\rangle |x^a \pmod{n}\rangle. \tag{3.49}$$

This step can be done reversibly since all the a's were kept in Register-1.

[4] [Perform a quantum FFT] Apply FFT on Register-1. The FFT maps each state $|a\rangle$ to

$$\frac{1}{\sqrt{q}} \sum_{c=0}^{q-1} \exp(2\pi iac/q) |c\rangle. \tag{3.50}$$

That is, we apply the unitary matrix with the (a,c) entry equal to $\frac{1}{\sqrt{q}}\exp(2\pi iac/q)$. This leaves the machine in the state $|\Psi_3\rangle$:

$$|\Psi_3\rangle = \frac{1}{q} \sum_{a=0}^{q-1}\sum_{c=0}^{q-1} \exp(2\pi iac/q)|c\rangle|x^a \ (\mathrm{mod}\ n)\rangle. \tag{3.51}$$

[5] [Detect periodicity in x^a] Observe the machine. For clarity, we observe both $|c\rangle$ in Register-1 and $|x^a \ (\mathrm{mod}\ n)\rangle$ in Register-2, measure both arguments of this superposition, obtaining the values of $|c\rangle$ in the first argument and some $|x^k \ (\mathrm{mod}\ n)\rangle$ as the answer for the second one $(0 < k < r)$.

[6] [Extract r] Finally extract the required value of r. Given the pure state $|\Psi_3\rangle$, the probabilities of different results for this measurement will be given by the probability distribution:

$$\mathrm{Prob}(c, x^k) = \left|\frac{1}{q}\sum_{a=0}^{q-1}\exp(2\pi iac/q)\right|^2 \tag{3.52}$$

where the sum is over all values of a such that

$$x^a \equiv x^k \ (\mathrm{mod}\ n). \tag{3.53}$$

Independent of k, $\mathrm{Prob}(c, x^k)$ is periodic in c with period q/r; but since q is known, we can deduce r with just a few trial executions (this can be accomplished by using a continued fraction expansion).

[7] [Resolution] Once r is found, the factors of n can be *possibly* obtained from computing $\gcd(x^{r/2}-1, n)$ and $\gcd(x^{r/2}+1, n)$, that is, the pair of integers (a, b) satisfying

$$(a, b) = \{\gcd(x^{r/2} \pm 1, n)$$

could be the nontrivial factors of n. If it fails to produce a nontrivial factor of N, goto step [1] to choose a new base.

Theorem 3.8.1. Shor's quantum algorithm for factoring can be performed in $\mathcal{O}((\log n)^{2+\epsilon})$ quantum steps.

Since the finding of the order of x modulo n is related to the computation of discrete logarithms, the above quantum factoring algorithm can also be used to computing discrete logarithms.

Example 3.8.1. On 19 December 2001, IBM made the first demonstration (an important breakthrough) of the quantum factoring algorithm [245], that correctly identified 3 and 5 as the factors of 15. Although the answer may appear to be trivial, it may have good potential and practical implication. In this example, we show how to factor 15 quantum-mechanically [164].

[1] Choose at random $x = 7$ such that $\gcd(x, N) = 1$. We aim to find $r = \text{order}_{15} 7$ such that $7^r \equiv 1 \pmod{15}$.

[2] Initialize two four-qubit registers to state 0:

$$| \Psi_0 \rangle = | 0 \rangle | 0 \rangle .$$

[3] Randomize the first register as follows:

$$| \Psi_0 \rangle \rightarrow | \Psi_1 \rangle = \frac{1}{\sqrt{2^t}} \sum_{k=0}^{2^t - 1} | k \rangle | 0 \rangle .$$

[4] Unitarily compute the function $f(a) \equiv 13^a \pmod{15}$ as follows:

$$
\begin{aligned}
| \Psi_1 \rangle \rightarrow | \Psi_2 \rangle &= \frac{1}{\sqrt{2^t}} \sum_{k=0}^{2^t - 1} | k \rangle | 13^k \ (\text{mod } 15) \rangle \\
&= \frac{1}{\sqrt{2^t}} [\ | 0 \rangle | 1 \rangle + | 1 \rangle | 7 \rangle + | 2 \rangle | 4 \rangle + | 3 \rangle | 13 \rangle + \\
&\quad\ | 4 \rangle | 1 \rangle + | 5 \rangle | 7 \rangle + | 6 \rangle | 4 \rangle + | 7 \rangle | 13 \rangle + \\
&\quad\ | 8 \rangle | 1 \rangle + | 9 \rangle | 7 \rangle + | 10 \rangle | 4 \rangle + | 11 \rangle | 13 \rangle + \\
&\quad\ + \cdots]
\end{aligned}
$$

[5] We now apply the FFT to the second register and measure it (it can be done in the first), obtaining a random result from $1, 7, 4, 13$. Suppose we incidently get 4, then the state input to FFT would be

$$\sqrt{\frac{4}{2^t}} \ [\ | 2 \rangle + | 6 \rangle + | 10 \rangle + | 14 \rangle + \cdots] .$$

After applying FFT, some state

$$\sum_{\lambda} \alpha_\lambda | \lambda \rangle$$

with the probability distribution for $q = 2^t = 2048$ (see [164]). The final measurement gives $0, 512, 1024, 2048$, each with probability almost exactly $1/4$. Suppose $\lambda = 1536$ was obtained from the measurement. Then we compute the continued fraction expansion

$$\frac{\lambda}{q} = \frac{1536}{2048} = \frac{1}{1+\frac{1}{3}}, \quad \text{with convergents} \quad \left[0, 1, \frac{3}{4}, \right]$$

Thus, $r = 4 = \text{order}_{15}(7)$. Therefore,

$$\gcd(x^{r/2} \pm 1, N) = \gcd(7^2 \pm 1, 15) = (5, 3).$$

Remark 3.8.1. Quantum factoring is still in its very earlier stage and will not threaten the security of RSA at least at present, as the current quantum computer can only factor a number with only 2 digits such as 15 which is essentially hopeless.

Problems for Section 3.8

Problem 3.8.1. Let $n = pq$ be an odd integer with p and q distinct prime factors. Show that for a randomly chosen number $x \in (\mathbb{Z}/n\mathbb{Z})^*$ with multiplicative order r modulo n, the probability that r is even and $x^{r/2} \not\equiv -1 \pmod{n}$ is at least $1/2$.

Problem 3.8.2. Let $n > 1$ be an odd integer with k distinct prime factors. Show that for a randomly chosen number $x \in (\mathbb{Z}/n\mathbb{Z})^*$ with multiplicative order r modulo n, the probability that r is even and $x^{r/2} \not\equiv -1 \pmod{n}$ is at least $1 - 1/2^{k-1}$.

Problem 3.8.3. Let $n = pq$ with p and q prime. Use the Chinese Remainder Theorem to show that with probability at least $3/4$, the order r of a modulo n is even. If r is even, show the probability that $x^{r/2} \equiv \pm 1 \pmod{n}$ is at most $1/2$.

Problem 3.8.4. In [93], Griffiths and Niu proposed a network of only one-qubit gates for performing Quantum Fourier Transforms. Construct a quantum circuit to perform the Quantum Fourier Transforms on two and four qubits, respectively.

Problem 3.8.5. There are currently many pseudo-simulations of Shor's quantum factoring algorithm; for example, the paper by Schneiderman, Stanley and Aravind [214] is one of the simulations. Consult [214] and construct a Java (C/C++, Mathematica or Maple) program to simulate Shor's algorithm.

3.9 Discrete Logarithms

The *discrete logarithm problem* (DLP) can be described as follows:

$$\left.\begin{array}{ll} \text{Input}: & a, b, n \in \mathbb{B}^+ \\[6pt] \text{Output}: & x \in \mathbb{N} \text{ with } a^x \equiv b \ (\mathrm{mod}\ n) \\ & \text{if such an } x \text{ exists,} \end{array}\right\} \qquad (3.54)$$

where the modulus n can either be a composite or a prime.

According to Adleman in 1979, the Russian mathematician Bouniakowsky developed a clever algorithm to solve the congruence $a^x \equiv b \ (\mathrm{mod}\ n)$, with the asymptotic complexity $\mathcal{O}(n)$ in 1870. Despite its long history, no efficient algorithm has ever emerged for the discrete logarithm problem. It is believed to be extremely hard, and harder than the integer factorization problem (IFP) even in the average case. The best known algorithm for DLP at present, using NFS and due to Gordon, requires an expected running time

$$\mathcal{O}\left(\exp\left(c(\log n)^{1/3}(\log\log n)^{2/3}\right)\right).$$

There are essentially three different categories of algorithms in use for computing discrete logarithms:

(1) Algorithms that work for arbitrary groups, that is, those that do not exploit any specific properties of groups; Shanks' baby-step giant-step method, Pollard's ρ-method (an analogue of Pollard's ρ-factoring method) and the λ-method (also known as wild and tame Kangaroos) are in this category.

(2) Algorithms that work well in finite groups for which the order of the groups has no large prime factors; more specifically, algorithms that work for groups with smooth orders. A positive integer is called *smooth* if it has no large prime factors; it is called y-smooth if it has no large prime factors exceeding y. The well-known Silver–Pohlig–Hellman algorithm based on the Chinese Remainder Theorem is in this category.

(3) Algorithms that exploit methods for representing group elements as products of elements from a relatively small set (also making use of the Chinese Remainder Theorem); the typical algorithms in this category are Adleman's index calculus algorithm and Gordon's NFS algorithm.

In the sections that follow, we shall introduce the basic ideas of each of these three categories; more specifically, we shall introduce Shanks' baby-step giant-step algorithm, the Silver–Pohlig–Hellman algorithm, Adleman's index calculus algorithm as well as Gordon's NFS algorithm for computing discrete logarithms.

Shanks' Baby-Step Giant-Step Algorithm. Let G be a finite cyclic group of order n, a a generator of G and $b \in G$. The *obvious* algorithm for computing successive powers of a until b is found takes $\mathcal{O}(n)$ group operations. For example, to compute $x = \log_2 15 \pmod{19}$, we compute $2^x \bmod 19$ for $x = 0, 1, 2, \ldots, 19 - 1$ until $2^x \bmod 19 = 15$ for some x is found, that is:

x	0	1	2	3	4	5	6	7	8	9	10	11
a^x	1	2	4	8	16	13	7	14	9	18	17	15

So $\log_2 15 \pmod{19} = 11$. It is clear that when n is large, the algorithm is inefficient. In this section, we introduce a type of square root algorithm, called the baby-step giant-step algorithm, for taking discrete logarithms, which is better than the above mentioned *obvious* algorithm. The algorithm, due to Daniel Shanks (1917–1996), works on arbitrary groups [222].

Let $m = \lfloor \sqrt{n} \rfloor$. The baby-step giant-step algorithm is based on the observation that if $x = \log_a b$, then we can uniquely write $x = i + jm$, where $0 \leq i, j < m$. For example, if $11 = \log_2 15 \bmod 19$, then $a = 2$, $b = 15$, $m = 5$, so we can write $11 = i + 5j$ for $0 \leq i, j < m$. Clearly here $i = 1$ and $j = 2$ so we have $11 = 1 + 5 \cdot 2$. Similarly, for $14 = \log_2 6 \bmod 19$ we can write $14 = 4 + 5 \cdot 2$, for $17 = \log_2 10 \bmod 19$ we can write $17 = 2 + 5 \cdot 3$, etc. The following is a description of the algorithm:

Algorithm 3.9.1 (Shanks' baby-step giant-step algorithm). This algorithm computes the discrete logarithm x of y to the base a, modulo n, such that $y = a^x \pmod{n}$:

[1] (Initialization) Computes $s = \lfloor \sqrt{n} \rfloor$.

[2] (Computing the baby step) Compute the first sequence (list), denoted by S, of pairs (ya^r, r), $r = 0, 1, 2, 3, \ldots, s - 1$:

$$S = \{(y, 0), (ya, 1), (ya^2, 2), (ya^3, 3), \ldots, (ya^{s-1}, s - 1) \bmod n\} \quad (3.55)$$

and sort S by ya^r, the first element of the pairs in S.

[3] (Computing the giant step) Compute the second sequence (list), denoted by T, of pairs (a^{ts}, ts), $t = 1, 2, 3, \ldots, s$:

$$T = \{(a^s, 1), (a^{2s}, 2), (a^{3s}, 3), \ldots, (a^{s^2}, s) \bmod n\} \quad (3.56)$$

and sort T by a^{ts}, the first element of the pairs in T.

[4] (Searching, comparing and computing) Search both lists S and T for a match $ya^r = a^{ts}$ with ya^r in S and a^{ts} in T, then compute $x = ts - r$. This x is the required value of $\log_a y \pmod{n}$.

This algorithm requires a table with $\mathcal{O}(m)$ entries ($m = \lfloor \sqrt{n} \rfloor$, where n is the modulus). Using a sorting algorithm, we can sort both the lists S and T in $\mathcal{O}(m \log m)$ operations. Thus this gives an algorithm for computing discrete logarithms that uses $\mathcal{O}(\sqrt{n}\log n)$ time and space for $\mathcal{O}(\sqrt{n})$ group elements. Note that Shanks' idea was originally for computing the order of a group element g in the group G, but here we use his idea to compute discrete logarithms. Note also that although this algorithm works on arbitrary groups, if the order of a group is larger than 10^{40}, it will be infeasible.

Example 3.9.1. Suppose we wish to compute the discrete logarithm $x = \log_2 6 \bmod 19$ such that $6 = 2^x \bmod 19$. According to Algorithm 3.9.1, we perform the following computations:

[1] $y = 6$, $a = 2$ and $n = 19$, $s = \lfloor \sqrt{19} \rfloor = 4$.

[2] Computing the baby step:

$$
\begin{aligned}
S &= \{(y,0),(ya,1),(ya^2,2),(ya^3,3) \bmod 19\} \\
&= \{(6,0),(6\cdot 2,1),(6\cdot 2^2,2),(6\cdot 2^3,3) \bmod 19\} \\
&= \{(6,0),(12,1),(5,2),(10,3)\} \\
&= \{(5,2),(6,0),(10,3),(12,1)\}.
\end{aligned}
$$

[3] Computing the giant step:

$$
\begin{aligned}
T &= \{(a^s,s),(a^{2s},2s),(a^{3s},3s),(a^{4s},4s) \bmod 19\} \\
&= \{(2^4,4),(2^8,8),(2^{12},12),(2^{16},16) \bmod 19\} \\
&= \{(16,4),(9,8),(11,12),(5,16)\} \\
&= \{(5,16),(9,8),(11,12),(16,4)\}.
\end{aligned}
$$

[4] Matching and computing: The number 5 is the common value of the first element in pairs of both lists S and T with $r = 2$ and $st = 16$, so $x = st - r = 16 - 2 = 14$. That is, $\log_2 6 \ (\bmod\ 19) = 14$, or equivalently, $2^{14} \ (\bmod\ 19) = 6$.

Example 3.9.2. Suppose now we wish to find the discrete logarithm $x = \log_{59} 67 \bmod 113$, such that $67 = 59^x \bmod 113$. Again by Algorithm 3.9.1, we have:

[1] $y = 67$, $a = 59$ and $n = 113$, $s = \lfloor \sqrt{113} \rfloor = 10$.

[2] Computing the baby step:

$$
\begin{aligned}
S &= \{(y,0),(ya,1),(ya^2,2),(ya^3,3),\ldots,(ya^9,9) \bmod 113\} \\
&= \{(67,0),(67\cdot 59,1),(67\cdot 59^2,2),(67\cdot 59^3,3),(67\cdot 59^4,4), \\
&\qquad (67\cdot 59^5,5),(67\cdot 59^6,6),(67\cdot 59^7,7),(67\cdot 59^8,8), \\
&\qquad (67\cdot 59^9,9) \bmod 113\} \\
&= \{(67,0),(111,1),(108,2),(44,3),(110,4),(49,5),(66,6), \\
&\qquad (52,7),(17,8),(99,9)\} \\
&= \{(17,8),(44,3),(49,5),(52,7),(66,6),(67,0),(99,9), \\
&\qquad (108,2),(110,4),(111,1)\}.
\end{aligned}
$$

[3] Computing the giant-step:

$$
\begin{aligned}
T &= \{(a^s, s), (a^{2s}, ss), (a^{3s}, 3s), \ldots (a^{10s}, 10s) \mod 113\} \\
&= \{(59^{10}, 10), (59^{2 \cdot 10}, 2 \cdot 10), (59^{3 \cdot 10}, 3 \cdot 10), (59^{4 \cdot 10}, 4 \cdot 10), \\
&\qquad (59^{5 \cdot 10}, 5 \cdot 10), (59^{6 \cdot 10}, 6 \cdot 10), (59^{7 \cdot 10}, 7 \cdot 10), (59^{8 \cdot 10}, 8 \cdot 10), \\
&\qquad (59^{9 \cdot 10}, 9 \cdot 10) \mod 113\} \\
&= \{(72, 10), (99, 20), (9, 30), (83, 40), (100, 50), (81, 60), \\
&\qquad (69, 70), (109, 80), (51, 90), (56, 100)\} \\
&= \{(9, 30), (51, 90), (56, 100), (69, 70), (72, 10), (81, 60), (83, 40), \\
&\qquad (99, 20), (100, 50), (109, 80)\}.
\end{aligned}
$$

[4] Matching and computing: The number 99 is the common value of the first element in pairs of both lists S and T with $r = 9$ and $st = 20$, so $x = st - r = 20 - 9 = 11$. That is, $\log_{59} 67 \pmod{113} = 11$, or equivalently, $59^{11} \pmod{113} = 67$.

Shanks' baby-step giant-step algorithm is a type of *square root method* for computing discrete logarithms. In 1978 Pollard also gave two other types of square root methods, namely the ρ-method and the λ-method for taking discrete logarithms. Pollard's methods are probabilistic but remove the necessity of precomputing the lists S and T, as with Shanks' baby-step giant-step method. Again, Pollard's algorithm requires $\mathcal{O}(n)$ group operations and hence is infeasible if the order of the group G is larger than 10^{40}.

Silver–Pohlig–Hellman Algorithm. In 1978, Pohlig and Hellman proposed an important special algorithm, now widely known as the Silver–Pohlig–Hellman algorithm for computing discrete logarithms over GF(q) with $\mathcal{O}(\sqrt{p})$ operations and a comparable amount of storage, where p is the largest prime factor of $q - 1$. Pohlig and Hellman showed that if

$$
q - 1 = \prod_{i=1}^{k} p_i^{\alpha_i}, \tag{3.57}
$$

where the p_i are distinct primes and the α_i are natural numbers, and if $r_1, \ldots, r_k$ are any real numbers with $0 \le r_i \le 1$, then logarithms over GF(q) can be computed in

$$
\mathcal{O}\left(\sum_{i=1}^{k} \left(\log q + p_i^{1 - r_i} (1 + \log p_i^{r_i}) \right) \right) \tag{3.58}
$$

field operations, using

$$
\mathcal{O}\left(\log q \sum_{i=1}^{k} (1 + p_i^{r_i}) \right) \tag{3.59}
$$

bits of memory, provided that a precomputation requiring

$$\mathcal{O}\left(\sum_{i=1}^{k} p_i^{r_i} \log p_i^{r_i} + \log q\right) \qquad (3.60)$$

field operations is performed first. This algorithm is very efficient if q is "smooth", i.e., all the prime factors of $q - 1$ are small. We shall give a brief description of the algorithm as follows:

Algorithm 3.9.2 (Silver–Pohlig–Hellman Algorithm). This algorithm computes the discrete logarithm $x = \log_a b \bmod q$:

[1] Factor $q - 1$ into its prime factorization form:

$$q - 1 = \prod_{i=1}^{k} p_1^{\alpha_1} p_2^{\alpha_2} \cdots p_k^{\alpha_k}.$$

[2] Precompute the table $r_{p_i,j}$ for a given field:

$$r_{p_i,j} = a^{j(q-1)/p_i} \bmod q, \quad 0 \le j < p_i. \qquad (3.61)$$

This only needs to be done once for any given field.

[3] Compute the discrete logarithm of b to the base a modulo q, i.e., compute $x = \log_a b \bmod q$:

[3-1] Use an idea similar to that in the baby-step giant-step algorithm to find the individual discrete logarithms $x \bmod p_i^{\alpha_i}$: To compute $x \bmod p_i^{\alpha_i}$, we consider the representation of this number to the base p_i:

$$x \bmod p_i^{\alpha_i} = x_0 + x_1 p_i + \cdots + x_{\alpha_i - 1} p_i^{\alpha_i - 1}, \qquad (3.62)$$

where $0 \le x_n < p_i - 1$.

(a) To find x_0, we compute $b^{(q-1)/p_i}$ which equals $r_{p_i,j}$ for some j, and set $x_0 = j$ for which

$$b^{(q-1)/p_i} \bmod q = r_{p_i,j}.$$

This is possible because

$$b^{(q-1)/p_i} \equiv a^{x(q-1)/p} \equiv a^{x_0(q-1)/p} \bmod q = r_{p_i,x_0}.$$

(b) To find x_1, compute $b_1 = ba^{-x_0}$. If

$$b_1^{(q-1)/p_i^2} \bmod q = r_{p_i,j},$$

then set $x_1 = j$. This is possible because

$$b_1^{(q-1)/p_i^2} \equiv a^{(x-x_0)(q-1)/p_i^2} \equiv a^{(x_1+x_2p_i+\cdots)(q-1)/p_i}$$
$$\equiv a^{x_1(q-1)/p} \bmod q = r_{p_i,x_1}.$$

(c) To obtain x_2, consider the number $b_2 = ba^{-x_0-x_1p_i}$ and compute

$$b_2^{(q-1)/p_i^3} \bmod q.$$

The procedure is carried on inductively to find all $x_0, x_1, \ldots, x_{\alpha_i-1}$.

[3-2] Use the Chinese Remainder Theorem to find the unique value of x from the congruences $x \bmod p_i^{\alpha_i}$.

We now give an example of how the above algorithm works:

Example 3.9.3. Suppose we wish to compute the discrete logarithm $x = \log_2 62 \bmod 181$. Now we have $a = 2$, $b = 62$ and $q = 181$ (2 is a generator of $\mathbb{F}_{181}^*$). We follow the computation steps described in the above algorithm:

[1] Factor $q - 1$ into its prime factorization form:

$$180 = 2^2 \cdot 3^2 \cdot 5.$$

[2] Use the following formula to precompute the table $r_{p_i,j}$ for the given field $\mathbb{F}_{181}^*$:

$$r_{p_i,j} = a^{j(q-1)/p_i} \bmod q, \quad 0 \le j < p_i.$$

This only needs to be done once for this field.

(a) Compute

$$r_{p_1,j} = a^{j(q-1)/p_1} \bmod q = 2^{90j} \bmod 181 \text{ for } 0 \le j < p_1 = 2 :$$

$$r_{2,0} = 2^{90\cdot0} \bmod 181 = 1,$$
$$r_{2,1} = 2^{90\cdot1} \bmod 181 = 180.$$

(b) Compute

$$r_{p_2,j} = a^{j(q-1)/p_2} \bmod q = 2^{60j} \bmod 181 \text{ for } 0 \le j < p_2 = 3 :$$

$$r_{3,0} = 2^{60\cdot0} \bmod 181 = 1,$$
$$r_{3,1} = 2^{60\cdot1} \bmod 181 = 48,$$
$$r_{3,2} = 2^{60\cdot2} \bmod 181 = 132.$$

(c) Compute

$$r_{p_3,j} = a^{j(q-1)/p_3} \bmod q = 2^{36j} \bmod 181 \text{ for } 0 \le j < p_3 = 5 :$$

$$r_{5,0} = 2^{36 \cdot 0} \bmod 181 = 1,$$

$$r_{5,1} = 2^{36 \cdot 1} \bmod 181 = 59,$$

$$r_{5,2} = 2^{36 \cdot 2} \bmod 181 = 42,$$

$$r_{5,3} = 2^{36 \cdot 3} \bmod 181 = 125,$$

$$r_{5,4} = 2^{36 \cdot 4} \bmod 181 = 135.$$

Construct the $r_{p_i,j}$ table as follows:

p_i	j				
	0	1	2	3	4
2	1	180			
3	1	48	132		
5	1	59	42	125	135

This table is manageable if all p_i are small.

[3] Compute the discrete logarithm of 62 to the base 2 modulo 181, that is, compute $x = \log_2 62 \bmod 181$. Here $a = 2$ and $b = 62$:

[3-1] Find the individual discrete logarithms $x \bmod p_i^{\alpha_i}$ using

$$x \bmod p_i^{\alpha_i} = x_0 + x_1 p_i + \cdots + x_{\alpha_i - 1} p_i^{\alpha_i - 1}, \quad 0 \le x_n < p_i - 1.$$

(a-1) Find the discrete logarithms $x \bmod p_1^{\alpha_1}$, i.e., $x \bmod 2^2$:

$$x \bmod 181 \iff x \bmod 2^2 = x_0 + 2x_1.$$

(i) To find x_0, we compute

$$b^{(q-1)/p_1} \bmod q = 62^{180/2} \bmod 181 = 1 = r_{p_1,j} = r_{2,0}$$

hence $x_0 = 0$.

(ii) To find x_1, compute first $b_1 = ba^{-x_0} = b = 62$, then compute

$$b_1^{(q-1)/p_1^2} \bmod q = 62^{180/4} \bmod 181 = 1 = r_{p_1,j} = r_{2,0}$$

hence $x_1 = 0$. So

$$x \bmod 2^2 = x_0 + 2x_1 \implies x \bmod 4 = 0.$$

(a-2) Find the discrete logarithms $x \bmod p_2^{\alpha_2}$, that is, $x \bmod 3^2$:

$$x \bmod 181 \iff x \bmod 3^2 = x_0 + 2x_1.$$

(i) To find x_0, we compute

$$b^{(q-1)/p_2} \bmod q = 62^{180/3} \bmod 181 = 48 = r_{p_2,j} = r_{3,1}$$

hence $x_0 = 1$.

(ii) To find x_1, compute first $b_1 = ba^{-x_0} = 62 \cdot 2^{-1} = 31$, then compute

$$b_1^{(q-1)/p_2^2} \bmod q = 31^{180/3^2} \bmod 181 = 1 = r_{p_2,j} = r_{3,0}$$

hence $x_1 = 0$. So

$$x \bmod 3^2 = x_0 + 2x_1 \implies x \bmod 9 = 1.$$

(a-3) Find the discrete logarithms $x \bmod p_3^{\alpha_3}$, that is, $x \bmod 5^1$:

$$x \bmod 181 \iff x \bmod 5^1 = x_0.$$

To find x_0, we compute

$$b^{(q-1)/p_3} \bmod q = 62^{180/5} \bmod 181 = 1 = r_{p_3,j} = r_{5,0}$$

hence $x_0 = 0$. So we conclude that

$$x \bmod 5 = x_0 \implies x \bmod 5 = 0.$$

[3-2] Find the x in

$$x \bmod 181,$$

such that

$$\begin{cases} x \bmod 4 = 0, \\ x \bmod 9 = 1, \\ x \bmod 5 = 0. \end{cases}$$

To do this, we just use the Chinese Remainder Theorem to solve the following system of congruences:

$$\begin{cases} x \equiv 0 \ (\mathrm{mod}\ 4), \\ x \equiv 1 \ (\mathrm{mod}\ 9), \\ x \equiv 0 \ (\mathrm{mod}\ 5). \end{cases}$$

The unique value of x for this system of congruences is $x = 100$. (This can be easily done by using, for example, the Maple function chrem([0, 1, 0], [4,9, 5]).) So the value of x in the congruence $x \bmod 181$ is 100. Hence $x = \log_2 62 = 100$.

Index Calculus Algorithm. In 1979, Adleman [1] proposed a general purpose, subexponential algorithm for computing discrete logarithms, called the *index calculus method*, with the following expected running time:

$$\mathcal{O}\left(\exp\left(c\sqrt{\log n \log\log n}\right)\right).$$

The index calculus is, in fact, a wide range of methods, including CFRAC, QS and NFS for IFP. In what follows, we discuss a variant of Adleman's index calculus for DLP in $(\mathbb{Z}/p\mathbb{Z})^*$.

Algorithm 3.9.3 (Index calculus for DLP). This algorithm tries to find an integer k such that

$$k \equiv \log_\beta \alpha \pmod{p} \quad \text{or} \quad \alpha \equiv \beta^k \pmod{p}.$$

[1] Precomputation

[1-1] (Choose Factor Base) Select a factor base Γ, consisting of the first m prime numbers,

$$\Gamma = \{p_1, p_2, \ldots, p_m\},$$

with $p_m \leq B$, the bound of the factor base.

[1-2] (Compute $\beta^e \bmod p$) Randomly choose a set of exponent $e \leq p-2$, compute $\beta^e \bmod p$, and factor it as a product of prime powers.

[1-3] (Smoothness) Collect only those relations $\beta^e \bmod p$ that are smooth with respect to B. That is,

$$\beta^e \bmod p = \prod_{i=1}^{m} p_i{}^{e_i}, e_i \geq 0. \tag{3.63}$$

When such relations exist, get

$$e \equiv \sum_{j=1}^{m} e_j \log_\beta p_j \pmod{p-1}. \tag{3.64}$$

[1-4] (Repeat) Repeat [1-3] to find at least m such e in order to find m relations as in (3.64) and solve $\log_\beta p_j$ for $j = 1, 2, \ldots, m$.

[2] Compute $k \equiv \log_\beta \alpha \pmod{p}$

[2-1] For each e in (3.64), determine the value of $\log_\beta p_j$ for $j = 1, 2, \ldots, m$ by solving the m modular linear equations with unknown $\log_\beta p_j$.

[2-2] (Compute $\alpha\beta^r \bmod p$) Randomly choose exponent $r \leq p-2$ and compute $\alpha\beta^r \bmod p$.

[2-3] (Factor $\alpha\beta^r \bmod p$ over Γ)

$$\alpha\beta^r \bmod p = \prod_{j=1}^{m} p_j{}^{r_i}, r_j \geq 0. \tag{3.65}$$

If (3.65) is unsuccessful, go back to to Step [2-2]. If it is successful, then

$$\log_\beta \alpha \equiv -r + \sum_{j=1}^{m} r_j \log_\beta p_j. \tag{3.66}$$

Example 3.9.4 (Index calculus for DLP). Find

$$x \equiv \log_{22} 4 \pmod{3361}$$

such that

$$4 \equiv 22^x \pmod{3361}.$$

[1] Precomputation

    [1-1] (Choose Factor Base) Select a factor base Γ, consisting of the first 4 prime numbers,

$$\Gamma = \{2, 3, 5, 7\},$$

    with $p_4 \leq 7$, the bound of the factor base.

    [1-2] (Compute $22^e \bmod 3361$) Randomly choose a set of exponent $e \leq 3359$, compute $22^e \bmod 3361$, and factor it as a product of prime powers:

$$22^{48} \equiv 2^5 \cdot 3^2 \pmod{3361},$$
$$22^{100} \equiv 2^6 \cdot 7 \pmod{3361},$$
$$22^{186} \equiv 2^9 \cdot 5 \pmod{3361},$$
$$22^{2986} \equiv 2^3 \cdot 3 \cdot 5^2 \pmod{3361}.$$

    [1-3] (Smoothness) The above four relations are smooth with respect to $B = 7$. Thus

$$48 \equiv 5\log_{22} 2 + 2\log_{22} 3 \pmod{3360},$$
$$100 \equiv 6\log_{22} 2 + \log_{22} 7 \pmod{3360},$$
$$186 \equiv 9\log_{22} 2 + \log_{22} 5 \pmod{3360},$$
$$2986 \equiv 3\log_{22} 2 + \log_{22} 3 + 2\log_{22} 5 \pmod{3360}.$$

[2] Compute $k \equiv \log_\beta \alpha \pmod{p}$

    [2-1] Compute

$$\log_{22} 2 \equiv 1100 \pmod{3360},$$
$$\log_{22} 3 \equiv 2314 \pmod{3360},$$
$$\log_{22} 5 \equiv 366 \pmod{3360},$$
$$\log_{22} 7 \equiv 220 \pmod{3360}.$$

[2-2] (Compute $4 \cdot 22^r \bmod p$) Randomly choose exponent $r = 754 \leq 3659$ and compute $4 \cdot 22^{754} \bmod 3361$.

[2-3] (Factor $4 \cdot 22^{754} \bmod 3361$ over Γ)

$$4 \cdot 22^{754} \equiv 2 \cdot 3^2 \cdot 5 \cdot 7 \pmod{3361}.$$

Thus,

$$
\begin{aligned}
\log_{22} 4 &\equiv -754 + \log_{22} 2 + 2 \log_{22} 3 + \log_{22} 5 + \log_{22} 7 \\
&\equiv 2200.
\end{aligned}
$$

That is,

$$22^{2200} \equiv 4 \pmod{3361}.$$

Example 3.9.5. Find $k \equiv \log_{11} 7 \pmod{29}$ such that $\beta^k \equiv 11 \pmod{29}$.

[1] (Factor Base) Let the factor base $\Gamma = \{2, 3, 5\}$.

[2] (Compute and Factor $\beta^e \bmod p$) Randomly choose $e < p$, compute and factor $\beta^e \bmod p = 11^e \bmod 29$ as follows:

    (1) $11^2 \equiv 5 \pmod{29}$ (success),

    (2) $11^3 \equiv 2 \cdot 13 \pmod{29}$ (fail),

    (3) $11^5 \equiv 2 \cdot 7 \pmod{29}$ (fail)

    (4) $11^6 \equiv 3^2 \pmod{29}$ (success),

    (5) $11^7 \equiv 2^3 \cdot 3 \pmod{29}$ (success),

    (6) $11^9 \equiv 2 \cdot 7 \pmod{29}$ (success).

[3] (Solve the systems of congruences for the quantities $\log_\beta p_i$)

    (1) $\log_{11} 5 \equiv 2 \pmod{28}$,

    (4) $\log_{11} 3 \equiv 3 \pmod{28}$,

    (6) $\log_{11} 2 \equiv 9 \pmod{28}$,

    (5) $2 \cdot \log_{11} 2 + \log_{11} 3 \equiv 7 \pmod{28}$,

        $\log_{11} 3 \equiv 17 \pmod{28}$.

[4] (Compute and Factor $\alpha\beta^e \bmod p$) Randomly choose $e < p$, compute and factor $\alpha\beta^e \bmod p = 7 \cdot 11^e \bmod 29$ as follows:

    $7 \cdot 11 \equiv 19 \pmod{29}$ (fail),

    $7 \cdot 11^2 \equiv 2 \cdot 3 \pmod{29}$ (success).

Thus
$$\log_{11} 7 \equiv \log_{11} 2 + \log_{11} 3 - 2 \equiv 24 \ (\text{mod } 28).$$
This is true since
$$11^{24} \equiv 7 \ (\text{mod } 29).$$

For more than ten years since its invention, Adleman's method and its variants were the fastest algorithms for computing discrete logarithms. But the situation changed when Gordon [86] in 1993 proposed an algorithm for computing discrete logarithms in GF(p). Gordon's algorithm is based on the Number Field Sieve (NFS) for integer factorization, with the heuristic expected running time

$$\mathcal{O}\left(\exp\left(c(\log p)^{1/3}(\log\log p)^{2/3}\right)\right),$$

the same as that used in factoring. The algorithm can be briefly described as follows:

Algorithm 3.9.4 (Gordon's NFS). This algorithm computes the discrete logarithm x such that $a^x \equiv b \ (\text{mod } p)$ with input a, b, p, where a and b are generators and p is prime:

[1] (Precomputation): Find the discrete logarithms of a factor base of small rational primes, which must only be done once for a given p.

[2] (Compute individual logarithms): Find the logarithm for each $b \in \mathbb{F}_p$ by finding the logarithms of a number of "medium-sized" primes.

[3] (Compute the final logarithm): Combine all the individual logarithms (by using the Chinese Remainder Theorem) to find the logarithm of b.

Interested readers are referred to Gordon's paper [86] for more detailed information. Note also that Gordon, with co-author McCurley [85], discussed some implementation issues of massively parallel computations of discrete logarithms over GF(2^n).

Finally, we give a brief description of the quantum algorithm for solving the DLP problem in polynomial time (provided that there is a practical quantum computer). It is interesting to note that this quantum algorithm for DLP is in fact a modified version of the quantum algorithm for IFP, introduced earlier. It will certainly generate more interest if this algorithm can be extended to the elliptic curve discrete logarithms problem (ECDLP), but at present, we do not know whether or not the quantum algorithm can be modified to solve the ECDLP problem.

Algorithm 3.9.5 (Quantum algorithm for discrete logarithms).
Given $g, x \in \mathbb{N}$ and p prime. This algorithm will find the integer r such that $g^r \equiv x \ (\text{mod } p)$ if r exists. It uses three quantum registers.

[1] Find q a power of 2 such that q is close to p, that is, $p < q < 2p$.

[2] Put in the first two registers of the quantum computer the uniform super-position of all $|a\rangle$ and $|b\rangle$ (mod $p-1$), and compute $g^a x^{-b}$ (mod p) in the third register. This leaves the quantum computer in the state $|\Psi_1\rangle$:

$$|\Psi_1\rangle = \frac{1}{p-1} \sum_{a=0}^{p-2} \sum_{b=0}^{p-2} |a,\ b,\ g^a x^{-b} \ (\text{mod } p)\rangle. \qquad (3.67)$$

[3] Use the Fourier transform A_q to map $|a\rangle \rightarrow |c\rangle$ and $|b\rangle \rightarrow |d\rangle$ with probability amplitude

$$\frac{1}{q} \exp\left(\frac{2\pi i}{q}(ac+bd)\right).$$

Thus, the state $|a,b\rangle$ will be changed to the state:

$$\frac{1}{q} \sum_{c=0}^{q-1} \sum_{d=0}^{q-1} \exp\left(\frac{2\pi i}{q}(ac+bd)\right) |c,\ d\rangle. \qquad (3.68)$$

This leaves the machine in the state $|\Psi_2\rangle$:

$$|\Psi_2\rangle = \frac{1}{(p-1)q} \sum_{a,b=0}^{p-2} \sum_{c,d=0}^{q-1} \exp\left(\frac{2\pi i}{q}(ac+bd)\right) |c,\ d,\ g^a x^{-b} \ (\text{mod } p)\rangle. \qquad (3.69)$$

[4] Observe the state of the quantum computer and extract the required information. The probability of observing a state $|c,\ d,\ g^k \ (\text{mod } p)\rangle$ is

$$\left| \frac{1}{(p-1)q} \sum_{a,b} \exp\left(\frac{2\pi i}{q}(ac+bd)\right) \right|^2 \qquad (3.70)$$

where the sum is over all (a,b) such that

$$a - rb \equiv k \ (\text{mod } p-1). \qquad (3.71)$$

The better outputs (observed states) we get, the more chance of deducing r we will have; readers are referred to [224] for a justification.

Problems for Section 3.9

Problem 3.9.1. Let the group $G = (\mathbb{Z}/p\mathbb{Z})^*$ with p prime, let also $x, y \in G$ such that $y \equiv x^k$ (mod p). Use smooth numbers to show that the discrete logarithm $k \equiv \log_x y$ (mod p) can be computed in expect time $L(p)^{\sqrt{2}+o(1)}$.

Problem 3.9.2. Use the baby-step giant-step method to solve the DLP problem $k \equiv \log_5 57105961 \pmod{58231351}$.

Problem 3.9.3. Use Silver-Pohliq-Hellman method to solve the DLP problem $k \equiv \log_5 57105961 \pmod{58231351}$.

Problem 3.9.4. Use the index calculus with FB $= (2, 3, 5, 7, 11)$ to solve the DLP problem $k \equiv \log_7 13 \pmod{2039}$.

Problem 3.9.5. Let

$$
\begin{aligned}
p &= 31415926535897932384626433832795028841971693993751058209749 \\
  &= 45923078164062862089986280348253421170679821480865132823066 4 \\
  &= 70938446095505822317253594081284812 37299,
\end{aligned}
$$

$$
x = 2,
$$

$$
\begin{aligned}
y &= 27182818284590452353602874713526624977572470936999595749669 6 \\
  &= 76277240766303535475945713821785251664274274663919320030599 2 \\
  &= 18174135966290435729003342952605956307 38.
\end{aligned}
$$

Use Gordon's index calculus method (Algorithm 3.9.4) to compute the k such that

$$
y \equiv x^k \pmod{p}.
$$

Problem 3.9.6. Give a heuristic argument for the expected running time

$$
\mathcal{O}\left(\exp\left(c(\log p)^{1/3}(\log\log p)^{2/3} \right) \right)
$$

of Gordon's index calculus method (based on NFS) for DLP.

Problem 3.9.7. Give a complexity analysis of Shor's quantum discrete logarithm:

$$
\mathcal{O}((\log p)^{2+\epsilon}).
$$

3.10 kth Roots

There are six closely related computational number theoretic problems concerning the congruence

$$
y \equiv x^k \pmod{n} \tag{3.72}
$$

(1) The Modular Exponentiation Problem (MEP): Given the triple (k, x, n), compute $y \equiv x^k \pmod{n}$. This problem is easy because it can be performed in polynomial time.

(2) The Discrete Logarithm Problem (DLP): Given the triple (x, y, n), find an exponent k such that $y \equiv x^k \pmod{n}$. That is,

$$k \equiv \log_x y \pmod{n} \tag{3.73}$$

This problem is hard, since no polynomial time algorithm has been found yet for this problem.

(3) The kth Root Problem (kRTP): Given the triple (k, y, n), find an x, the kth root of y modulo n, such that $y \equiv x^k \pmod{n}$. That is,

$$x \equiv \sqrt[k]{y} \pmod{n} \tag{3.74}$$

This problem is slightly easier than the discrete logarithm problem, since there are efficient *randomized* algorithms for it, provided that n is a prime power. However, for general n, even the problem for finding *square roots* modulo n (see next item) is as hard as the well-known integer factorization problem (IFP). It should be noted that if the value for $\phi(n)$ is known, then the kth root of y modulo n can be found fairly easily.

(4) The SQuare RooT Problem (SQRTP): Given the pair (a, n), find an x, the square root of y modulo n, such that $y \equiv x^2 \pmod{n}$. That is,

$$x \equiv \sqrt{y} \pmod{n} \tag{3.75}$$

When n is prime, x is easy to find, however when n is composite, x is difficult to find. Clearly, the SQRTP problem is a special case of the kRTP problem for $k = 2$.

(5) The Quadratic Residuosity Problem (QRP): An integer y is a quadratic residue modulo n, denoted by $y \in Q_n$, if $\gcd(y, n) = 1$ and if there exists a solution x to the congruence

$$y \equiv x^2 \pmod{n}, \tag{3.76}$$

otherwise, y is a quadratic non-residue modulo n, denoted by $y \in \overline{Q}_n$. Thus the QRP problem becomes

Given (y, n), decide whether or not $y \in Q_n$.
It is believed that solving QRP is equivalent to computing the prime factorization of n, so, same as the SQRTP, it is computationally infeasible. The quadratic residuosity problem can be extended to the kth power residuosity problem, in the same manner as to extend the square root problem to the kth root problem.

(6) The kth Power Residuosity Problem (kPRP): The QRP problem can be extended to the kPRP as follows:

Given (y, n), decide whether or not $y \in P_n$.

In this section, we shall discuss some algorithms for solving the kth root and the square root problems.

kth Root Problem. First of all, present an efficient and practical algorithm for computing the kth roots modulo n, provided $\phi(n)$ is known.

Algorithm 3.10.1 (kth roots modulo n). Given (k, y, n), this algorithm tries to find an integer x, the kth root of y modulo n, in the congruence $y \equiv x^k \pmod{n}$, that is,

$$x \equiv \sqrt[k]{y} \pmod{n},$$

provided that (k, y, n) is known.

[1] Compute $\phi(n)$.

[2] Find positive integers u and v such that $ku - \phi(n)v = 1$.

[3] Compute $x = y^u \pmod{n}$; this x is the required value.

How do we know $x = y^u$ is a solution to the congruence $x^k \equiv y \pmod{n}$? This can be verified by

$$
\begin{aligned}
x^k &= (y^u)^k \\
&= y^{ku} \\
&= y^{1+\phi(n)v} && \text{(by step [2])} \\
&= y \cdot (y^{\phi(n)})^v \\
&\equiv y && \text{(by Euler's Theorem).}
\end{aligned}
$$

Example 3.10.1. Find the 131th root of 758 modulo 1073

$$x \equiv \sqrt[131]{758} \pmod{1073}.$$

That is, find an x such that $x^{131} \equiv 758 \pmod{1073}$. We follow the steps in Algorithm 3.10.1.

[1] Compute $\phi(n)$: Since $1073 = 29 \cdot 37, \quad \phi(1073) = 28 \cdot 36 = 1008$.

[2] Solve the linear Diophantine equation:

$$131u - 1008v = 1.$$

Since $131/1008$ can be expanded as a finite continued fraction with convergents:

$$\left[0, \frac{1}{7}, \frac{1}{8}, \frac{3}{23}, \frac{10}{77}, \frac{13}{100}, \frac{23}{177}, \frac{36}{277}, \frac{131}{1008} \right],$$

we have

$$u = (-1)^{n-1} q_{n-1} = (-1)^7 \cdot 277 = -277,$$
$$v = (-1)^{n-1} p_{n-1} = (-1)^7 \cdot 36 = -36.$$

Therefore,

$$131 \cdot (-277) - 1008 \cdot (-36) = -36287 + 35288 = 1.$$

Thus, $u = -277$ and $v = -36$.

[3] Finally, compute

$$
\begin{aligned}
x &\equiv y^u \ (\text{mod } 1073) \\
&\equiv (x^{131})^{-277} \ (\text{mod } 1073) \\
&\equiv 758^{-277} \ (\text{mod } 1073) \\
&\equiv 1/(758^{277}) \ (\text{mod } 1073) \\
&\equiv 1/875 \ (\text{mod } 1073) \\
&\equiv 905 \ (\text{mod } 1073).
\end{aligned}
$$

Clearly, $x = 905$ is the required solution to

$$x^{131} \equiv 758 \ (\text{mod } 1073),$$

since

$$905^{131} \equiv 758 \ (\text{mod } 1073).$$

When n is a prime, the computation of the kth root problem is easy, because we do not need to factor n. The following is such an example.

Example 3.10.2. Find the cube root of 2 modulo 17

$$x \equiv \sqrt[3]{2} \ (\text{mod } 17).$$

That is, find an x such that $x^3 \equiv 2 \ (\text{mod } 17)$.

We first calculate $\phi(17) = 16$. Then we try to solve the linear Diophantine equation:

$$3u - 16v = 1.$$

Since $3/16$ can be expanded as a finite continued fraction with convergents:

$$\left[0, \frac{1}{5}, \frac{3}{16} \right],$$

we have

$$u = (-1)^{n-1}q_{n-1} = (-1)^{2-1} \cdot 5 = -5,$$
$$v = (-1)^{n-1}p_{n-1} = (-1)^{2-1} \cdot 1 = -1.$$

Thus

$$3u - 16v = 3(-5) - 16(-1) == 3 \cdot 11 - 16 \cdot 2 = 1.$$

So, $u = 11$ and $v = 2$. Finally, we have

$$x \equiv y^u \pmod{17}$$
$$\equiv (x^3)^{11} \pmod{17}$$
$$\equiv 2^{11} \pmod{17}$$
$$\equiv 8 \pmod{17}.$$

Clearly, $8^3 \bmod 17 = 2$.

Note that to solve $3u - 16v = 1$, we can directly use Euclid's algorithm as follows:

$$16 = 3 \cdot 5 + 1 \qquad 1 = 16 - 3 \cdot 5$$
$$3 = 1 \cdot 3 + 0 \qquad 1 = 3(-5) + 16(-1).$$

So, $u = -5$ and $v = -1$.

Exercise 3.10.1. Find an x, such that $x^5 \equiv 128 \pmod{171}$. That is, find a 5th root of 128 modulo 171:

$$x \equiv \sqrt[5]{128} \pmod{171}.$$

Remark 3.10.1. The above method works only when $\phi(n)$ is known. It will not work if we cannot calculate $\phi(n)$, but it is exactly this *weakness* (unreasonable effectiveness) which is used by Rivest, Shamir, and Adleman in 1978 to construct their *unbreakable* cryptosystem.

Square Root Problem. Recall that the square root problem (SQRTP) in $\mathbb{Z}/n\mathbb{Z}$ is to find an x such that

$$y \equiv x^2 \pmod{n} \tag{3.77}$$

given (y, n). That is, we need to find such an x

$$x \equiv \sqrt{y} \pmod{n}. \tag{3.78}$$

If n is prime, the square roots modulo n are easy to find, however, they will be difficult to find if n is a large composite whose prime factors are unknown. In what follows, we shall first introduce an algorithm for computing x when $n = p$ is prime.

Algorithm 3.10.2 (Square root modulo p). Let p be an odd prime and $y \in \mathbb{Z}$. This algorithm tries to find an x such that

$$x \equiv \sqrt{y} \pmod{p}$$

or says that *such an x does not exist* (i.e., that y is a quadratic non-residue modulo p).

[1] (Distinguish three cases of p) If $p \equiv 3 \pmod 4$, then

$$x = y^{(p+1)/4} \pmod{p}. \tag{3.79}$$

else if $p \equiv 5 \pmod 8$, then

$$\begin{cases} x = y^{(p+3)/8} \pmod{p}, & \text{if } y^{(p-1)/4} = 1, \\ x = 2y \cdot (4y)^{(p-5)/8} \pmod{p}, & \text{otherwise} \end{cases} \tag{3.80}$$

In either case, terminates the algorithm and output x. If $p \equiv 1 \pmod 8$, then write $p - 1 = 2^e \cdot q$ with q odd, and perform the following steps.

[2] (Find generator) Choose a random number n such that $\left(\frac{n}{p}\right) = -1$. Then set $z \leftarrow n^q \pmod{p}$.

[3] (Initialize) Set $w \leftarrow z$, $r \leftarrow e$, $x \leftarrow y^{(q-1)/2} \pmod{p}$, $b \leftarrow yx^2 \pmod{p}$, and $x \leftarrow yx \pmod{p}$.

[4] (Find exponent) If $b \equiv 1 \pmod{p}$, output x and terminate the algorithm. Otherwise, find the smallest $m \geq 1$ such that $b^{2^m} \equiv 1 \pmod{p}$. If $m = r$, output a message saying that y *is a quadratic non-residue modulo p*, and terminates the algorithm.

[5] (Reduce exponent) Set $t \leftarrow w^{2^{r-m-1}}$, $w \leftarrow t^2$, $r \leftarrow m$, $x \leftarrow xt$, $b \leftarrow bw$ (all modulo p), and go to Step [4].

The algorithm has an expected running time of $\mathcal{O}(\log p)^4$.

Example 3.10.3. Find a square root of 35 modulo 3329. That is, find an integer x such that

$$x \equiv \sqrt{35} \pmod{3329}, \quad \text{or equivalently,} \quad x^2 \equiv 35 \pmod{3329}.$$

We follow the steps in Algorithm 3.10.2.

[1] (Determine the types of p) First we notice that $3329 \equiv 1 \pmod 8$, so we write $3329 - 1 = 2^8 \cdot 13$, thus we have $e = 8$ and $q = 13$. Notice also that $y = 35$.

[2] (Find generator) Choose a random number $n = 15$, so that $\left(\frac{n}{p}\right) = \left(\frac{15}{3329}\right) = -1$. Then set $z \equiv n^q \equiv 15^{13} \equiv 2594 \pmod{3329}$.

[3] (Initialize)

$$
\begin{aligned}
w &= z = 2594, \\
r &= e = 8, \\
x &\equiv y^{(q-1)/2} \equiv 35^{(13-1)/2} \equiv 1812 \pmod{3329}, \\
b &\equiv yx^2 \equiv 35 \cdot 1812^2 \equiv 3289 \pmod{3329}, \\
x &= yx \equiv 35 \cdot 1812 \equiv 169 \pmod{3329}.
\end{aligned}
$$

[4] (Find exponent) Now $b = 3289 \not\equiv 1 \pmod{3329}$, we find that $m = 3$ is the smallest positive integer that makes $3289^{2^3} \equiv 1 \pmod{3329}$. Note that $(m = 3) \neq (r = 8)$, so y is a quadratic residue.

[5] (Reduce exponent) Compute

$$
\begin{aligned}
t &\equiv w^{2^{r-m-1}} \equiv 2594^{2^{8-3-1}} \equiv 2594^{2^4} \equiv 1432 \pmod{3329}, \\
w &= t^2 = 1432^2 \equiv 3289 \pmod{3329}, \\
r &= m = 3, \\
x &= xt = 169 \cdot 1432 \equiv 2320 \pmod{3329}, \\
b &= bw = 3289 \cdot 3289 \equiv 1600 \pmod{3329}.
\end{aligned}
$$

and go to Step [4] to repeat.

[4] (Find exponent) Now $b = 1600 \not\equiv 1 \pmod{3329}$, we find that $m = 2$ is the smallest positive integer that makes $1600^{2^2} \equiv 1 \pmod{3329}$. Note that now $(m = 2) \neq (r = 3)$, so y is a quadratic residue.

[5] (Reduce exponent) Compute

$$
\begin{aligned}
t &\equiv w^{2^{r-m-1}} \equiv 3289^{2^{3-2-1}} \equiv 3289 \pmod{3329}, \\
w &= t^2 = 3289^2 \equiv 1600 \pmod{3329}, \\
r &= m = 2, \\
x &= xt = 2320 \cdot 3289 \equiv 412 \pmod{3329}, \\
b &= bw = 1600 \cdot 1600 \equiv 3328 \pmod{3329}.
\end{aligned}
$$

and go to Step [4] to repeat.

[4] (Find exponent) Now $b = 3328 \not\equiv 1 \pmod{3329}$, we find that $m = 1$ is the smallest positive integer that makes $3328^{2^1} \equiv 1 \pmod{3329}$. Note that now $(m = 1) \neq (r = 2)$, so y is a quadratic residue.

[5] (Reduce exponent) Compute

$$
\begin{aligned}
t &\equiv w^{2^{r-m-1}} \equiv 1600^{2^{2-1-1}} \equiv 1600 \pmod{3329}, \\
w &= t^2 = 1600^2 \equiv 3328 \pmod{3329}, \\
r &= m = 1, \\
x &= xt = 412 \cdot 1600 \equiv 58 \pmod{3329}, \\
b &= bw = 3328 \cdot 3328 \equiv 1 \pmod{3329},
\end{aligned}
$$

and go to Step [4] to repeat.

[4] (Find exponent) Now $b = 1 \pmod{3329}$, so $x = 58$ is the required solution to
$$x \equiv \sqrt{35} \pmod{3329}.$$

In fact, $58^2 \equiv 35 \pmod{3329}$. Terminate the algorithm.

Now we move on to the case when $n = pq$ is a composite, with p, q primes.

Algorithm 3.10.3 (Square root modulo *n*). Let $n = pq$ be a composite with p, q primes and $a \in Q_n$. This program finds the two square roots $\pm r$ of a modulo n.

[1] Use Algorithm 3.10.2 to find the square root r of a modulo q.

[2] Use Algorithm 3.10.2 to find the square root s of a modulo q.

[3] Use the extended Euclid's algorithms to find integers c and d such that $cp + dq = 1$.

[4] Set $x \leftarrow (rdq \pm scp) \pmod{n}$ and output x.

The algorithm has an expected running time of $\mathcal{O}((\log p)^4)$.

Problems for Section 3.10

Problem 3.10.1. Show that computing the kth roots

$$y \equiv \sqrt[k]{x} \pmod{n}$$

is as hard as factoring n.

Problem 3.10.2. Show that computing the square roots

$$y \equiv \sqrt{x} \ (\text{mod} \ n)$$

is as hard as factoring n.

Problem 3.10.3. Use Algorithm 3.10.1 to solve the following 13th root problem:

$$y \equiv \sqrt[13]{237} \ (\text{mod} \ 712983765311).$$

Problem 3.10.4. Use Algorithm 3.10.2 to solve the following square root problem:

$$y \equiv \sqrt{37} \ (\text{mod} \ 3329)$$

where 3329 is a prime.

Problem 3.10.5. Use Algorithm 3.10.3 to solve the following square root problem:

$$y \equiv \sqrt{36} \ (\text{mod} \ 332917)$$

where $332917 = 13 \cdot 25609$.

3.11 Elliptic Curve Discrete Logarithms

The elliptic curve discrete logarithm problem (ECDLP): Let E be an elliptic curve over the finite field $\mathbb{F}_p$, say, given by a Weierstrass equation

$$E : \ y^2 \equiv x^3 + ax + b \ (\text{mod} \ p), \tag{3.81}$$

S and T the two points in the elliptic curve group $E(\mathbb{F}_p)$. Then the ECDLP is to find the integer k (assuming that such an integer k exists)

$$k = \log_T S \in \mathbb{Z}, \quad \text{or} \quad k \equiv \log_T S \ (\text{mod} \ p) \tag{3.82}$$

such that

$$S = kT \in E(\mathbb{F}_p), \quad \text{or} \quad S \equiv kT \ (\text{mod} \ p). \tag{3.83}$$

The ECDLP is a more difficult problem than the DLP, on which the elliptic curve digital signature algorithm (ECDSA) is based on. Clearly, the ECDLP is the generalization of DLP, which extends the multiplicative group $\mathbb{F}_p^*$ to the elliptic curve group $E(\mathbb{F}_p)$.

The xedni calculus was first proposed by Joseph Silverman in 1998 [230], and analyzed in [112] [125] and [231]. It is called *xedni calculus* because it "stands index calculus on its head". The xedni calculus is a new method that *might* be used to solve the ECDLP, although it has not yet been tested in practice. It can be described as follows [230]:

[1] Choose points in $E(\mathbb{F}_p)$ and lift them to points in $\mathbb{Z}^2$.

[2] Choose a curve $E(\mathbb{Q})$ containing the lift points; use Mestre's method [147] (in reverse) to make rank $E(\mathbb{Q})$ small.

Whilst the index calculus works in reverse:

[1] Lift $E/\mathbb{F}_p$ to $E(\mathbb{Q})$; use Mestre's method to make rank $E(\mathbb{Q})$ large.

[2] Choose points in $E(\mathbb{F}_p)$ and try to lift them to points in $E(\mathbb{Q})$.

A brief description of the xedni algorithm is as follows (a complete description and justification of the algorithm can be found in [230]).

Algorithm 3.11.1 (Xedni calculus for the ECDLP). Let $\mathbb{F}_p$ be a finite field with p elements (p prime), $E/\mathbb{F}_p$ an elliptic curve over $\mathbb{F}_p$, say, given by

$$E: \quad y^2 + a_{p,1}xy + a_{p,3}y = x^3 + a_{p,2}x^2 + a_{p,4}x + a_{p,6}.$$

N_p the number of points in $E(\mathbb{F}_p)$, S and T the two points in $E(\mathbb{F}_p)$. This algorithm tries to find an integer k

$$k = \log_T S$$

such that

$$S = kT \quad \text{in } E(\mathbb{F}_p).$$

[1] Fix an integer $4 \leq r \leq 9$ and an integer M which is a product of small primes.

[2] Choose r points:

$$P_{M,i} = [x_{M,i}, y_{M,i}, z_{M,i}], \quad 1 \leq i \leq r \tag{3.84}$$

having integer coefficients and satisfying

[a] the first 4 points are $[1, 0, 0]$, $[0, 1, 0]$, $[0, 0, 1]$ and $[1, 1, 1]$.

[b] For every prime $l \mid M$, the matrix $\mathbf{B}(P_{M,1}, \ldots, P_{M,r})$ has maximal rank modulo l.

Further choose coefficients $u_{M,1}, \ldots, u_{M,10}$ such that the points $P_{M,1}, \ldots, P_{M,r}$ satisfy the congruence:

$$u_{M,1}x^3 + u_{M,2}x^2y + u_{M,3}xy^2 + u_{M,4}y^3 + u_{M,5}x^2z + u_{M,6}xyz + u_{M,7}y^2z$$
$$+ u_{M,8}xz^2 + u_{M,9}yz^2 + u_{M,10}z^3 \equiv 0 \pmod{M}. \tag{3.85}$$

[3] Choose r random pair of integers (s_i, t_i) satisfying $1 \le s_i, t_i < N_p$, and for each $1 \le i \le r$, compute the point $P_{p,i} = (x_{p,i}, y_{p,i})$ defined by

$$P_{p,i} = s_i S - t_i T \quad \text{in } E(\mathbb{F}_p). \tag{3.86}$$

[4] Make a change of variables in $\mathbb{P}^2$ of the form

$$\begin{pmatrix} X' \\ Y' \\ Z' \end{pmatrix} = \begin{pmatrix} a_{11} & a_{12} & a_{13} \\ a_{21} & a_{22} & a_{23} \\ a_{31} & a_{32} & a_{33} \end{pmatrix} \begin{pmatrix} X \\ Y \\ Z \end{pmatrix} \tag{3.87}$$

so that the first four points become

$$P_{p,1} = [1,0,0], \;\; P_{p,2} = [0,1,0], \;\; P_{p,3} = [0,0,1], \;\; P_{p,4} = [1,1,1].$$

The equation for E will then have the form:

$$u_{p,1}x^3 + u_{p,2}x^2y + u_{p,3}xy^2 + u_{p,4}y^3 + u_{p,5}x^2z + u_{p,6}xyz$$
$$+u_{p,7}y^2z + u_{p,8}xz^2 + u_{p,9}yz^2 + u_{p,10}z^3 = 0. \tag{3.88}$$

[5] Use the Chinese Remainder Theorem to find integers $u'_1, \ldots, u'_{10}$ satisfying

$$u'_i \equiv u_{p,i} \;(\text{mod } p) \text{ and } u'_i \equiv u_{M,i} \;(\text{mod } M) \text{ for all } 1 \le i \le 10. \tag{3.89}$$

[6] Lift the chosen points to $\mathbb{P}^2(\mathbb{Q})$. That is, choose points

$$P_i = [x_i, y_i, z_i], \quad 1 \le i \le r, \tag{3.90}$$

with integer coordinates satisfying

$$P_i \equiv P_{p,i} \;(\text{mod } p) \text{ and } P_i \equiv P_{M,i} \;(\text{mod } M) \text{ for all } 1 \le i \le r. \tag{3.91}$$

In particular, take $P_1 = [1,0,0], P_2 = [0,1,0], P_3 = [0,0,1], P_4 = [1,1,1]$.

[7] Let $\mathbf{B} = \mathbf{B}(P_1, \ldots, P_r)$ be the matrix of cubic monomials defined earlier. Consider the system of linear equations:

$$\mathbf{B}\mathbf{u} = 0. \tag{3.92}$$

Find a small integer solution $\mathbf{u} = [u_1, \ldots, u_{10}]$ to (3.92) which has the additional property

$$\mathbf{u} \equiv [u'_1, \ldots, u'_{10}] \;(\text{mod } M_p), \tag{3.93}$$

where $u'_1, \ldots, u'_{10}$ are the coefficients computed in Step [5]. Let $C_{\mathbf{u}}$ denote the associated cubic curve:

$$C_{\mathbf{u}} : \; u_1x^3 + u_2x^2y + u_3xy^2 + u_4y^3 + u_5x^2z + u_6xyz$$
$$+u_7y^2z + u_8xz^2 + u_9yz^2 + u_{10}z^3 = 0. \tag{3.94}$$

[8] Make a change of coordinates to put $C_{\mathbf{u}}$ into standard minimal Weierstrass form with the point $P_1 = [1, 0, 0]$ the point at infinity, $\mathcal{O}$. Write the resulting equation as

$$E_{\mathbf{u}} : \quad y^2 + a_1 xy + a_3 y = x^3 + a_2 x^2 + a_4 x + a_6 \qquad (3.95)$$

with $a_1, \ldots, a_6 \in \mathbb{Z}$, and let $Q_1, Q_2, \ldots, Q_r$ denote the images of $P_1, P_2, \ldots, P_r$ under this change of coordinates (so in particular, $Q_1 = \mathcal{O}$). Let $c_4(\mathbf{u})$, $c_6(\mathbf{u})$, and $\Delta(\mathbf{u})$ be the usual quantities in [230] associated to the equation (3.95).

[9] Check if the points $Q_1, Q_2, \ldots, Q_r \in E_{\mathbf{u}}(\mathbb{Q})$ are independent. If they are, return to Step [2] or [3]. Otherwise compute a relation of dependence

$$n_2 Q_2 + n_3 Q_3 + \cdots + n_r Q_r = \mathcal{O}, \qquad (3.96)$$

set

$$n_1 = -n_2 - n_3 - \cdots - n_r, \qquad (3.97)$$

and continue with the next step.

[10] Compute

$$s = \sum_{i=1}^{r} n_i s_i \quad \text{and} \quad t = \sum_{i=1}^{r} n_i t_i. \qquad (3.98)$$

If $\gcd(s, n_p) > 1$, go to Step [2] or [3]. Otherwise compute an inverse $ss' \equiv 1 \pmod{N_p}$. Then

$$\log_T S \equiv s't \pmod{N_p}, \qquad (3.99)$$

and the ECDLP is solved.

As can be seen, the basic idea in the above algorithm is that we first choose points $P_1, P_2, \ldots, P_r$ in $E(\mathbb{F}_p)$ and lift them to points $Q_1, Q_2, \ldots, Q_r$ having integer coordinates, then we choose an elliptic curve $E(\mathbb{Q})$ that goes through the points $Q_1, Q_2, \ldots, Q_r$, finally, check if the points $Q_1, Q_2, \ldots, Q_r$ are *dependent*. If they are, the ECDLP is almost solved. Thus, the goal of the xedni calculus is to find an instance where an elliptic curve has *smaller* than expected rank. Unfortunately, a set of points $Q_1, Q_2, \ldots, Q_r$ as constructed above will usually be *independent*. So, it will not work. To make it work, a congruence method, due to Mestre [147], is used *in reverse* to produce the lifted curve E having smaller than expected rank[1]. Again unfortunately, Mestre's method is based on some deep ideas and unproved conjectures in analytic number theory and arithmetic algebraic geometry, it is not possible for us at present to give even a rough estimate of the algorithm's running time.

---

[1] Mestre's original method is to produce elliptic curves of large rank.

So, virtually we know nothing about the complexity of the xedni calculus. We also do not know if the xedni calculus will be practically useful; it may be completely useless from a practical point of view. Much needs to be done before we can have a better understanding of the xedni calculus.

The index calculus is probabilistic, subexponential-time algorithm applicable for both the integer factorization problem (IFP) and the finite field discrete logarithm problem (DLP). However, there is no known subexponential-time algorithm for the elliptic curve discrete logarithm (ECDLP); the index calculus will not work for the ECDLP. The *xedni calculus*, on the other hand, is applicable to ECDLP (it is in fact also applicable to IFP and DLP), but unfortunately its complexity is essentially unknown. From a computability point of view, xedni calculus is applicable to IFP, DLP, and ECDLP, but from a complexity point of view, the xedni calculus may turn out to be useless (i.e., not at all practical). As for quantum algorithms, we now know that IFP, DLP, and ECDLP can all be solved in polynomial time if a quantum computer is available for use. However, the problem with quantum algorithms is that a practical quantum computer is out of reach in today's technology. We summarise various algorithms for IFP, DLP and ECDLP in Table 3.1.

IFP	DLP	ECDLP
Trial Divisions	Baby-Step Giant-Step	
Pollard's ρ-method	Pollard's λ-method	
CFRAC/MPQS	Index Calculus	
NFS	NFS	
Xedni Calculus	Xedni Calculus	Xedni Calculus
Quantum Algorithms	Quantum Algorithms	Quantum Algorithms

Table 3.1. Algorithms for IFP, DLP and ECDLP

Finally, we conclude that we do have algorithms to solve the IFP, DLP and ECDLP; the only problem is that we do not have an efficient algorithm, nor does any one proved that no such an efficient algorithm exists. From a computational complexity point of view, a $\mathcal{P}$-type problem is easy to solve, whereas an $\mathcal{NP}$-type problem is easy to verify [79], so the IFP, DLP and ECDLP are clearly in $\mathcal{NP}$. For example, it might be difficult (indeed, it is difficult at present) to factor a large integer, but it is easy to verify whether or not a given factorization is correct. If $\mathcal{P} = \mathcal{NP}$, then the two types of the problems are the same, the factorization is difficult only because no one has been clever enough to find an easy/efficient algorithm yet (it may turn out that the integer factorization problem is indeed $\mathcal{NP}$-hard, regardless of the cleverness of the human beings). Whether or not $\mathcal{P} = \mathcal{NP}$ is one of the biggest open problems in both mathematics and computer science, and it is listed in the first of the seven Millennium Prize Problems by the Clay Mathematics Institute in Boston on 24 May 2000 [53]. The struggle continues and more research needs to be done before we can say anything about whether or not $\mathcal{P} = \mathcal{NP}$!

Problems for Section 3.11

Problem 3.11.1. As Shank's Baby-Step Giant-Step method works for arbitrary groups, it can be extended of course to elliptic curve groups.

(1) Develop an elliptic curve analog of Shank's algorithm to solve the ECDLP problem.

(2) Use the analog algorithm to solve the following ECDLP problem, that is, to find k such that
$$Q \equiv kP \pmod{p}$$
where $E/\mathbb{F}_{41} : y^2 \equiv x^3 + 2x + 1 \pmod{41}$, $P = (0,1)$ and $Q = (30,40)$.

Problem 3.11.2. Poland ρ and λ methods [177] for IFP/DLP can also be extended to ECDLP.

(1) Develop an elliptic curve analog of Poland ρ algorithm to solve the ECDLP problem.

(2) Use the analog algorithm to solve the following ECDLP problem: find k such that
$$Q \equiv kP \pmod{p}$$
where $E\mathbb{F}_{1093} : y^2 \equiv x^3 + x + 1 \pmod{1093}$, $P = (0,1)$ and $Q = (413,959)$.

Problem 3.11.3. (Extend the Silver-Pohlig-Hellman method)

(1) Develop an elliptic curve analog of Silver-Pohlig-Hellman method for ECDLP.

(2) Use this analog method to solve the following ECDLP problem: find k such that
$$Q \equiv kP \pmod{p}$$
where $E\mathbb{F}_{599} : y^2 \equiv x^3 + 1 \pmod{1093}$, $P = (60,19)$ and $Q = (277,239)$.

Problem 3.11.4. (ECDLP Challenges) In November 1997, Certicom, a computer security company in Waterloo, Canada, introduced the Elliptic Curve Cryptosystem (ECC) Challenge,
    http://www.certicom.com/index.php?action=ecc,ecc_challenge
developed to increase industry understanding and appreciation for the difficulty of ECDLP and to encourage and stimulate further research in the

security analysis of ECC. The challenge is to compute the ECC private keys from the given list of ECC public keys and associated system parameters. It is the type of problem facing an adversary who wishes to attack ECC. These problems are defined on curves either over $\mathbb{F}_{2^m}$ or over $\mathbb{F}_p$ with p prime (see Table 3.2 and Table 3.3). Also there are three levels of difficulty associated to the curves: exercise level (with bits less than 109), rather easy level (with bits in 109-131), and very hard level (with bits in 163-359). Readers who are interested in solving real-world ECDLP problems are suggested to try to solve the problems listed in Table 3.2 and Table 3.3, particularly those with the question marks "?" as they are still open.

Curve	Field size (in bits)	Estimated number of machine days	Prize in US dollars	Status
ECC2K-95	97	8637	$5,000	May 1998
ECC2-97	97	180448	$5,000	Sept 1999
ECC2K-108	109	1.3×10^6	$10,000	April 2000
ECC2-109	109	2.1×10^7	$10,000	April 2004
ECC2K-130	131	2.7×10^9	$20,000	?
ECC2-131	131	6.6×10^{10}	$20,000	?
ECC2-163	163	2.9×10^{15}	$30,000	?
ECC2K-163	163	4.6×10^{14}	$30,000	?
ECC2-191	191	1.4×10^{20}	$40,000	?
ECC2-238	239	3.0×10^{27}	$50,000	?
ECC2K-238	239	1.3×10^{26}	$50,000	?
ECC2-353	359	1.4×10^{45}	$100,000	?
ECC2K-358	359	2.8×10^{44}	$100,000	?

Table 3.2. Elliptic curves over $\mathbb{F}_{2^m}$

Curve	Field size (in bits)	Estimated number of machine days	Prize in US dollars	Status
ECCp-97	97	71982	$5,000	March 1998
ECCp-109	109	9×10^7	$10,000	Nov 2002
ECCp-131	131	2.3×10^{10}	$20,000	?
ECCp-163	163	2.3×10^{15}	$30,000	?
ECCp-191	191	4.8×10^{19}	$40,000	?
ECCp-239	239	1.4×10^{27}	$50,000	?
ECC2p-359	359	3.7×10^{45}	$100,000	?

Table 3.3. Elliptic curves over $\mathbb{F}_p$

3.12 Chapter Notes and Further Reading

As pointed out by the eminent computational number theorist Hugh Williams [260] *"of all the problems in the theory of numbers to which computers have been applied, probably none has been influenced more than of factoring"*. At present, we still do not have an efficient algorithm for factoring large integers, the security of the famous RSA cryptosystem (will be discussed in next chapter) is based on the intractability of the integer factorization problem. There are many books and papers on the Integer Factorization Problem (IFP). Some of the most popular books on IFP include Bressoud [39], Cohen [50], Crandall and Pomerance [62], Knuth [119], Wagstaff [251], and Yan [268]; these books all provide a good introduction to various algorithms for attacking the integer factorization problem and the related problems. Brent [37] and Montgomery [159] present a very good survey of some modern factoring algorithms.

More references on integer factorization and the related problems such as the kth root problem, the square root problem, the discrete logarithm problem, and the elliptic curve discrete logarithm problem, etc. can be found in the Bibliography section at the end of the book, see, e.g, Adleman ([1] and [2]), Bach, Giesbrecht and McInnes [11], Bach and Shallit [12], Buhler [41], Burr [42], Cormen, Ceiserson and Rivest [59], Gordon and McCurley [85], Huizing [107], Jacobson [112], Knuth [119], Lenstra [131], Lenstra and Lenstra [132], McCurley [139], Odlyzko[168], Pohlig and Hellman [175], Pomerance [182], Riesel [200], Robert Silverman ([232] and [234]), and Williams [262].

For some recent advanced techniques on factoring, particularly from an *analytic number theoretic* point of view, see Pomerance ([183], Pomerance:1995, [186] and [187]), and Granville ([91] and [63]). For some recent advanced techniques for attacking the Elliptic Curve Discrete Logarithm Problem (ECDLP), see Blake, Seroussi and Smart [25], Cohen and Frey [52], Silverman and Suzuki [231], Joseph Silverman ([229] and [230]), and Washington [254].

4. Number-Theoretic Cryptography

Cryptography relies heavily on number-theoretic tools. In particular, systems based on (assumed) hardness of problems in number theory, such as factoring and discrete log, form an important part of modern cryptography.

MOTWANI AND RAGHAVAN [161]

Cryptography was concerned initially with providing secrecy for written messages. Its principles apply equally well to securing data flow between computers, to digitized speech, and to encrypting facsimile and television signals. For example, most satellites routinely encrypt the data flow to and from ground stations to provide both privacy and security for their subscribers. In this chapter, we shall introduce some basic concepts and techniques in public-key cryptography based on primality testing/prime number generation, integer factorization, discrete logarithms, quadratic residuosity, and elliptic curve discrete logarithms, etc.

4.1 Public-Key Cryptography

Cryptography (from the Greek *Kryptós*, "hidden", and *gráphein*, "to write") is the study of the principles and techniques by which information can be concealed in cipher-texts and later revealed by legitimate users employing the secret key, but in which it is either impossible or computationally infeasible for an unauthorized person to do so. Cryptanalysis (from the Greek *Kryptós* and *analýein*, "to loosen") is the science (and art) of recovering information from cipher-texts without knowledge of the key. Both terms are subordinate to the more general term *cryptology* (from the Greek *Kryptós* and *lógos*, "word"). That is,

$$\text{Cryptology} \stackrel{\text{def}}{=} \text{Cryptography} + \text{Cryptanalysis},$$

and

$$\text{Cryptography} \overset{\text{def}}{=} \text{Encryption} + \text{Decryption}.$$

Modern cryptography, however, is the study of "mathematical" systems for solving the following two main types of security problems:

(1) privacy,

(2) authentication.

A privacy system prevents the extraction of information by unauthorized parties from messages transmitted over a public and often insecure channel, thus assuring the sender of a message that it will only be read by the intended receiver. An authentication system prevents the unauthorized injection of messages into a public channel, assuring the receiver of a message of the legitimacy of its sender. It is interesting to note that the computational engine, designed and built by a British group led by Alan Turing at Bletchley Park, Milton Keynes to crack the German ENIGMA code is considered to be among the very first real electronic computers; thus one could argue that modern cryptography is the mother (or at least the midwife) of modern computer science.

Definition 4.1.1. A cryptosystem is a 6-tuple mathematical system

$$T = (M, C, \mathcal{E}, \mathcal{D}, e, d),$$

where

(1) M is a set of plain-texts;

(2) C is a set of cipher-text;

(3) e is a set of encryption keys;

(4) d is a set of decryption keys;

(5) $\mathcal{E}$ is the encryption process using e

$$\mathcal{E}_e : M \to C;$$

(6) $\mathcal{D}$ is the decryption process using d

$$\mathcal{D}_d : C \to M.$$

Remark 4.1.1. Notice that

(1) $\mathcal{E}$ and $\mathcal{D}$ are a pair of invertible functions, and must satisfy

$$\mathcal{D}_{d_i}(\mathcal{E}_{e_i}(m)) = m, \quad m \in M, e_i \in e, d_i \in d.$$

(2) If $e = d$, then T is a symmetric (secret) key cryptosystem, otherwise, it is an asymmetric (public) key cryptosystem.

(3) In the case that $e \neq d$, both e and $\mathcal{E}_e(M)$ should be easy to calculate, whereas $\mathcal{D}_d(C)$, particularly d must be difficult to calculate, at least for the unauthorized parties.

In their seminal paper [70] in 1976, "New Directions in Cryptography", Diffie and Hellman, then both in the Department of Electrical Engineering at Stanford University, first proposed the idea and the concept of public-key cryptography as well as digital signatures; in addition to that they also proposed in the same time a key-exchange protocol, based on the hard *discrete logarithm problem*, for two parties to form a common private key over the insecure channel. It should be noted that Ralph Merkle deserves equal credit with Diffie and Hellman for the invention of public-key cryptography [101]. Although his paper [146] *Secure Communication Over Insecure Channels* was published in 1978, two years later than Diffie and Hellman's 1976 paper it was submitted in August 1975. Also, his conception of *public-key distribution* occurred in the Fall of 1974, again before Diffie and Hellman conceived of *public-key cryptosystems*[1].

Remarkably enough, just about one or two years later, three MIT computer scientists, Rivest, Shamir and Adleman, proposed in 1978 a practical public-key cryptosystem based on primality testing and integer factorization, now widely known as RSA cryptosystem (see Section 4.2). More specifically, they based their encryption and decryption on mod-n arithmetic, where n is the product of two large prime numbers p and q. A special case based on mod-p arithmetic with p prime, now known as exponential cipher, had already been studied by Pohlig and Hellman [166] before RSA.

It is also interesting to note that in December 1997 the Communication-Electronics Security Group (CESG) in the British Government Communications Headquarters (GCHQ) claimed that public-key cryptography was conceived by James H. Ellis in 1970, about six years earlier than Diffie, Hellman and Merkle, by releasing the following five papers by Ellis and his two colleagues Cocks and Williamson:

[1] James H. Ellis, *The Possibility of Non-Secret Encryption*, January 1970, 9 pages.

---

[1] The Association for Computing Machinery (ACM) presented its first Paris Kanellakis Theory and Practice Award in 1997 to Diffie, Hellman and Merkle for their conception of public-key cryptography (along with Adleman, Rivest and Shamir for their first effective realization of public-key cryptography). As mentioned in the citation of the award, "the idea of a public-key cryptosystem was a major conceptual breakthrough that continues to stimulate research to this day, and without it today's rapid growth of electronic commerce would have been impossible". Prof. Peter Wegner, the chair of the ACM's Paris Kanellakis Theory and Practice Award committee even stressed that "the only practical way to maintain privacy and integrity of information is by using public-key cryptography."

Figure 4.1. Merkle, Hellman and Diffie (Photo by courtesy of Dr. Simon Singh)

[2] Clifford C. Cocks, *A Note on Non-Secret Encryption*, 20 November 1973, 2 pages.

[3] Malcolm J. Williamson, *Non-Secret Encryption Using a Finite Field*, 21 January 1974, 2 pages.

[4] Malcolm Williamson, *Thoughts on Cheaper Non-Secret Encryption*, 10 August 1976, 3 pages.

[5] James Ellis, *The Story of Non-Secret Encryption*, 1987, 9 pages.

The US Government's National Security Agency (NSA) also made a similar claim that they had public-key cryptography a decade earlier. There are apparently two parallel universes in cryptography, the public and secret worlds [101]. The CESG and NSA people, on the secret world side, certainly deserve some kind of credit, but according to the "first to publish" rule, the full credit for the invention of public-key cryptography goes to Diffie, Hellman and Merkle.

In a public-key cryptosystem (see Figure 4.2), the encryption key e_k and decryption key d_k are different, that is, $e_k \neq d_k$ (this is why we call public-key cryptosystems *asymmetric key cryptosystems*). Since e_k is only used for encryption, it can be made public; only d_k must be kept a secret for decryption. To distinguish public-key cryptosystems from secret-key cryptosystems,

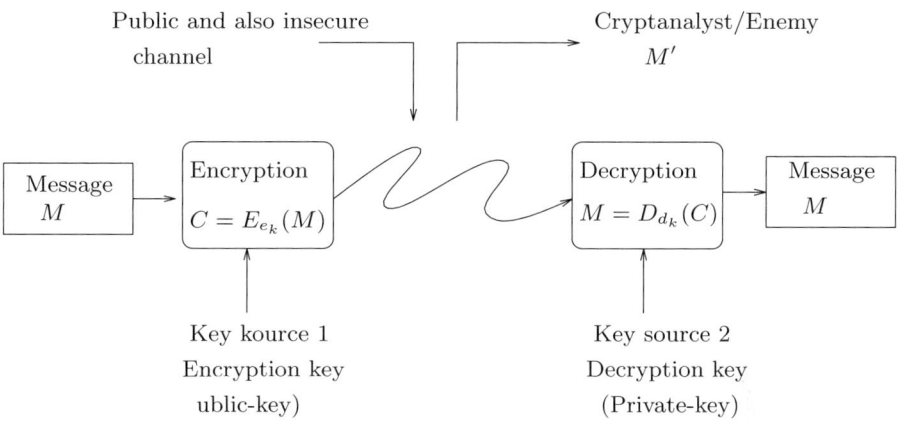

Figure 4.2. Modern public-key cryptosystems ($e_k \neq d_k$)

e_k is called the *public-key*, and d_k the *private key*; only the key used in secret-key cryptosystems is called the *secret-key*. The implementation of public-key cryptosystems is based on *trapdoor one-way functions*.

Remark 4.1.2. Public-key cryptosystems have some important advantages over secret-key cryptosystems in the distribution of the keys. However, when a large amount of information has to be communicated, it may be that the use of public-key cryptography would be too slow whereas the use of secret-key cryptography could be impossible due to lack of a shared secret key. In practice, it is better to combine the secret-key and public-key cryptography into a single cryptosystem for secure communications. Such a combined system is often called a *hybrid cryptosystem*. A hybrid cryptosystem uses a public-key cryptosystem once at the beginning of the communication to share a short piece of information that is then used as the key for encryption and decryption by means of a "conventional" secret-key cryptosystem in later stages. Such a cryptosystem is essentially a secret-key cryptosystem but still enjoys the advantages of the public-key cryptosystems.

Problems for Section 4.1

Problem 4.1.1. Explain the following basic concepts in cryptology:

(1) Cryptography;

(2) Cryptanalysis;

(3) Cryptology;

(4) Public-key cryptography;

(5) Secret-key cryptography;

(6) Encryption;

(7) Decryption;

(8) Digital Signatures;

(9) Digital certificate;

(10) Intrusion Detection;

(11) Secure Hashing Functions.

Problem 4.1.2. Explain the following basic concepts in information security:

(1) Confidentiality/Privacy;

(2) Integrity;

(3) Authentication;

(4) authorization;

(5) Nonrepudiation;

(6) Access Control.

Problem 4.1.3. Write an essay on the history and the development of public-key cryptography.

4.2 RSA Cryptosystem

In 1978, just shortly after Diffie and Hellman proposed the first public-key exchange protocol at Stanford, three MIT researchers Ronald Rivest, Adi Shamir and Leonard Adleman (See Figure 4.3)

proposed the first practical public-key cryptosystem, now widely known as the RSA public-key cryptosystem[2]. The RSA cryptosystem is based on the following assumption:

---

[2] The Association for Computing Machinery (ACM) has offered the Year 2002 A. M. Turing Award (the equivalent of the Nobel prize in the physical sciences) to Adleman, Rivest and Shamir for their contributions to the theory and practical application of public-key cryptography, particularly the invention of the RSA cryptosystem, because the RSA system now "has become the foundation for an entire generation of technology security products and has also inspired important work in both theoretical computer science and mathematics."

Figure 4.3. Shamir, Rivest and Adleman (Photo by courtesy of Prof. Adleman)

RSA Assumption: It is not too difficult to find two large prime numbers, but it is very difficult to factor a large composite into its prime factorization form.

The RSA cryptosystem (See Figure 4.4) works as follows:

$$\left. \begin{array}{l} C \equiv M^e \pmod{n} \\ M \equiv C^d \pmod{n} \end{array} \right\} \tag{4.1}$$

where

(1) M is the plain-text;

(2) C is the cipher-text;

(3) $n = pq$ is the modulus, with p and q large and distinct primes;

(4) e is the *public* encryption exponent (key) and d the *private* decryption exponent (key) , with $ed \equiv 1 \pmod{\phi(n)}$. $\langle n, e \rangle$ should be made public, but d (as well as $\phi(n)$) should be kept secret.

Clearly, the function $f : M \to C$ is a one-way trap-door function, since it is easy to compute by the fast exponentiation method, but its inverse $f^{-1} : C \to M$ is difficult to compute, because for those who do not know the private decryption key (the trap-door information) d, they will have to factor

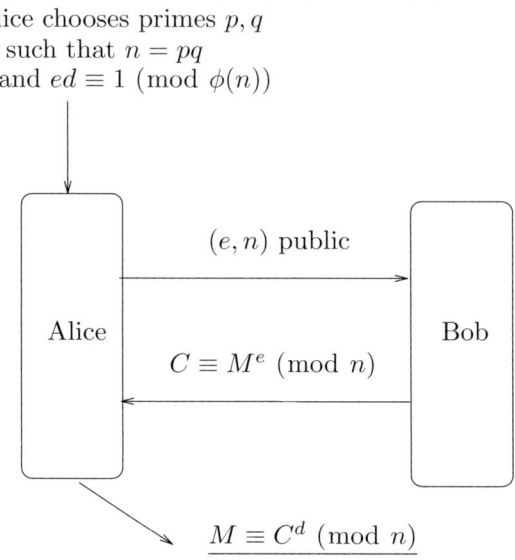

Alice chooses primes p, q
such that $n = pq$
and $ed \equiv 1 \pmod{\phi(n)}$

(e, n) public

Alice   Bob

$C \equiv M^e \pmod{n}$

$M \equiv C^d \pmod{n}$

Figure 4.4. RSA Cryptosystem

n and to compute $\phi(n)$ in order to find d. However, for those who know d, then the computation of f^{-1} is as easy as of f. This is exactly the idea of RSA cryptography.

Suppose now the sender, say, for example, Alice wants to send a message M to the receiver, say, for example, Bob. Bob will have already chosen a one-way trapdoor function f described above, and published his *public-key* (e, n), so we can assume that both Alice and any potential adversary know (e, n). Alice splits the message M into blocks of $\lfloor \log n \rfloor$ bits or less (padded on the right with zeros for the last block), and treats each block as an integer $x \in \{0, 1, 2, \ldots, n - 1\}$. Alice computes

$$y \equiv x^e \pmod{n} \tag{4.2}$$

and transmits y to Bob. Bob, who knows the private key d, computes

$$x \equiv y^d \pmod{n} \tag{4.3}$$

where $ed \equiv 1 \pmod{\phi(n)}$. An adversary who intercepts the encrypted message should not be able to decrypt it without knowledge of d. There is no known way of cracking the RSA system without essentially factoring N, so it is clear that the security of the RSA system depends on the difficulty of factoring N. Some authors, for example, Woll in 1987 observed that finding the RSA decryption key d is random polynomial-time equivalent to factorization, and Pinch in 1997 showed that an algorithm $A(n, e)$ for obtaining d given n and e can be turned into an algorithm which obtains p and q with positive probability.

Example 4.2.1. Suppose the message to be encrypted is "PLEASE WAIT FOR ME". Let $n = 5515596313 = 71593 \cdot 77041$. Let also $e = 1757316971$ with $\gcd(e, n) = 1$. Then $d \equiv 1/1757316971 \equiv 2674607171 \pmod{(71593 - 1)(77041 - 1)}$. To encrypt the message, we first translate the message into its numerical equivalent by the letter-digit encoding scheme defined by $A \rightarrow 01, B \rightarrow 02, \ldots, Z \rightarrow 26$ and space $\rightarrow 00$:

$$M = 1612050119050023010920061518001305.$$

Then we split it into 4 blocks, each with 10 digits, padded on the right with zeros for the last block:

$$M = (M_1, M_2, M_3, M_4) = (1612050119 \ 0500230109 \ 2000061518 \ 0013050000).$$

Now, we have

$$C_1 \equiv 1612050119^{1757316971} \equiv 763222127 \pmod{5515596313},$$
$$C_2 \equiv 0500230109^{1757316971} \equiv 1991534528 \pmod{5515596313},$$
$$C_3 \equiv 2000061518^{1757316971} \equiv 74882553 \pmod{5515596313},$$
$$C_4 \equiv 0013050000^{1757316971} \equiv 3895624854 \pmod{5515596313}.$$

That is,

$$C = (C_1, C_2, C_3, C_4) = (763222127, 1991534528, 74882553, 3895624854).$$

To decrypt the cipher text, we perform:

$$M_1 \equiv 763222127^{2674607171} \equiv 1612050119 \pmod{5515596313},$$
$$M_2 \equiv 1991534528^{2674607171} \equiv 500230109 \pmod{5515596313},$$
$$M_3 \equiv 74882553^{2674607171} \equiv 2000061518 \pmod{5515596313},$$
$$M_4 \equiv 3895624854^{2674607171} \equiv 13050000 \pmod{5515596313}.$$

By padding the necessary zeros on the left of some blocks, we get

$$M = (M_1, M_2, M_3, M_4) = (1612050119 \ 0500230109 \ 2000061518 \ 0013050000)$$

which is "Please wait for me", the original plain-text message.

Example 4.2.2. We now give a reasonably large RSA example. In one of his series of Mathematical Games, Martin Gardner [78] reported an RSA challenge with US\$100 to decrypt the following message C:

9686961375462206147714092225435588290575999112457431987469512093_
0816298225145708356931476622883989628013391990551829945157815154.

The public-key consists of a pair of integers (e, n), where $e = 9007$ and N is a "random" 129-digit number (called RSA-129):

114381625757888867669235779976146612010218296721242362562561 8429_
35706935245733897830597123563958705058989075147599290026879543541.

The RSA-129 was factored by Derek Atkins, Michael Graff, Arjen K. Lenstra, Paul Leyland et al. on 2 April 1994 to win the \$100 prize offered by RSA in 1977. Its two prime factors are as follows:

3490529510847650949147849619903898133417764638493387843990820577,

32769132993266709549961988190834461413177642967992942539798288533.

They used the double large prime variation of the Multiple Polynomial Quadratic Sieve (MPQS) factoring method. The sieving step took approximately 5000 mips years, and was carried out in 8 months by about 600 volunteers from more than 20 countries, on all continents except Antarctica. As we have explained in the previous example, to encrypt an RSA-encrypted message, we only need to use the public-key (n, e) to compute

$$x^e \equiv y \pmod{n}.$$

But decrypting an RSA-message requires factorization of N if one does not know the secret decryption key. This means that if we can factor n, then we can compute the secret key d, and get back the original message by calculating

$$y^d \equiv x \pmod{n}.$$

Since now we know the prime factorization of n, it is trivial to compute the secret key $d = 1/e \bmod \phi(n)$, which in fact is

106698614368578024442868771328920154780709906633937862801226 2244_
9663106312591177447087334016859746230655396854451327710905360609 5.

So we shall be able to compute

$$C^d \equiv M \pmod{n}$$

without any problem. To use the fast exponential method to compute $C^d \bmod N$, we first write d in its binary form $d_1 d_2 \cdots d_{\text{size}}$ (where size is the number of the bits of d) as follows:

$d = d_1 d_2 \cdots d_{426} =$
    1001110110011111100101001100100010000010000011101001111001001 10_
    0100111101001110000000000000011111110100001101010110001011101111_
    0101000011111011000000100000111011010101011110101010011111110110_
    1101000011111101000000111101001100010110010110011010010100011 00_
    1001110101100001011101001010110100000111000000011100011101010 10_
    0110111010001111010011100011010110101010100100111010100010011 11_
    0000001001110100110001101111101011001000110011111

and perform the following computation:

> $M \leftarrow 1$
> for i from 1 to 426 do
> $\quad M \leftarrow M^2 \bmod n$
> $\quad$ if $d_i = 1$ then $M \leftarrow M \cdot C \bmod n$
> print M

which gives the plain-text M:

2008050013010709030023151804190001180500191721050113091908001519_
19090618010705

and hence the original message:

THE MAGIC WORDS ARE SQUEAMISH OSSIFRAGE

via the encoding alphabet ⊔ $= 00, A = 01, B = 02, \dots, Z = 26$. Of course, by
the public encryption key $e = 9007$, we can compute $M^e \equiv C \pmod{n}$; first
write e in the binary form $e = e_1 e_2 \cdots e_{14} = 10001100101111$, then perform
the following procedure:

> $C \leftarrow 1$
> for i from 1 to 14 do
> $\quad C \leftarrow C^2 \bmod n$
> $\quad$ if $e_i = 1$ then $C \leftarrow C \cdot M \bmod n$
> print C

which gives the encrypted text C at the beginning of this example:

9686961375462206147714092225435588290575999112457431987469512093_
08162982251457083569314766228839896280133919905518299445157815154.

Exercise 4.2.1. Let the ciphertexts $C_1 \equiv M_1^e \pmod{n}$ and $C_2 \equiv M_2^e \pmod{n}$ be as follows, where $e = 9137$ and $n =$ RSA-129:

4660490643506009639239112238711202373603916347008276824341038329_
6685073462027217982000297925067088337283567804532383891140719579,

6506409693851106974152831334247539664897855173581383677796350373_
8147209287793861787878189741574391857183608196124160093438830158

Find M_1 and M_2.

Remark 4.2.1. In fact, anyone who can factor the integer RSA-129 can
decrypt the message. Thus, decrypting the message is essentially factoring
the 129-digit integer. The factorization of RSA-129 implies that it is possible
to factor any random 129-digit integer. It should be also noted that, as we
mentioned in Chapter 3, the current best know general factoring record is the
RSA-155. It follows that the composite number (i.e., the modulus) n used in
the RSA cryptosystem should have more than 155 decimal digits.

A better example [33] of a trap-door one-way function of the form used in the RSA cryptosystem would use Carmichael's λ-function rather than Euler's ϕ-function, and is as follows:

$$y = f(x) \equiv x^e \pmod{n} \tag{4.4}$$

where
$$
\left.
\begin{aligned}
&n = pq \quad (p \text{ and } q \text{ are two large primes}), \\
&e > 1, \quad \gcd(e, \lambda) = 1, \\
&\lambda(n) = \text{lcm}(p-1, \ q-1) = \frac{(p-1)(q-1)}{\gcd(p-1, \ q-1)}.
\end{aligned}
\right\} \tag{4.5}
$$

We assume that e and n are publicly known but p, q and $\lambda(n)$ are not. The inverse function of $f(x)$ is defined by

$$x = f^{-1}(y) \equiv y^d \pmod{n} \quad \text{with} \ \ ed \equiv 1 \pmod{\lambda}. \tag{4.6}$$

To show it works, we see

$$
\begin{aligned}
x &\equiv y^d \equiv (x^e)^d \equiv x^{ed} \equiv x^{k\lambda(n)+1} \\
&\equiv (x^{\lambda(n)})^k \cdot x \equiv 1^k \cdot x \quad \text{(by Carmichael's theorem)} \\
&\equiv x.
\end{aligned}
$$

It should be easy to compute $f^{-1}(y) \equiv y^d \pmod{n}$ if d is known, provided that $f^{-1}(y)$ exists (note that $f^{-1}(y)$ may not exist). The assumption underlying the RSA cryptosystem is that it is hard to compute $f^{-1}(y)$ without knowing d. However, the knowledge of p, q or $\lambda(n)$ makes it easy to compute d.

Algorithm 4.2.1 (Construction of the above trapdoor function).
This algorithm constructs the trapdoor function and generates both the public and the secret keys suitable for RSA cryptography:

[1] Use Algorithm 2.11.2 or Algorithm 2.11.1 to find two large primes p and q, each with at least 100 digits such that:

[a] $|p - q|$ is large;

[b] $p \equiv -1 \pmod{12}, q \equiv -1 \pmod{12}$;

[c] The following values of p', p'', q' and q'' are all primes:

$$
\begin{aligned}
p' &= (p-1)/2, \\
p'' &= (p+1)/12, \\
q' &= (q-1)/2, \\
q'' &= (q+1)/12.
\end{aligned}
$$

[2] Compute $n = pq$ and $\lambda = 2p'q'$.

[3] Choose a random integer e relatively prime to λ such that $e - 1$ is not a multiple of p' or q'.

[4] Apply the extended Euclidean algorithm to e and λ to find d and λ' such that $0 < d < \lambda$ and
$$ed + \lambda\lambda' = 1.$$

[5] Destroy all evidence of p, q, λ and λ'.

[6] Make (e, n) public but keep d secret.

Remark 4.2.2. Since the primality testing or prime generation can be done in polynomial time, the expected running time required to construct the trapdoor function is of course also in polynomial time.

Problems for Section 4.2

Problem 4.2.1. The RSA function $M \mapsto C \bmod n$ is a trap-door one-way, as it is computationally intractable to invert the function if the prime factorization $n = pq$ is unknown. Give any other *five* trap-door one-way functions that can be used to construct public-key cryptosystems. Justify your answer.

Problem 4.2.2. Implement Algorithm 4.2.1 and generate a triple of information (e, n, d) to be used for RSA encryption and decryption. (Note that n should have about 200 digits.)

Problem 4.2.3. Use the information (e, n, d) generated in Problem 4.2.2 to encrypt the following message M:

the winning city for the year 2012 olympic games is london

and then to decrypt your C back to M again. (Use the following letter-number coding scheme $a \leftrightarrow 01, b \leftrightarrow 01, \ldots, z \leftrightarrow 26, \text{space} \leftrightarrow 00$.)

Problem 4.2.4. Let

e = 65537,

n = 2519590847565789349402718324004839857142928212620403202777_
13783604366202070759555626401852588078440691829064124951508_
21892985591491761845028084891200728449926873928072877767359_
71418347270261896375014971824691165077613337985909570009733O_
45974880842840179742910064245869181719511874612151517265463_
22822168699875491824224336372590851418654620435767984233871_
84774447920739934236584823824281198163815010674810451660377_
30605620161967625613384414360383390441495263443219011465754_
44541784240209246165157233507787077498171257724679629263863_
56373289912154831438167899885040445364023527381951378636564_
39121201039712282212072O357,

M = 1905031805200009142005121209070514030500070922051900200805O_
0071522051814130514200001002209200112000504O705.

Find C such that $C \equiv M^e \pmod{n}$.

Problem 4.2.5. Let e and n be the same as given in Problem 4.2.4, and C as follows:

21859805614455554930240193896291771597538111447285434229215O0499254_
18121103256208767902225983106799128610119089769511935775476540852 2_
69795663824292287063708323169440487394769407843277578199861497994 2_
06436166946261408885274160021723305205957488066846353603028794423 5_
82262770813499706106470077169306460071262980916541699844999292531 3_
37428138732590332878186320959546870156074276759915720731486943230 5_
89265183618950810376467872168336018311899427370639870779548080069 8_
50187887587515053212373800623567195852763946133986860441037844981 8_
38391305986458712839620011281598913455842775066742715153760973671 2_
0464775711605903168458 7.

Find M such that $C \equiv M^e \pmod{n}$.

4.3 Security and Cryptanalysis of RSA

As can be seen from the previous section, the whole idea of the RSA encryption and decryption is as follows:

$$\left. \begin{array}{rcl} C & \equiv & M^e \ (\mathrm{mod} \ n) \\ M & \equiv & C^d \ (\mathrm{mod} \ n) \end{array} \right\} \tag{4.7}$$

where

$$\left. \begin{array}{rcl} ed & \equiv & 1 \ (\mathrm{mod} \ \phi(n)) \\ n & = & pq \ \text{with} \ p, q \in \ \mathrm{Primes.} \end{array} \right\} \tag{4.8}$$

Thus, the *RSA function* may be defined by

$$f_{\mathrm{RSA}} : M \mapsto M^e \ \mathrm{mod} \ n. \tag{4.9}$$

The *inverse of the RSA function* is then defined by

$$f_{\mathrm{RSA}}^{-1} : M^e \mapsto M \ \mathrm{mod} \ n. \tag{4.10}$$

Clearly, the RSA function is a *one-way trap-door function*, with

$$\{d, p, q, \phi(n)\} \tag{4.11}$$

the RSA *trap-door information*mitrap-door information. For security purposes, this set of information must be kept as a secret and should never be disclosed in anyway even in part. Now suppose that Bob sends C to Alive, but Eve intercepts it and wants to understand it. Since Eve only has (e, n, C) and does not have any piece of the trap-door information in (4.11), then it should be infeasible/intractable for her to recover M from C:

$$\{e, n, C \equiv M^e \ (\mathrm{mod} \ n)\} \xrightarrow{\text{hard}} \{M \equiv C^d \ (\mathrm{mod} \ n)\}. \tag{4.12}$$

On the other hand, for Alice, since she knows d, which implies that she knows all the pieces of trap-door information in (4.11), since

$$\{d\} \overset{\mathcal{P}}{\Longleftrightarrow} \{p\} \overset{\mathcal{P}}{\Longleftrightarrow} \{q\} \overset{\mathcal{P}}{\Longleftrightarrow} \{\phi(n)\}. \tag{4.13}$$

(We shall show (4.13) later in this section.) Thus, it is easy for Alice to recover M from C:

$$\{N, C \equiv M^e \ (\mathrm{mod} \ n)\} \xrightarrow[\text{easy}]{\{d, p, q, \phi(n)\}} \{M \equiv C^d \ (\mathrm{mod} \ n)\}. \tag{4.14}$$

Why is it hard for Eve to recover M from C? This is because Eve is facing a hard computational problem, namely, the *RSA problem* [203]:

The RSA problem: Given the RSA public-key (e, n) and the RSA ciphertext C, find the corresponding RSA plaintext M. That is,

$$\{e, n, C\} \longrightarrow \{M\}.$$

It is conjectured although it has never been proved or disproved that:

The RSA conjecture: Given the RSA public-key (e, N) and the RSA ciphertext C, it is hard to find the corresponding RSA plaintext M. That is,

$$\{e, n, C\} \xrightarrow{\text{hard}} \{M\}.$$

But how hard is it for Alice to recover M from C? This is another version of the RSA conjecture, often called the *RSA assumption*, which again has never been proved or disproved:

The RSA assumption: Given the RSA public-key (e, n) and the RSA ciphertext C, then finding M is as hard as factoring the RSA modulus n. That is,

$$\text{IFP}(n) \Longleftrightarrow \text{RSA}(M)$$

provided that n is sufficiently large and randomly generated, and M and C are random integers between 0 and $n - 1$. More precisely, it is conjectured (or assumed) that

$$\text{IFP}(n) \overset{\mathcal{P}}{\Longleftrightarrow} \text{RSA}(M).$$

That is, if n can be factorized in polynomial-time $(\mathcal{P})$, then M can be recovered from C in polynomial-time as well. In other words, cryptoanalyzing RSA must be as difficult as solving the IFP problem. But the problem is that no-one knows whether or not IFP can be solved in polynomial-time, so RSA is only assumed to be secure, not proved to be secure:

$$\text{IFP}(n) \text{ is hard} \longrightarrow \text{RSA}(M) \text{ is secure}.$$

The real situtaion is that

$$\text{IFP}(n) \overset{\checkmark}{\Longrightarrow} \text{RSA}(M),$$

$$\text{IFP}(n) \overset{?}{\Longleftarrow} \text{RSA}(M).$$

Now re return to show the relations expressed (4.13).

Theorem 4.3.1. (The equivalence of $\phi(n)$ and $\text{IFP}(n)$)

$$\phi(n) \overset{\mathcal{P}}{\Longleftrightarrow} \text{IFP}(n). \tag{4.15}$$

Proof. Note first that if $(n, \phi(n))$ is known and n is assumed to be the product of two primes p and q, then n can be easily factored. Assume

$$n = pq,$$

then

$$\phi(n) = (p - 1)(q - 1),$$

thus

$$pq - p - q + 1 - \phi(n) = 0 \qquad (4.16)$$

substituting $q = n/p$ into (4.16) gives

$$p^2 - (n - \phi(N) + 1)p + n = 0. \qquad (4.17)$$

Let $A = n - \phi(n) + 1$, then

$$(p, q) = \frac{A \pm \sqrt{A^2 - 4n}}{2}$$

will be the two roots of (4.17), and hence, the two prime factors of n. On the other hand, if the two prime factors p and q of n are known, then $\phi(n) = (p - 1)(q - 1)$ immediately from

$$\phi(n) = n \prod_{i=1}^{k} p_i^{\alpha_i}$$

if

$$n = \prod_{i=1}^{k} p_i.$$

$\square$

What this theorem says is that if an enemy cryptanalyst could compute $\phi(n)$ then he could break RSA by computing d as the multiplicative inverse of e modulo $\phi(n)$. That is, $d \equiv 1/e \pmod{\phi(n)}$. On the other hand, the knowledge of $\phi(n)$ can lead to an easy way of factoring n, since

$$
\begin{aligned}
p + q &= n - \phi(n) + 1, \\
(p - q)^2 &= (p + q)^2 - 4n, \\
p &= \frac{(p + q) + (p - q)}{2}, \\
q &= \frac{(p + q) - (p - q)}{2}.
\end{aligned}
$$

In other words, computing $\phi(n)$ is no easier than factoring n.

Example 4.3.1. Let

$$N = 74153950911911911.$$

Suppose the cryptanalyst knows by guessing, interception or whatever that

$$\phi(N) = 74153950339832712.$$

Then

$$
\begin{aligned}
A &= n - \phi(n) + 1 \\
&= 74153950911911911 - 74153950339832712 + 1 \\
&= 572079200
\end{aligned}
$$

Thus

$$p^2 - 572079200p + 74153950911911911 = 0.$$

Solving this equation gives the two roots

$$\{p, q\} = \{198491317, \ 373587883\},$$

and hence the complete prime factorization of n

$$
\begin{aligned}
n &= 74153950911911911 \\
&= 198491317 \cdot 373587883.
\end{aligned}
$$

Theorem 4.3.2. The RSA encryption is breakable in polynomial-time if the cryptanalyst knows $\phi(n)$. That is,

$$\phi(n) \quad \overset{\mathcal{P}}{\Longrightarrow} \quad \mathrm{RSA}(M). \tag{4.18}$$

Proof. If $\phi(n)$ is known, then $d \equiv 1/e \pmod{\phi(n)}$, hence recovers M from C: $M \equiv C^d \pmod{n}$. □

Theorem 4.3.3. If the prime factorization of the RSA modulus n is known, then d can be calculated in polynomial-time. That is,

$$\mathrm{IFP}(n) \overset{\mathcal{P}}{\Longrightarrow} \{d\}. \tag{4.19}$$

Proof. If $n = pq$ is known, then

$$d \equiv 1/e \pmod{(p-1)(q-1)}$$

since e is also known. As the modular inverse $1/e$ can be done in polynomial-time by the extended Euclid's algorithm, d can be calculated in polynomial-time. □

The next theorem shows that given the RSA private exponent d, the prime factorization of N can be done in polynomial time:

$$\{d\} \overset{\mathcal{P}}{\Longrightarrow} \mathrm{IFP}(n). \tag{4.20}$$

Theorem 4.3.4 (Coron and May [60]). Let $n = pq$ with p and q prime numbers. Let also e and d be the public and private exponent, respectively, satisfying $ed \equiv 1 \pmod{\phi(n)}$.

(1) If p and q are with the same bit size and $1 < ed \leq n^{3/2}$, then given (n, e, d), the prime factorization of n can be computed deterministically in time $\mathcal{O}((\log n)^2)$.

(2) If p and q are with the same bit size and $1 < ed \leq n^2$, then given (n, e, d), the prime factorization of N can be computed deterministically in time $\mathcal{O}((\log n)^9)$.

(3) Let β and $0 < \delta \leq 1/2$ be real numbers such that $2\beta\delta(1 - \delta) \leq 1$. Let $n = pq$ with p and q primes such that $p < n^\delta$ and $q < 2N^{1-\delta}$. Let $1 < ed \leq n^\beta$. Then given (n, e, d), the prime factorization of N can be computed deterministically in time $\mathcal{O}((\log n)^9)$.

Proof. The results follow by applying Coppersmith's technique [55] of finding small solutions to the univariable modular polynomial equations using lattice reduction [133]. For more details, see [60]. $\qquad\qquad\square$

Corollary 4.3.1. If d is known, then the prime factorization N can be found in deterministic polynomial-time. That is,

$$\{d\} \overset{\mathcal{P}}{\Longrightarrow} \mathrm{IFP}(n). \tag{4.21}$$

Combining Theorem 4.3.3 and Corollary 4.3.1, we have

Theorem 4.3.5 (The equivalence of $\mathrm{RSA}(d)$ and $\mathrm{IFP}(n)$). Computing the private exponent d by giving the prime factorization N and computing the prime factorization of N by giving the private exponent d are deterministic polynomial-time equivalent. That is,

$$\{d\} \overset{\mathcal{P}}{\Longleftrightarrow} \mathrm{IFP}(n). \tag{4.22}$$

It is evident that the most direct and straightforward cryptanalytic attacks on RSA is by factoring the modulus n. There are, however, many other possible cryptanalytic attacks on RSA, that do not rely on factoring. In what follows, we shall just discuss some of these attacks; more attacks can be found in Boneh [28] and Yan [270].

Short d attack. If the RSA private exponent d is chosen to be too small, e.g., $d < n^{0.25}$, then by *Weiner's Diophantine attack* [255], d can be efficiently recovered (in polynomial-time) from the public-key (e, n). That is,

$$\{e, n\} \xrightarrow[d < N^{0.25}]{\mathcal{P}} \{d\} \tag{4.23}$$

Lemma 4.3.1. Suppose that $\gcd(e, n) = \gcd(k, d) = 1$ and

$$\left| \frac{e}{n} - \frac{k}{d} \right| < \frac{1}{2d^2}.$$

Then k/d is one of the convergents of the continued fraction expansion of e/n.

Proof. Let $n = pq$ with p, q prime and $1 < p < 2q$. Let also the private exponent d is small, say, e.g.,

$$d < \frac{1}{3} \sqrt[4]{n}.$$

Then, as we shall show, d will be the denominator of a convergent to the continued fraction expansion of e/n.   □

Theorem 4.3.6 (Weiner). Let $n = pq$ with p and q primes such that

$$\begin{cases} q < p < 2q \\ d < \frac{1}{3} \sqrt[4]{n} \end{cases} \tag{4.24}$$

then given e with $ed \equiv 1 \pmod{\phi(n)}$, d can be efficiently calculated.

Proof. Since $ed \equiv 1 \pmod{\phi(n)}$,

$$ed - k\phi(n) = 1$$

for some $k \in \mathbb{Z}$. Therefore

$$\left| \frac{e}{\phi(n)} - \frac{k}{d} \right| = \frac{1}{d\phi(n)}.$$

Since $n = pq > q^2$, we have $q < \sqrt{N}$. Also since $\phi(n) = N - p - q + 1$,

$$0 < n - \phi(n) = p + q - 1 < 2q + q - 1 < 3q < 3\sqrt{n}.$$

Now,

$$\left| \frac{e}{n} - \frac{k}{d} \right| = \left| \frac{ed - kn}{dn} \right|$$

$$= \left| \frac{ed - kn + k\phi(n) - k\phi(n)}{dn} \right|$$

$$= \left| \frac{1 - k(n - \phi(n))}{dn} \right|$$

$$< \frac{3k\sqrt{n}}{dn}$$

$$= \frac{3k}{d\sqrt{n}}$$

$$< \frac{1}{2d^2}.$$

Thus, by Lemma 4.3.1, k/d must be one of convergents of the simple continued fraction e/n. Therefore, if $d < \frac{1}{3}\sqrt[4]{n}$, the d can be computed via the elementary task of computing a few convergent of e/n, which can be done in polynomial-time. $\square$

To defend Wiener's Diophantine attack, it is important that the private-key d should be large (nearly as many bits as the modulus n), or otherwise, by the properties of continued fractions, the private-key d can be found in time polynomial in the length of the modulus n, and hence decrypt RSA(M).

Example 4.3.2. Suppose that $n = 160523347$ and $e = 60728973$. Then the continued fraction expansion of e/n is as follows:

$$\frac{e}{n} = 0 + \cfrac{1}{2 + \cfrac{1}{1 + \cfrac{1}{1 + \cfrac{1}{1 + \cfrac{1}{4 + \cfrac{1}{12 + \cfrac{1}{102 + \cfrac{1}{1 + \cfrac{1}{1 + \cfrac{1}{2 + \cfrac{1}{3 + \cfrac{1}{2 + \cfrac{1}{2 + \cfrac{1}{36}}}}}}}}}}}}}$$

$$= [0, 2, 1, 1, 1, 4, 12, 102, 1, 1, 2, 3, 2, 2, 36]$$

and the convergents of the continued fraction are as follows:

$$\Big[0, \frac{1}{2}, \frac{1}{3}, \frac{2}{5}, \frac{3}{8}, \frac{14}{37}, \frac{171}{452}, \frac{17456}{46141}, \frac{17627}{46593}, \frac{35083}{92734}, \frac{87793}{232061}, \frac{298462}{788917},$$

$$\frac{684717}{1809895}, \frac{1667896}{4408707}, \frac{60728973}{160523347}\Big].$$

If condition (4.24) is satisfied, then the unknown fraction k/d is a close approximation to the known fraction of e/N. Lemma 4.3.1 tells us that k/d must be one of the covergents of the continued fraction expansion of e/n. By just a few trials of computation, we find that $d = 37$.

Iterated encryption or fixed-point attack. Suppose e has order r in the multiplicative group modulo $\lambda(n)$. Then $e^r \equiv 1 \pmod{\lambda(N)}$, so $M^{e^r} \equiv M \pmod n$. This is just the rth iterate of the encryption of M. So To defend the fixed-point attack, we must ensure that r is large. We illustrate the fixed-point attack by selecting an encryption example from the original RSA paper (Page 124 of [201]), where

$$
\begin{aligned}
e &= 17, \\
n &= 2773, \\
C &= 2342.
\end{aligned}
$$

We construct the sequence $C^{e^k} \bmod n$ for $k = 1, 2, 3, \ldots$ as follows:

2365	1157	2018	985	1421	2101	1664	2047	1539	980
1310	1103	1893	1629	2608	218	1185	1039	602	513
772	744	720	2755	890	2160	2549	926	536	449
2667	2578	182	2278	248	454	1480	1393	2313	2637
2247	1688	1900	2342						

$$\qquad\qquad\quad \Uparrow \qquad \Downarrow$$
$$\qquad\qquad\quad M \qquad C$$

Indeed, 1900 is the plaintext of 2342, since one can easily verify that

$$1900^{17} \bmod 2773 = 2342.$$

Common modulus attack. Suppose that Bob sends Alice two ciphertexts C_1 and C_2 as follows:

$$C_1 \equiv M^{e_1} \pmod n,$$

$$C_2 \equiv M^{e_2} \pmod n,$$

where $\gcd(e_1, e_2) = 1$. Then as the following theorem shows, Eve can recover the plaintext M without factoring n or without using any of the trap-door information $\{d, p, q, \phi(n)\}$.

Theorem 4.3.7. Let $n = n_1 = n_2$, $M_1 = M_2$, $e_1 \neq e_2$ and $\gcd(e_1, e_2) = 1$ such that

$$C_1 \equiv M^{e_1} \pmod{n},$$
$$C_2 \equiv M^{e_2} \pmod{n}.$$

Then M can be recovered easily. That is,

$$\{[C_1, e_1, n], [C_2, e_2, n]\} \overset{\mathcal{P}}{\Longrightarrow} \{M\}. \tag{4.25}$$

Proof. Since $\gcd(e_1, e_2) = 1$, then $e_1 x + e_2 y = 1$ with $x, y \in \mathbb{Z}$, which can be done by the extended Euclid's algorithm (or the equivalent continued fraction algorithm) in polynomial-time. Thus,

$$
\begin{aligned}
C_1^x C_2^y &\equiv (M_1^{e_1})^x (M_2^{e_2})^y \\
&\equiv M^{e_1 x + e_2 y} \\
&\equiv M \pmod{n}.
\end{aligned}
$$

$\square$

Example 4.3.3. Let

$$
\begin{aligned}
e_1 &= 9007, \\
e_2 &= 65537, \\
M &= 19050321180920251905182209030519, \\
n &= 114381625757888867669235779976146612010218296721242362\_ \\
&\quad 562561842935706935245733897830597123563958705058989075\_ \\
&\quad 147599290026879543541.
\end{aligned}
$$

Then

$$
\begin{aligned}
C_1 &\equiv M^{e_1} \bmod n \\
&\equiv 104202250941196238413638382607974125774449084724929\\
&\quad 59\_1257433745889265297771717182413024642938078351979089\\
&\quad 9\_4534340746416137797212, \\
C_2 &\equiv M^{e_2} \bmod n \\
&\equiv 764527507291887001807199705175445747109447573179098\\
&\quad 96\_041340987488285573190280783480309084978021563396490\\
&\quad 75\_9750600519496071304348.
\end{aligned}
$$

Now we determine x and y in

$$9007x + 65537y = 1.$$

First, we get the continued fraction expansion of $9007/65537$ as follows:

$$9007/65537 = [0, 7, 3, 1, 1, 1, 1, 1, 2, 1, 1, 1, 1, 2, 7].$$

Then we get the convergents of the continued fraction of $9007/65537$ as follows:

$$\left[0, \frac{1}{7}, \frac{3}{22}, \frac{4}{29}, \frac{7}{51}, \frac{11}{80}, \frac{18}{131}, \frac{29}{211}, \frac{76}{553}, \frac{105}{764}, \frac{181}{1317}, \frac{286}{2081}, \frac{467}{3398}, \frac{1220}{8877}, \frac{9007}{65537}\right]$$

Thus,

$$\begin{cases} x = (-1)^{n-1}q_{n-1} = (-1)^{13}8877 = -8877, \\ y = (-1)^{n}p_{n-1} = (-1)^{14}1220 = 1220. \end{cases}$$

Therefore,

$$\begin{aligned}
M &\equiv C_1^x C_2^y \\
&\equiv 104202250941196238413638382607974125774449084724929 59\_ \\
&\quad 125743374588926529777171718241302464293807835197908 99\_ \\
&\quad 453434074641613779772 12^{-8877} \cdot \\
&\quad 764527507291887001807199705175445747109447573179098 96\_ \\
&\quad 041340987488285573190280783480309084978021563396490 75\_ \\
&\quad 9750600519496071304348^{1220} \\
&\equiv 19050321180920251905182209030519 \pmod{n}.
\end{aligned}$$

So, we can recover the plaintext M without factoring n and/or using any of the trap-door information $d, p, q, \phi(n)$.

This attack suggests that to defend RSA, one should never use common modulus in RSA encryption.

Short e attacks. We consider one more attack, in this section, on short e encryption for the related messages using the same modulus, due to Coppersmith, Franklin and Reiter [58].

Theorem 4.3.8. Let M_1 and M_2 are messages related to the following known relation:

$$M_2 = \alpha M_1 + \beta. \qquad (4.26)$$

Suppose the RSA M^3 encryption uses the same modulus n as follows:

$$\left. \begin{array}{l} C_1 \equiv M_1^3 \pmod{n}, \\ C_2 \equiv (\alpha M_1 + \beta)^3 \pmod{n}. \end{array} \right\} \qquad (4.27)$$

Then

$$M_1 \equiv \frac{\beta(C_2 + 2\alpha^3 C_1 - \beta^3)}{\alpha(C_2 - \alpha^3 C_1 + 2\beta^3)}$$

$$\equiv \frac{3\alpha^3\beta M_1^3 + 3\alpha^2\beta^2 M_1^2 + 3\alpha\beta^3 M_1}{3\alpha^3\beta M_1^2 + 3\alpha^2\beta^2 M_1 + 3\alpha\beta^3} \pmod{n}. \qquad (4.28)$$

Corollary 4.3.2. Let $M_1 = M$, $M_2 = M + 1$, and $\alpha = \beta = 1$, then (4.28) reduces to the following special case:

$$M \equiv \frac{(M+1)^3 + 2M^3 - 1}{(M+1)^3 - M3 + 2}$$

$$\equiv \frac{3M^3 + 3M^2 + 3M}{3M^2 + 3M + 3} \pmod{n}. \qquad (4.29)$$

Example 4.3.4. Give the relation $M_2 = \alpha M_1 + \beta$ and let

$\alpha = 3,$

$\beta = 5,$

$n = 77903022885101595423624756547055783624857676209739839\underline{4}$
1084402222135728725117099985850483876481319443405109\underline{3}2\underline{}
265136815168574119934775586854274094225644500087912723\underline{}
2585749337061853958340278434058208881085485078737,

$C_1 = 13205758404493740923120838932339899687881248694981155\underline{8}$
7242149830720913809890543081612779597338248650686875\underline{9}4\underline{}
2131398266220555437000745522936935039403511872032667\underline{4}0\underline{}
9110568061708806799784622122282312925753339240\underline{0}6,

$C_2 = 35655547692133100492426265117317729157279371476449120\underline{9}$
0997362096862084038123081043744658925329430451812652\underline{0}8\underline{}
1858712220905928591327874274888835176225741122966452\underline{9}9\underline{}
2998335410453929161733393892204730002674838955287.

Then

$$M_1 \equiv \frac{5(C_2 + 2 \cdot 3^3 \cdot C_1 - 5^3)}{3(C_2 - 3^3 \cdot C_1 + 2 \cdot 5^3)}$$

$$\equiv 200850013010709030023151804190001180500191721050113\underline{0}9\underline{}$$
$$19080015191909061801070\underline{5} \pmod{n},$$

$$M_2 \equiv 3M_1 + 5 \pmod{n},$$

$$\equiv 6024150039032127090069455412570003541500575163150339\underline{2}7\underline{}$$
$$57240045575727185403212\underline{0} \pmod{n}.$$

It is easy to verify that M_1 and M_2 are the correct plaintexts of C_1 and C_2, since

$$
\begin{aligned}
C_1 &\equiv M_1^3 \bmod n \\
&= 1320575840449374092312083893233989968788124869498115558\_ \\
&\quad 72421498307209138098905430816127795973382486506868687594\_ \\
&\quad 21313982662205554370007455229369350394035118720326674 0\_ \\
&\quad 91105680617088067997846221222823129257533392 4006, \\
C_2 &\equiv M_2^3 \bmod n \\
&= 3565554769213310049242626511731772915727937147644912 09\_ \\
&\quad 09973620968620840381230810437446589253294304518126520 8\_ \\
&\quad 18587122209059285913278742748888351762257411229664529 9\_ \\
&\quad 29983354104539291617333938922047300026748389552 87.
\end{aligned}
$$

as required. Note that to recover M_1 and M_2 from C_1 and C_2 in the present case, it is no need to factor n, the 211 digit modulus, and also no need to find d.

Problems for Section 4.3

Problem 4.3.1. Given the RSA decryption exponent d satisfying

$$ed \equiv 1 \ (\bmod \ \phi(n))$$

where $e \geq 3$ and $\phi(n) = (p - 1)(q - 1)$, show that n can be factored in polynomial-time.

Problem 4.3.2. Prove or disprove the conjecture that factoring the RSA modulo n is equivalent to breaking the RSA system.

Problem 4.3.3. Show that computing the eth root of C (the RSA cipher-text)

$$M \equiv \sqrt[e]{C} \ (\bmod \ n)$$

is as hard as factoring n.

Problem 4.3.4. Let

$$e_1 = 9007,$$

$$e_2 = 65537,$$

$$M = 1905032118092025190518220903519.$$

$$n = 114381625757888867669235779976146612010218296721242362\_$$
$$56256184293570693524573389783059712356395870558989075\_$$
$$147599290026879543541,$$

$$C_1 \equiv M^{e_1} \bmod n$$
$$\equiv 1042022509411962384136383826079741257744490847249295 9\_$$
$$1257433745889265297771717182413024642938078351979089 9\_$$
$$45343407464161377977212,$$

$$C_2 \equiv M^{e_2} \bmod n$$
$$\equiv 76452750729188700180719970517544574710944757317909896\_$$
$$04134098748828557319028078348030908497802156339649075\_$$
$$9750600519496071304348.$$

Find the plain-text M.

Problem 4.3.5. (Coppersmith) Let $n = pq$ be a β bit RSA modulo such that p has about $\beta/2$ bits. Show that if either $\beta/4$ least (or most) significant bits of p is given, then n can be efficiently factored, and hence, RSA can be efficiently broken.

Problem 4.3.6. (Boneh, et al) Show that if $d/4$ least significant bits of d is given, then d can be computed in time $\mathcal{O}(e \log e)$, where e is the RSA public exponent.

Problem 4.3.7. (Knuth Decoding Problem [119]) Let

$$
\begin{aligned}
n \ = \ & 7790302288510159542362475654705578362485767620973983941\,0844\_\\
& 0222213572872511709998585048387648131944340510932265136\,8151\_\\
& 6857411993477558685427409422564450008791272325857493370\,6185\_\\
& 39583402784340582088810854850787\,37,
\end{aligned}
$$

$$
\begin{aligned}
C_1 \ = \ & 6875028364370892898789953506044079907168981402585834430\,3553\_\\
& 5588237479271080090293049630566651268112334056274332612\,1428\_\\
& 2318720373118151963944261656899892436827122751237714587\,9737\_\\
& 2299204125753023665954875641382\,171,
\end{aligned}
$$

$$
\begin{aligned}
C_2 \ = \ & 7130139886169274645420466503586462247282166640137557785\,6722\_\\
& 3219797011593220849557864249703775331317377532696534879\,7392\_\\
& 0186888756782951903268163268881275006025182238844628661\,5758\_\\
& 360493162805668669968333451929\,4663.
\end{aligned}
$$

Find M_1 and M_2 such that $C_1 \equiv M_1^3 \pmod{n}$ $C_2 \equiv M_2^3 \pmod{n}$.

4.4 Rabin Cryptography

As can be seen from the previous sections, RSA uses M^e for encryption, with $e \geq 3$ (3 is the smallest possible public exponent), we might call RSA encryption M^e encryption. In 1979, Michael Rabin proposed a scheme based on M^2 encryption. rather than the M^e for $e \geq 3$ encryption used in RSA. A brief description of the Rabin system is as follows (see also Figure 4.5).

[1] Key generation: Let $n = pq$ with p, q odd primes satisfying

$$
p \equiv q \equiv 3 \pmod 4. \tag{4.30}
$$

[2] Encryption:

$$
C \equiv M^2 \pmod n. \tag{4.31}
$$

[3] Decryption: Use the Chinese Remainder Theorem to solve the system of congruences:

$$
\begin{cases}
M_p \equiv \sqrt{C} \pmod p \\
M_q \equiv \sqrt{C} \pmod q
\end{cases} \tag{4.32}
$$

to get the four solutions: $\{\pm M_p, \pm M_q\}$. The true plaintext M will be one of these four values.

[4] Cryptanalysis: A cryptanalyst who can factor n can compute the four square roots of C modulo n, and hence can recover M from C. Thus, breaking the Rabin system is equivalent to factoring n.

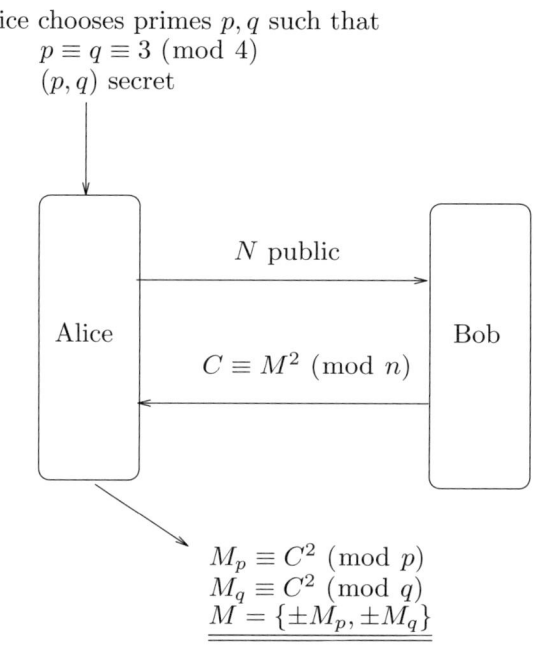

Alice chooses primes p, q such that
$$p \equiv q \equiv 3 \pmod{4}$$
(p, q) secret

N public

Alice

Bob

$C \equiv M^2 \pmod{n}$

$$M_p \equiv C^2 \pmod{p}$$
$$M_q \equiv C^2 \pmod{q}$$
$$M = \{\pm M_p, \pm M_q\}$$

Figure 4.5. Rabin cryptosystem

Unlike the RSA cryptosystem whose security was only conjectured to be equivalent to the intractability of IFP, the security of Rabin-Williams is proved to be equivalent to the intractability of IFP. First notice that there is a fast algorithm to compute the square roots modulo N if $n = pq$ is known.

Consider the following quadratic congruence

$$x^2 \equiv y \pmod{p} \tag{4.33}$$

there are essentially three cases for the prime p:

(1) $p \equiv 3 \pmod{4}$,

(2) $p \equiv 5 \pmod{8}$,

(3) $p \equiv 1 \pmod{8}$.

All three cases may be solved by the following process:

$$
\begin{cases}
\text{if } p \equiv 3 \ (\text{mod } 4), \quad x \equiv \pm y^{\frac{p+1}{4}} \ (\text{mod } p), \\[2mm]
\text{if } p \equiv 5 \ (\text{mod } 8), \quad
\begin{cases}
\text{if } y^{\frac{p+1}{4}} = 1, \quad x \equiv \pm y^{\frac{p+3}{8}} \ (\text{mod } p) \\[2mm]
\text{if } y^{\frac{p+1}{4}} \neq 1, \quad x \equiv \pm 2y(4y)^{\frac{p-5}{8}} \ (\text{mod } p).
\end{cases}
\end{cases}
\tag{4.34}
$$

Algorithm 4.4.1 (Computing Square Roots Modulo pq). Let $n = pq$ with p and q odd prime and $y \in \mathrm{QR}_n$. This algorithm will find all the four solutions in x to congruence $x^2 \equiv y \ (\text{mod } pq)$ in time $\mathcal{O}((\log p)^4)$.

[1] Use (4.34) to find a solution r to $x^2 \equiv y \ (\text{mod } p)$.

[2] Use (4.34) to find a solution s to $x^2 \equiv y \ (\text{mod } q)$.

[3] Use the Extended Euclid's algorithm to find integers c and d such that $cp + dq = 1$.

[4] Compute $x \equiv \pm(rdq \pm scp) \ (\text{mod } pq)$.

On the other hand, if there exists an algorithm to find the four solutions in x to $x^2 \equiv y \ (\text{mod } n)$, then there exists an algorithm to find the prime factorization of n. The following is the algorithm.

Algorithm 4.4.2 (Factoring via Square Roots). This algorithm seeks to find a factor of n by using an existing square root finding algorithm (namely, Algorithm 4.4.1).

[1] Choose at random an integer x such that $\gcd(x, n) = 1$, and compute $x^2 \equiv a \ (\text{mod } n)$.

[2] Use Algorithm 4.4.1 to find four solutions in x to $x^2 \equiv a \ (\text{mod } n)$.

[3] Choose one of the four solutions, say y such that $y \not\equiv \pm x \ (\text{mod } n)$, then compute $\gcd(x \pm y, n)$.

[4] If $\gcd(x \pm y, n)$ reveals p or q, then go to Step [5], or otherwise, go to Step [1].

[5] Exit.

Theorem 4.4.1. Let $N = pq$ with p, q odd prime. If there exists a polynomial-time algorithm A to factor $n = pq$, then there exists an algorithm B to find a solution to $x^2 \equiv y \ (\text{mod } n)$, for any $y \in \mathrm{QR}_N$.

Proof. If there exists an algorithm A to factor $n = pq$, then there exists an algorithm (in fact, Algorithm 4.4.1), which determines $x = \pm(rdq \pm scp) \ (\text{mod } pq)$, as defined in Algorithm 4.4.1, for $x^2 \equiv y \ (\text{mod } n)$. Clearly, Algorithm 4.4.1 runs in polynomial-time. $\square$

Theorem 4.4.2. Let $n = pq$ with p, q odd prime. If there exists a polynomial-time algorithm A to find a solution to $x^2 \equiv a \ (\text{mod } n)$, for any $a \in \mathrm{QR}_n$, then there exists a probabilistic polynomial time algorithm B to find a factor of n.

Proof. First note that for n composite, x and y integer, if $x^2 \equiv y^2 \pmod{n}$ but $x \not\equiv \pm y \pmod{n}$, then $\gcd(x+y, n)$ are proper factors of n. If there exists an algorithm A to find a solution to $x^2 \equiv a \pmod{n}$ for any $a \in \mathrm{QR}_n$, then there exists an algorithm (in fact, Algorithm 4.4.2), which uses algorithm A to find four solutions in x to $x^2 \equiv a \pmod{n}$ for a random x with $\gcd(x, n) = 1$. Select one of the solutions, say, $y \not\equiv \pm x \pmod{n}$, then by computing $\gcd(x \pm y, n)$, the probability of finding a factor of N will be $\geq 1/2$. If Algorithm 4.4.2 runs for k times and each time randomly chooses a different x, then the probability of not factoring n is $\leq 1/2^k$. □

So, finally, we have

Theorem 4.4.3. Factoring integers, computing the modular square roots, and breaking the Rabin cryptosystem are computationally (deterministic polynomial-time) equivalent. That is,

$$\mathrm{IFP}(n) \overset{\mathcal{P}}{\Longleftrightarrow} \mathrm{Rabin}(M). \tag{4.35}$$

Williams [259] proposed a modified version of the RSA cryptographic system, particularly the Rabin's M^2 system in order to make it suitable as a public-key encryption scheme (Rabin's original system was intended to be used as a digital signature scheme). A description of *Williams' M^2 encryption* is as follows (suppose Bob wishes to send Alice a ciphertext $C \equiv M^2 \pmod{n}$):

[1] Key generation: Let $n = pq$ with q and q primes such that

$$\begin{cases} p \equiv 3 \pmod{8} \\ q \equiv 7 \pmod{8}. \end{cases}$$

So, $n \equiv 5 \pmod{8}$ and $(n, 2)$ is used public-key. The private-key d is defined by

$$d = \frac{(p-1)(q-1)}{4} + 1.$$

[2] Encryption: Let $\mathcal{M}$ be plaintext space containing all possible plaintexts M such that

$$2(2M + 1) < n \text{ if the Jacobi symbol } \left(\frac{2M+1}{n}\right) = -1,$$

$$4(2M + 1) < n \text{ if the Jacobi symbol } \left(\frac{2M+1}{n}\right) = 1.$$

The first step in encryption is for all $M \in \mathcal{M}$, put

$$M' = E_1(M) = \begin{cases} 2(2M+1) & \text{if the Jacobi symbol } \left(\frac{2M+1}{n}\right) = -1, \\ 4(2M+1) & \text{if the Jacobi symbol } \left(\frac{2M+1}{n}\right) = 1. \end{cases}$$

The last step in encryption is just the same as Rabin's encryption:

$$C \equiv (M')^2 \ (\text{mod } n).$$

[3] Decryption: On the reverse order of the encryption, the first step in decryption is as follows:

$$C' = D_2(C) \equiv C^d \ (\text{mod } n)$$

and the last step in decryption is defined by:

$$M = D_1(C') = \begin{cases} \frac{\frac{M'}{4}-1}{2} & \text{if } M' \equiv 0 \ (\text{mod } 4) \\ \frac{\frac{N-M'}{4}-1}{2} & \text{if } M' \equiv 1 \ (\text{mod } 4) \\ \frac{\frac{M'}{2}-1}{2} & \text{if } M' \equiv 2 \ (\text{mod } 4) \\ \frac{\frac{N-M'}{2}-1}{2} & \text{if } M' \equiv 3 \ (\text{mod } 4). \end{cases}$$

The whole process of encryption and decryption is as follows:

$$M \xrightarrow{E_1} M' \xrightarrow{E_2} C \xrightarrow{D_2} M' \xrightarrow{D_1} M.$$

[4] Cryptanalysis: A cryptanalyst who can factor n can find d, and hence can recover M from C. Thus, breaking the Williams' system is equivalent to factoring n.

Theorem 4.4.4 (Correctness of Williams' M^2 encryption). Let $M \in \mathcal{M}$. Then
$$M = D_1(D_2(E_2(E_1(M)))).$$

Theorem 4.4.5 (Equivalence of Williams(M) and IFP(n)). Breaking Williams' M^2 encryption (i.e., finding M from C) is equivalent to factoring the modulus n. That is,

$$\text{IFP(n)} \xleftrightarrow{\mathcal{P}} \text{Williams}(M). \tag{4.36}$$

For the justification of the above two theorems, see Williams [259].

Just the same as Rabin's system, Williams' M^2 encryption is also provably secure, as breaking the Williams' M^2 mod n encryption is equivalent to factoring n, where the N is a special form of $N = pq$, with p, q primes and $p \equiv 3 \ (\text{mod } 8)$ and $q \equiv 7 \ (\text{mod } 8)$. Note that this special integer factorization problem is not the same as the general IFP problem, although there is no any known reason to believe this special factoring problem is any easier than the general factoring problem. But unlike Rabin's system, Williams' M^2 encryption can be easily generalized to the general M^e encryption with $e > 2$, as in RSA. Thus, Williams' M^2 encryption is not just a variant of Rabin system, but also a variant of the general RSA system. In fact, Williams' original paper [259] discussed the general case that

$$ed \equiv \frac{\frac{(p-1)(q-1)}{4} + 1}{2} \ (\mathrm{mod}\,\lambda(n)).$$

Williams' M^2 encryption improved Rabin's M^2 encryption by eliminating the $4:1$ ciphertext ambiguity problem in decryption without adding extra information for removing the ambiguity. Williams in [263] also proposed a M^3 *encryption* variant to Rabin but eliminated the $9:1$ ciphertext ambiguity problem. The encryption is also proved to be as hard as factoring, although it is again still not the general IFP problem, since $n = pq$ was chosen to be

$$p \equiv q \equiv 1 \ (\mathrm{mod}\ 3)$$

and

$$\frac{(p-1)(q-1)}{9} \equiv -1 \ (\mathrm{mod}\ 3).$$

Problems for Section 4.4

Problem 4.4.1. Show that breaking the Rabin encryption is equivalent to factoring the Rabin modulo n.

Problem 4.4.2. Give a method to eliminate the $4:1$ ciphertext ambiguity problem in deciphering Rabin's codes without adding extra information.

Problem 4.4.3. Show that breaking the Williams M^2 encryption is equivalent to factoring the Williams modulo n.

Problem 4.4.4. Generalize Williams' M^2 encryption to M^e ($e \geq 3$) encryption.

Problem 4.4.5. Let

$$
\begin{aligned}
n \ = \ & 21290246318258757547497882016271517497806703963277216278233 \\
& 38321538470570413250102890108976982548192582551350925260960 \\
& 0236998394024335907529, \\
C \ \equiv \ & M^2 \ (\mathrm{mod}\ n) \\
= \ & 51285205060243481188122109876540661122140906807437327290641 \\
& 60633920242479741450841196687149365272035106423411648279363 \\
& 93204288427165138923 4.
\end{aligned}
$$

Find the plain-text M.

4.5 Quadratic Residuosity Cryptography

The RSA cryptosystem discussed in the previous section is *deterministic* in the sense that under a fixed public-key, a particular plain-text M is always encrypted to the same cipher-text C. Some of the drawbacks of a deterministic scheme are:

[1] It is not secure for all probability distributions of the message space. For example, in RSA encryption, the messages 0 and 1 always get encrypted to themselves, and hence are easy to detect.

[2] It is easy to obtain some partial information of the secret key (p, q) from the public modulus n (assume that $n = pq$). For example, when the least-significant digit of n is 3, then it is easy to obtain the partial information that the least-significant digits of p and q are either 1 and 3 or 7 and 9, and indicated as follows:

$$183 = 3 \cdot 61 \qquad 253 = 11 \cdot 23$$
$$203 = 7 \cdot 29 \qquad 303 = 3 \cdot 101$$
$$213 = 3 \cdot 71 \qquad 323 = 17 \cdot 19.$$

[3] It is sometimes easy to compute partial information about the plain-text M from the cipher-text C. For example, given (C, e, n), the Jacobi symbol of M over n can be easily deduced from C:

$$\left(\frac{C}{n} \right) = \left(\frac{M^e}{n} \right) = \left(\frac{M}{n} \right)^e = \left(\frac{M}{n} \right).$$

[4] It is easy to detect when the same message is sent twice.

Probabilistic encryption, or randomized encryption, however, utilizes randomness to attain a strong level of security, namely, the *polynomial security* and *semantic security*, defined as follows:

Definition 4.5.1. A public-key encryption scheme is said to be *polynomially secure* if no passive adversary can, in expected polynomial time, select two plain-texts M_1 and M_2 and then correctly distinguish between encryptions of M_1 and M_2 with probability significantly greater that $1/2$.

Definition 4.5.2. A public-key encryption scheme is said to be *semantically secure* if, for all probability distributions over the message space, whatever a passive adversary can compute in expected polynomial time about the plain-text given the cipher-text, it can also be computed in expected polynomial time without the cipher-text.

Intuitively, a public-key encryption scheme is semantically secure if the cipher-text does not leak any partial information whatsoever about the plain-text that can be computed in expected polynomial time. That is, given (C, e, n), it should be intractable to recover any information about M. Clearly, a public-key encryption scheme is semantically secure if and only if it is polynomially secure.

In this section, we shall introduce a semantically secure cryptosystem based on the *quadratic residuosity problem*. Recall that an integer a is a quadratic residue modulo n, denoted by $a \in Q_n$, if $\gcd(a, n) = 1$ and there exists a solution x to the congruence $x^2 \equiv a \pmod{n}$, otherwise a is a quadratic non-residue modulo n, denoted by $a \in \overline{Q}_n$. The Quadratic Residuosity Problem may be stated as:

> Given positive integers a and n, decide whether or not $a \in Q_n$.

It is believed that solving QRP is equivalent to computing the prime factorization of n, so it is computationally infeasible. If n is prime then

$$a \in Q_n \iff \left(\frac{a}{n} \right) = 1, \tag{4.37}$$

and if n is composite, then

$$a \in Q_n \implies \left(\frac{a}{n} \right) = 1, \tag{4.38}$$

but

$$a \in Q_n \;\not\!\!\impliedby\; \left(\frac{a}{n} \right) = 1, \tag{4.39}$$

however

$$a \in \overline{Q}_n \impliedby \left(\frac{a}{n} \right) = -1. \tag{4.40}$$

Let $J_n = \{a \in (\mathbb{Z}/n\mathbb{Z})^* : \left(\frac{a}{n} \right) = 1\}$, then $\tilde{Q}_n = J_n - Q_n$. Thus, $\tilde{Q}_n$ is the set of all pseudosquares modulo n; it contains those elements of J_n that do not belong to Q_n. Readers may wish to compare this result to Fermat's little theorem, namely (assuming $\gcd(a, n) = 1$),

$$n \text{ is prime} \implies a^{n-1} \equiv 1 \pmod{n}, \tag{4.41}$$

but

$$n \text{ is prime} \;\not\!\!\impliedby\; a^{n-1} \equiv 1 \pmod{n}, \tag{4.42}$$

however

$$n \text{ is composite} \impliedby a^{n-1} \not\equiv 1 \pmod{n}. \tag{4.43}$$

The Quadratic Residuosity Problem can then be further restricted to:

> Given a composite n and an integer $a \in J_n$, decide whether or not $a \in Q_n$.

For example, when $n = 21$, we have $J_{21} = \{1, 4, 5, 16, 17, 20\}$ and $Q_{21} = \{1, 4, 16\}$, thus $\tilde{Q}_{21} = \{5, 17, 20\}$. So, the QRP problem for $n = 21$ is actually to distinguish squares $\{1, 4, 16\}$ from pseudosquares $\{5, 17, 20\}$. The only method we know for distinguishing squares from pseudosquares is to factor n; since integer factorization is computationally infeasible, the QRP problem is computationally infeasible. In what follows, we shall present a cryptosystem whose security is based on the infeasibility of the Quadratic Residuosity Problem; it was first proposed by Goldwasser and Micali in 1984, under the term *probabilistic encryption*.

Algorithm 4.5.1 (Quadratic residuosity based cryptography). This algorithm uses the randomized method to encrypt messages and is based on the quadratic residuosity problem (QRP). The algorithm divides into three parts: key generation, message encryption and decryption.

[1] Key generation: Both Alice and Bob should do the following to generate their public and secret keys:

   [a] Select two large distinct primes p and q, each with roughly the same size, say, each with β bits.

   [b] Compute $n = pq$.
   Select a $y \in \mathbb{Z}/n\mathbb{Z}$, such that $y \in \overline{Q}_n$ and $\left(\dfrac{y}{n}\right) = 1$. ($y$ is thus a pseudosquare modulo n).

   [c] Make (n, y) public, but keep (p, q) secret.

[2] Encryption: To send a message to Alice, Bob should do the following:
   [a] Obtain Alice's public-key (n, y).

   [c] Represent the message m as a binary string $m = m_1 m_2 \cdots m_k$ of length k.

   [d] For i from 1 to k do
   [d-1] Choose at random an $x \in (\mathbb{Z}/n\mathbb{Z})^*$ and call it x_i.

   [d-2] Compute c_i:

$$
c_i = \begin{cases} x_i^2 \bmod n, & \text{if } m_i = 0, \quad \text{(r.s.)} \\ yx_i^2 \bmod n, & \text{if } m_i = 1, \quad \text{(r.p.s.)}, \end{cases} \qquad (4.44)
$$

   where r.s. and r.p.s. represent random square and random pseudosquare, respectively.
   Send the k-tuple $c = (c_1, c_2, \ldots, c_k)$ to Alice. (Note first that each c_i is an integer with $1 \leq c_i < n$. Note also that since n is a 2β-bit integer, it is clear that the cipher-text c is a much longer string than the original plain-text m.)

[3] Decryption: To decrypt Bob's message, Alice should do the following:
  [a] For i from 1 to k do

  [a-1] Evaluate the Legendre symbol:

$$e_i' = \left(\frac{c_i}{p}\right).$$

  [a-2] Compute m_i:

$$m_i = \begin{cases} 0, & \text{if } e_i' = 1 \\ 1, & \text{if otherwise.} \end{cases} \tag{4.45}$$

  That is, $m_i = 0$ if $c_i \in Q_n$, otherwise, $m_i = 1$.

  Finally, get the decrypted message $m = m_1 m_2 \cdots m_k$.

Remark 4.5.1. The above encryption scheme has the following interesting features:

1) The encryption is random in the sense that the same bit is transformed into different strings depending on the choice of the random number x. For this reason, it is called *probabilistic* (or *randomized*) encryption.
2) Each bit is encrypted as an integer modulo n, and hence is transformed into a 2β-bit string.
3) It is semantically secure against any threat from a polynomially bounded attacker, provided that the QRP is hard.

Example 4.5.1. In what follows we shall give an example of how Bob can send the message "HELP ME" to Alice using the above cryptographic method. We use the binary equivalents of letters as defined in Table 4.1. Now both Alice and Bob proceed as follows:

[1] Key Generation:
  − Alice chooses $(n, y) = (21, \ 17)$ as a public-key, where $n = 21 = 3 \cdot 7$ is a composite, and $y = 17 \in \tilde{Q}_{21}$ (since $17 \in J_{21}$ but $17 \notin Q_{21}$), so that Bob can use the public-key to encrypt his message and send it to Alice.
  − Alice keeps the prime factorization $(3, 7)$ of 21 as a secret; since $(3, 7)$ will be used as a private decryption key. (Of course, here we just show an example; in practice, the prime factors p and q should be at last 100 digits.)

Letter	Binary Code	Letter	Binary Code	Letter	Binary Code
A	00000	B	00001	C	00010
D	00011	E	00100	F	00101
G	00110	H	00111	I	01000
J	01001	K	01010	L	01011
J	01001	K	01010	L	01011
M	01100	N	01101	O	01110
P	01111	Q	10000	R	10001
S	10010	T	10011	U	10100
V	10101	W	10110	X	10111
Y	11000	Z	11001	␣	11010

Table 4.1. The binary equivalents of letters

[2] Encryption:
  - Bob converts his plain-text HELP ME to the binary stream $M = m_1 m_2 \cdots m_{35}$:

$$00111\ 00100\ 01011\ 01111\ 11010\ 01100\ 00100.$$

(To save space, we only consider how to encrypt and decrypt $m_2 = 0$ and $m_3 = 1$; readers are suggested to encrypt and decrypt the whole binary stream).

  - Bob randomly chooses integers $x_i \in (\mathbb{Z}/21\mathbb{Z})^*$. Suppose he chooses $x_2 = 10$ and $x_3 = 19$ which are elements of $(\mathbb{Z}/21\mathbb{Z})^*$.

  - Bob computes the encrypted message $C = c_1 c_2 \cdots c_k$ from the plain-text $M = m_1 m_2 \cdots m_k$ using Equation (4.44). To get, for example, c_2 and c_3, Bob performs:

$$c_2 = x_2^2 \bmod 21 = 10^2 \bmod 21 = 16, \qquad \text{since } m_2 = 0,$$
$$c_3 = y \cdot x_3^2 \bmod 21 = 17 \cdot 19^2 \bmod 21 = 5, \quad \text{since } m_3 = 1.$$

(Note that each c_i is an integer reduced to 21, i.e., m_i is a bit, but its corresponding c_i is not a bit but an integer, which is a string of bits, determined by Table 4.1.)

  - Bob then sends c_2 and c_3 along with all other c_i's to Alice.

[3] Decryption: To decrypt Bob's message, Alice evaluates the Legendre symbols $\left(\dfrac{c_i}{p}\right)$ and $\left(\dfrac{c_i}{q}\right)$. Since Alice knows the prime factorization (p, q)

of n, it should be easy for her to evaluate these Legendre symbols. For example, for c_2 and c_3, Alice first evaluates the Legendre symbols $\left(\dfrac{c_i}{p}\right)$:

$$e_2' = \left(\frac{c_2}{p}\right) = \left(\frac{16}{3}\right) = \left(\frac{4^2}{3}\right) = 1,$$

$$e_3' = \left(\frac{c_3}{p}\right) = \left(\frac{5}{3}\right) = \left(\frac{2}{3}\right) = -1.$$

then she gets

$$m_2 = 0, \quad \text{since } e_2' = 1,$$

$$m_3 = 1, \quad \text{since } e_3' = -1.$$

Remark 4.5.2. The scheme introduced above is a good extension of the public-key idea, but encrypts messages bit by bit. It is completely secure with respect to semantic security as well as bit security[3]. However, a major disadvantage of the scheme is the message expansion by a factor of $\log n$ bit. To improve the efficiency of the scheme, Blum and Goldwasser proposed in 1984 another randomized encryption scheme, in which the cipher-text is only longer than the plain-text by a constant number of bits; this scheme is comparable to the RSA scheme, both in terms of speed and message expansion.

Exercise 4.5.1. RSA encryption scheme is deterministic and not semantically secure, but it can be made semantically secure by adding randomness to the encryption process. Develop an RSA based probabilistic (randomized) encryption scheme that is semantically secure.

Several other cryptographic schemes, including digital signature schemes and authentication encryption schemes are based on the quadratic residuosity problem (QRP).

Problems for Section 4.5

Problem 4.5.1. The number a is a quadratic residue modulo n if $\gcd(a, n) = 1$ and there is a solution to the congruence $x^2 \equiv a \pmod{n}$. If p is an odd prime, then

$$a^{(p-1)/2} \equiv \left(\frac{a}{p}\right) \pmod{p}$$

where $\left(\dfrac{a}{p}\right)$ is the Legendre symbol. Based on the above facts, design an efficient algorithm that takes $\mathcal{O}((\log p)^3)$ bit operations to decide if a is a quadratic residue modulo p.

---

[3] Bit security is a special case of semantic security. Informally, bit security is concerned with not only that the whole message is not recoverable but also that individual bits of the message are not recoverable. The main drawback of the scheme is that the encrypted message is much longer than its original plain-text.

Problem 4.5.2. Show that the Jacobi symbol can be computed in $\mathcal{O}((\log p)^2)$ bit operations using Gauss' Quadratic Reciprocity Law.

Problem 4.5.3. Show that if the integer factorization problem can be solved in polynomial-time, then the quadratic residuosity problem can be solved in polynomial-time.

Problem 4.5.4. Use the binary-letter table in Exercise 4.5.1 to code the plain-text "HELP ME TO GO OUT THE DARK PLACE", then follow the steps in Algorithm 4.5.1 to encrypt the plain-text and decrypt the corresponding cipher-text. You need of course to choose the suitable values for p, q and y.

4.6 Discrete Logarithm Cryptography

The Diffie-Hellman-Merkle scheme, the first public-key cryptographic scheme, is based on the intractable discrete logarithm problem, which can be described as follows:

$$\text{Input}: \quad a, b, n \in \mathbb{Z}^+$$

$$\text{Output}: \quad x \in \mathbb{Z}^+ \text{ with } a^x \equiv b \pmod{n}$$

$$\text{if such a } x \text{ exists}$$

The Diffie-Hellman-Merkle scheme has found widespread use in practical cryptosystems, as for example in the optional security features of the NFS file system of SunOS operating system. In this section, we shall introduce some discrete logarithm based cryptosystems.

The Diffie-Hellman-Merkle Key-Exchange Protocol. Diffie and Hellman [70] in 1976 proposed for the first time a public-key cryptographic scheme based on the difficult discrete logarithm problem. Their scheme was not a public-key cryptographic system (first proposed in [70]), but rather a public-key distribution system as proposed by Merkle [146]. Such a public-key distribution scheme does not send secret messages directly, but rather allows the two parties to agree on a common private-key over public networks to be used later in exchanging messages through conventional cryptography. Thus, the Diffie-Hellman-Merkle scheme has the nice property that a very fast scheme such as DES or AES can be used for actual encryption, yet it still enjoys one of the main advantages of public-key cryptography. The Diffie-Hellman-Merkle key-exchange protocol works in the following way (see also Figure 4.6):

[1] A prime q and a generator g are made public (assume all users have agreed upon a finite group over a fixed finite field $\mathbb{F}_q$),

[2] Alice chooses a random number $a \in \{1, 2, \ldots, q-1\}$ and sends $g^a \bmod q$ to Bob,

[3] Bob chooses a random number $b \in \{1, 2, \ldots, q-1\}$ and sends $g^b \bmod q$ to Alice,

[4] Alice and Bob both compute $g^{ab} \bmod q$ and use this as a private key for future communications.

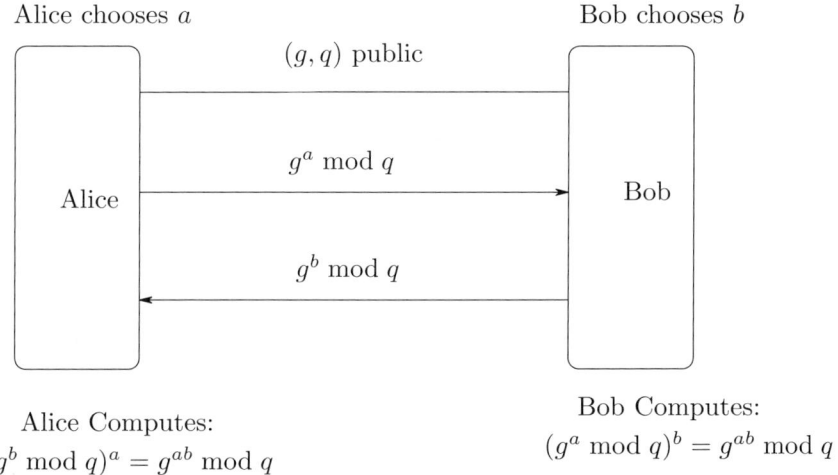

Figure 4.6. The Diffie-Hellman-Merkel key-exchange scheme

Clearly, an eavesdropper has g, q, $g^a \bmod q$ and $g^b \bmod q$, so if he can take discrete logarithms, he can calculate $g^{ab} \bmod q$ and understand the communications. That is, if the eavesdropper can use his knowledge of g, q, $g^a \bmod q$ and $g^b \bmod q$ to recover the integer a, then he can easily break the Diffie-Hellman-Merkle system. So, the security of the Diffie-Hellman-Merkle system is based on the following assumption:

Diffie-Hellman-Merkle Assumption: It is computationally infeasible to compute g^{ab} from g^a and g^b.

In theory, there could be a way to use knowledge of g^a and g^b to find g^{ab}. But at present we simply cannot imagine a way to go from g^a and g^b to g^{ab} without essentially solving the discrete logarithm problem.

Example 4.6.1. The following example [139], shows how the Diffie-Hellman-Merkle scheme works in a real situation:

[1] Let $q := (7^{149} - 1)/6$ and $p = 2 \cdot 739 \cdot q + 1$. (It can be shown that both p and q are primes.)

[2] Alice chooses a random number residue x modulo p, computes 7^x (mod p), and sends the result to Bob, keeping x secret.

[3] Bob receives

$7^x = 127402180119973946824269244334322849749382042586931621 65\_$
$45577352903229146790959986818609788130465951664554581442\_$
$80588076766033781.$

[4] Bob chooses a random number residue y modulo p, computes 7^y (mod p), and sends the result to Alice, keeping y secret.

[5] Alice receives

$7^y = 180162285287453102444782834836799895015967046695346697 31\_$
$30251217340599537720584759581769106253806921016518486623\_$
$62137934026803049.$

[6] Now both Alice and Bob can compute the private key 7^{xy} (mod p).

McCurley offered a prize of \$100 in 1989 to the first person to find the private key constructed from the above communication.

Remark 4.6.1. McCurley's 129-digit discrete logarithm challenge was actually solved on 25 January 1998 using the NFS method, by two German computer scientists, Damian Weber at the Institut für Techno -und Wirtschaftsmathematik in Kaiserslautern and Thomas F. Denny at the Debis IT Security Services in Bonn.

As we have already mentioned earlier the Diffie-Hellman-Merkle scheme is not intended to be used for actual secure communications, but for key-exchanges. There are, however, several other cryptosystems based on discrete logarithms, that can be used for secure message transmissions.

The ElGamal Cryptosystem for Secure Communications. In 1985, ElGamal proposed a public-key cryptosystem based on discrete logarithms:

[1] A prime q and a generator $g \in \mathbb{F}_q^*$ are made public.

[2] Alice chooses a private integer $a = a_A \in \{1, 2, \ldots, q-1\}$. This a is the private decryption key. The public encryption key is $g^a \in \mathbb{F}_q$.

[3] Suppose now Bob wishes to send a message to Alice. He chooses a random number $b \in \{1, 2, \ldots, q-1\}$ and sends Alice the following pair of elements of $\mathbb{F}_q$:
$$(g^b, \ Mg^{ab})$$
where M is the message.

[4] Since Alice knows the private decryption key a, she can recover M from this pair by computing $g^{ab} \pmod{q}$ and dividing this result into the second element, i.e., Mg^{ab}.

Remark 4.6.2. Someone who can solve the discrete logarithm problem in $\mathbb{F}_q$ breaks the cryptosystem by finding the secret decryption key a from the public encryption key g^a. In theory, there could be a way to use knowledge of g^a and g^b to find g^{ab} and hence break the cipher without solving the discrete logarithm problem. But as we have already seen in the Diffie-Hellman scheme, there is no known way to go from g^a and g^b to g^{ab} without essentially solving the discrete logarithm problem. So, the ElGamal cryptosystem is equivalent to the Diffie–Hellman key-exchange system.

The Massey–Omura Cryptosystem for Message Transmissions. This is another popular cryptosystem based on discrete logarithms; it works in the following way:

[1] All the users have agreed upon a finite group over a fixed finite field $\mathbb{F}_q$ with q a prime power.

[2] Each user secretly selects a random integer e between 0 and $q-1$ such that $\gcd(e, q-1) = 1$, and computes $d = e^{-1} \bmod (q-1)$ by using the extended Euclidean algorithm.

[3] Now suppose that user Alice wishes to send a secure message M to user Bob, then they follow the following procedure:

[a] Alice first sends M^{e_A} to Bob,

[b] On receiving Alice's message, Bob sends $M^{e_A e_B}$ back to Alice (note that at this point, Bob cannot read Alice's message M),

[c] Alice sends $M^{e_A e_B d_A} = M^{e_B}$ to Bob,

[d] Bob then computes $M^{d_B e_B} = M$, and hence recovers Alice's original message M.

Problems for Section 4.6

Problem 4.6.1. Let the DHM parameters be as follows:

$$p = 1000000000000000000000000000000000000000000000000000000000\_ \\ 00000000000000002047062703855328380597445351669742748036 08\_ \\ 39434012345969579867459152659137268522951065284733970579 7\_ \\ 62207550506983104348665168 2889,$$

$$13^x \equiv 10851945926748930321536897787511601536291411551215963735 7\_ \\ 97413754705002845778243766666788726776122805935695232661 4\_ \\ 81257320374720986213610649202854763331054158130244119857 3\_ \\ 77415713708744163529915144626 \pmod{p},$$

$$13^y \equiv 52200208400156523080484387248076760362198322255017014267 2\_ \\ 56873745866707749922777188091986977849828727835848382945 9\_ \\ 48956547764873325699997272322775368657123305830747697800 4\_ \\ 17855036551198719274264122371 \pmod{p}.$$

Find $13^{xy} \bmod p$.

Problem 4.6.2. In ElGamal cryptosystem, Alice makes (p, g, g^a) public with p prime p:

$$p = 1000000000000000000000000000000000000000000000000000000000\_ \\ 00000000000000002047062703855328380597445351669742748046 08\_ \\ 39434012345969579867459152659137268522951065284733970579 7\_ \\ 62207550506983104348665168 3281,$$

$$g = 137,$$

$$g^a \equiv 15219266397668101959283316151426320683674451858111063457 6\_ \\ 76905061579556925679355099442856564910069438554961438873 5\_ \\ 92866195042219679451267622593641925378022537537252639984 3\_ \\ 53500071774531090027331523676.$$

where $a \in \{1, 2, \ldots, p\}$ must be kept as a secret. Now Bob can send Alice an encrypted message $C = (g^b, Mg^{ab})$ to Alice by using her public-key information, where

$$
\begin{aligned}
g^b \equiv\ & 5954767560145832230236560413372022069605274694047335504600 \\
& 4974413791437414218363404323065365907081646746246663690438 \\
& 4382001528769925211730081006654249356412826389882146691842 \\
& 21777907261184240637405125 9,\\
Mg^{ab} \equiv\ & 4958786188281511383043041844766490753023726445360329447984 \\
& 9527736721533557707864314686330644624599660560087834147651 \\
& 1290381062014910855601264849526683408833232637420655255354 \\
& 9698164286521681700295976 0.
\end{aligned}
$$

Decode C. That is, recover M from C.

Problem 4.6.3. The Digital Signature Algorithm (DSA) [244] is a variant of ElGamal cryptosystem based on the intractability of the DLP program.

(1) Give a complete description of DSA.

(2) Write an assay on the cryptanalytic attacks on DSA.

4.7 Elliptic Curve Cryptography

Brief History of Elliptic Curve Cryptography. As can be seen, elliptic curve is quite ubiquitous in mathematics and computing, for example, we have seen that elliptic curves have novel applications to primality testing in Section 2.6 and to integer factorization in Section 3.3. (The basic ideas and concepts of elliptic curves can be found in Section 1.9.) However, the applications of elliptic curves to cryptography were not found until the following two seminal papers were published:

[1] Victor Miller, "Uses of Elliptic Curves in Cryptography", *Lecture Notes in Computer Science* **218**, Springer, 1986, pp 417–426.

[2] Neal Koblitz, "Elliptic Curve Cryptography", *Mathematics of Computation*, **48** (1987), pp 203–209.

Since then, elliptic curves have been studied extensively for the purpose of cryptography, and many practically more secure encryption and digital signature schemes have been developed based on elliptic curves. Now elliptic curve cryptography (ECC) is a standard term in the field and there is a text by

Menezes in 1993 that is solely devoted to elliptic curve cryptography. There is even a computer company in Canada, called Certicom, which is a leading provider of cryptographic technology based on elliptic curves. Now we move on to the discussion of the basic ideas and computational methods of elliptic curve cryptography.

Precomputations of Elliptic Curve Cryptography. To implement elliptic curve cryptography, we need to do the following precomputations:

[1] Embed Messages on Elliptic Curves: Our aim here is to do cryptography with elliptic curve groups in place of $\mathbb{F}_q$. More specifically, we wish to embed plain-text messages as points on an elliptic curve defined over a finite field $\mathbb{F}_q$, with $q = p^r$ and $p \in$ Primes. Let our message units m be integers $0 \leq m \leq M$, let also κ be a large enough integer for us to be satisfied with an error probability of $2^{-\kappa}$ when we attempt to embed a plain-text message m. In practice, $30 \leq \kappa \leq 50$. Now let us take $\kappa = 30$ and an elliptic curve $E : y^2 = x^3 + ax + b$ over $\mathbb{F}_q$. Given a message number m, we compute a set of values for x:

$$x = \{m\kappa + j, \ j = 0, 1, 2, \ldots\} = \{30m, \ 30m + 1, \ 30m + 2, \ \cdots\} \quad (4.46)$$

until we find $x^3 + ax + b$ is a square modulo p, giving us a point $(x, \sqrt{x^3 + ax + b})$ on E. To convert a point (x, y) on E back to a message number m, we just compute $m = \lfloor x/30 \rfloor$. Since $x^3 + ax + b$ is a square for approximately 50% of all x, there is only about a $2^{-\kappa}$ probability that this method will fail to produce a point on E over $\mathbb{F}_q$. In what follows, we shall give a simple example of how to embed a message number by a point on an elliptic curve. Let E be $y^2 = x^3 + 3x$, $m = 2174$ and $p = 4177$ (in practice, we select $p > 30m$). Then we calculate $x = \{30 \cdot 2174 + j, \ j = 0, 1, 2, \ldots\}$ until $x^3 + 3x$ is a square modulo 4177. We find that when $j = 15$:

$$
\begin{aligned}
x &= 30 \cdot 2174 + 15 \\[6pt]
&= 65235, \\[6pt]
x^3 + 3x &= (30 \cdot 2174 + 15)^3 + 3(30 \cdot 2174 + 15) \\[6pt]
&= 277614407048580 \\[6pt]
&\equiv 1444 \bmod 4177 \\[6pt]
&\equiv 38^2.
\end{aligned}
$$

So we get the message point for $m = 2174$:

$$(x, \ \sqrt{x^3 + ax + b}) = (65235, 38).$$

To convert the message point $(65235, 38)$ on E back to its original message number m, we just compute

$$m = \lfloor 65235/30 \rfloor = \lfloor 2174.5 \rfloor = 2174.$$

[2] Multiply Points on Elliptic Curves over $\mathbb{F}_q$: We have discussed the calculation of $kP \in E$ over $\mathbb{Z}/n\mathbb{Z}$. In elliptic curve public-key cryptography, we are now interested in the calculation of $kP \in E$ over $\mathbb{F}_q$, which can be done in $\mathcal{O}(\log k (\log q)^3)$ bit operations by the *repeated doubling method*. If we happen to know N, the number of points on our elliptic curve E and if $k > N$, then the coordinates of kP on E can be computed in $\mathcal{O}((\log q)^4)$ bit operations; recall that the number N of points on E satisfies $N \le q + 1 + 2\sqrt{q} = \mathcal{O}(q)$ and can be computed by René Schoof's algorithm [217] in $\mathcal{O}((\log q)^8)$ bit operations.

[3] Compute Elliptic Curve Discrete Logarithms: Let E be an elliptic curve over $\mathbb{F}_q$, and B a point on E. Then the *discrete logarithm* on E is the problem, given a point $P \in E$, find an integer $x \in \mathbb{Z}$ such that $xB = P$ if such an integer x exists. It is likely that the discrete logarithm problem on elliptic curves over $\mathbb{F}_q$ is more intractable than the discrete logarithm problem in $\mathbb{F}_q$. It is this feature that makes cryptographic systems based on elliptic curves even more secure than that based on the discrete logarithm problem. In the rest of this section, we shall discuss elliptic curve analogues of some important public-key cryptosystems.

Elliptic Curve Analogues of Cryptosystems

We are now in a position to discuss some elliptic curve cryptosystems. More specifically, we shall present elliptic curve analogues of four widely used public-key cryptosystems, namely the Diffie–Hellman key-exchange system, the Massey–Omura, the ElGamal, and the RSA public-key cryptosystems.

(1) EC Analogue of the Diffie–Hellman Key Exchange System:

[1] Alice and Bob publicly choose a finite field $\mathbb{F}_q$ with $q = p^r$ and $p \in$ Primes, an elliptic curve E over $\mathbb{F}_q$, and a random *base* point $P \in E$ such that P generates a large subgroup of E, preferably of the same size as that of E itself. All of this is public information.

[2] To agree on a secret key, Alice and Bob choose two secret random integers a and b. Alice computes $aP \in E$ and sends aP to Bob; Bob computes $bP \in E$ and sends bP to Alice. Both aP and bP are, of course, public but a and b are not.

[3] Now both Alice and Bob compute the secret key $abP \in E$, and use it for further secure communications.

There is no known fast way to compute abP if one only knows P, aP and bP – this is the discrete logarithm problem on E.

(2) Analogue of the Massey–Omura Cryptosystem:

[1] Alice and Bob publicly choose an elliptic curve E over $\mathbb{F}_q$ with q large, and we suppose also that the number of points (denoted by N) is publicly known.

[2] Alice chooses a secret pair of numbers $(e_A,\ d_A)$ such that $d_A e_A \equiv 1 \pmod{N}$. Similarly, Bob chooses $(e_B,\ d_B)$.

[3] If Alice wants to send a secret message-point $P \in E$ to Bob, the procedure is as follows:

    [a] Alice sends $e_A P$ to Bob,

    [b] Bob sends $e_B e_A P$ to Alice,

    [c] Alice sends $d_A e_B e_A P = e_B P$ to Bob,

    [d] Bob computes $d_B e_B P = P$ and hence recovers the original message number.

Note that an eavesdropper would know $e_A P$, $e_B e_A P$, and $e_B P$. So if he could solve the discrete logarithm problem on E, he could determine e_B from the first two points and then compute $d_B = e_B^{-1} \bmod N$ and hence get $P = d_B(e_B P)$.

(3) Analogue of the ElGamal Cryptosystem:

[1] Alice and Bob publicly choose an elliptic curve E over $\mathbb{F}_q$ with $q = p^r$ and $p \in \text{Primes}$, and a random *base* point $P \in E$.

[2] Alice chooses a random integer r_a and computes $r_a P$; Bob also chooses a random integer r_b and computes $r_b P$.

[3] To send a message-point M to Bob, Alice chooses a random integer k and sends the pair of points $(kP,\ M + k(r_b P))$.

[4] To read M, Bob computes

$$M + k(r_b P) - r_b(kP) = M. \qquad (4.47)$$

An eavesdropper who can solve the discrete logarithm problem on E can, of course, determine r_b from the publicly known information P and $r_b P$. But as everybody knows, there is no efficient way to compute discrete logarithms, so the system is secure.

(4) Analogue of the RSA Cryptosystem:

RSA, the most popular cryptosystem in use, also has the following elliptic curve analogue:

[1] $N = pq$ is a public-key which is the product of the two large secret primes p and q.

[2] Choose two random integers a and b such that $E : y^2 = x^3 + ax + b$ defines an elliptic curve both mod p and mod q.

[3] To encrypt a message-point P, just perform eP mod N, where e is the public (encryption) key. To decrypt, one needs to know the number of points on E modulo both p and q.

The above are some elliptic curve analogues of certain public-key cryptosystems. It should be noted that almost every public-key cryptosystem has an elliptic curve analogue; it is of course possible to develop new elliptic curve cryptosystems which do not rely on the existing cryptosystems.

It should be also noted that the digital signature schemes can also be analogued by elliptic curves over $\mathbb{F}_q$ or over $\mathbb{Z}/n\mathbb{Z}$ with $n = pq$ and $p, q \in$ Primes in exactly the same way as that for public-key cryptography; several elliptic curve analogues of digital signature schemes have already been proposed, say, e.g., [148].

(5) Menezes-Vanstone Elliptic Curve Cryptosystem

A serious problem with the above mentioned elliptic curve cryptosystems is that the plain-text message units m lie on the elliptic curve E, and there is no convenient method known of deterministically generating such points on E. Fortunately, Menezes and Vanstone had discovered a more efficient variation [144]; in this variation which we shall describe below, the elliptic curve is used for "masking", and the plain-text and cipher-text pairs are allowed to be in $\mathbb{F}_p^* \times \mathbb{F}_p^*$ rather than on the elliptic curve.

[1] Key generation: Alice and Bob publicly choose an elliptic curve E over $\mathbb{F}_p$ with $p > 3$ is prime and a random *base* point $P \in E(\mathbb{F}_p)$ such that P generates a large subgroup H of $E(\mathbb{F}_p)$, preferably of the same size as that of $E(\mathbb{F}_p)$ itself. Assume that randomly chosen $k \in \mathbb{Z}_{|H|}$ and $a \in \mathbb{N}$ are secret.

[2] Encryption: Suppose now Alice wants to sent message

$$m = (m_1, m_2) \in (\mathbb{Z}/p\mathbb{Z})^* \times (\mathbb{Z}/p\mathbb{Z})^* \tag{4.48}$$

to Bob, then she does the following:

[a] $\beta = aP$, where P and β are public;

[b] $(y_1, y_2) = k\beta$;

[c] $c_0 = kP$;

[d] $c_j \equiv y_j m_j \pmod{p}$ for $j = 1, 2$;

[e] Alice sends the encrypted message c of m to Bob:

$$c = (c_0, c_1, c_2). \tag{4.49}$$

[3] Decryption: Upon receiving Alice's encrypted message c, Bob calculates the following to recover m:

[a] $ac_0 = (y_1, y_2)$;

[b] $m = \left(c_1 y_1^{-1} \pmod{p}, \ c_2 y_2^{-1} \pmod{p}\right)$.

Example 4.7.1. The following is a nice example of Menezes-Vanstone cryptosystem [153].

[1] Key generation: Let E be the elliptic curve given by $y^2 = x^3 + 4x + 4$ over $\mathbb{F}_{13}$, and $P = (1, 3)$ be a point on E. Choose $E(\mathbb{F}_{13}) = H$ which is cyclic of order 15, generated by P. Let also the private keys $k = 5$ and $a = 2$, and the plain-text $m = (12, 7) = (m_1, m2)$.

[2] Encryption: Alice computes:

$$\beta = aP = 2(1, 3) = (12, 8),$$
$$(y_1, y_2) = k\beta = 5(12, 8) = (10, 11),$$
$$c_0 = kP = 5(1, 3) = (10, 2),$$
$$c_1 \equiv y_1 m_1 \equiv 10 \cdot 2 \equiv 3 \pmod{13},$$
$$c_2 \equiv y_2 m_2 \equiv 11 \cdot 7 \equiv 12 \pmod{13}.$$

Then Alice sends

$$c = (c_0, c_1, c_2) = ((10, 2), 3, 12)$$

to Bob.

[3] Decryption: Upon receiving Alice's message, Bob computes:

$$ac_0 = 2(10, 2) = (10, 11) = (y_1, y_2),$$
$$m_1 \equiv c_1 y_1^{-1} \equiv 12 \ (\text{mod } 13),$$
$$m_2 \equiv c_2 y_2^{-1} \equiv 7 \ (\text{mod } 13).$$

Thus, Bob recovers the message $m = (12, 7)$.

We have introduced so far the most popular public-key cryptosystems, such as Diffie-Hellman-Merkle, RSA, Elliptic curve and probabilistic cryptosystems. There are, of course, many other types of public-key cryptosystems in use, such as Rabin, McEliece and Knapsack cryptosystems. Readers who are interested in the cryptosystems which are not covered in this book are suggested to consult Menezes et al. [145].

Problems for Section 4.7

Problem 4.7.1. The exponential cipher, invented by Pohlig and Hellman in 1978 [166] and based on mod p arithmetic, is a secret-key cryptosystem, but very close to the RSA public-key cryptosystem, which is based on mod n arithmetic. In essence, the Pohlig-Hellman cryptosystem works as follows:

[1] Choose a large prime number p and the encryption key k such that $0 < k < p$ and $\gcd(k, p - 1) = 1$.

[2] Compute the decryption key k' such that $k \cdot k' \equiv 1 \ (\text{mod } p - 1)$.

[3] Encryption: $C \equiv M^k \ (\text{mod } p)$.

[4] Decryption: $M \equiv C^{k'} \ (\text{mod } p)$.

Clearly, if you change modulo p to modulo $n = pq$, then the Pohlig-Hellman cryptosystem is just the RSA cryptosystem (but invented earlier than RSA). Now you are asked to design an elliptic curve analog of the Pohlig-Hellman cryptosystem.

Problem 4.7.2. The Elliptic Curve Digital Signature Algorithm (ECDSA) (see [113] for more information) is the elliptic curve analog of the Digital Signature Algorithm (DSA).

(1) Give a complete description of ECDSA.

(2) What are the advantages of ECDSA over DSA.

(3) Write an assay on the cryptanalytic attacks on ECDSA.

4.8 Zero-Knowledge Techniques

Zero-knowledge is a technique, by which one can convince someone else that he has a certain knowledge without revealing any information about that knowledge. Let us look at an example of Zero-knowledge technique based on square root problem (SQRTP). Recall that finding square roots modulo n is hard and equivalent to factoring.

Algorithm 4.8.1 (Zero knowledge proof). Let $n = pq$ be product of two large prime numbers. Let also $y \equiv x^2 \pmod{n}$ with $\gcd(y, n) = 1$. Now suppose that Alice claims to know x, the square root of y, she does not want to reveal the x. Now Bob wants to verify this.

[1] Alice first chooses two random numbers r_1 and r_2 with

$$r_1 r_2 \equiv x \pmod{n}.$$

(She can do so by first choosing r_1 with $\gcd(r_1, n) = 1$ and then letting $r_2 \equiv x r_1^{-1} \pmod{n}$). She then computes

$$x_1 \equiv r_1^2, \quad x_2 \equiv r_2^2 \pmod{n}$$

and sends x_1 and x_2 to Bob.

[2] Bob checks that $x_1 x_2 \equiv y \pmod{n}$, then choose either x_1 or x_2 and asks Alice to supply a square root of it. He then checks that it is indeed a square root.

Example 4.8.1. Let $n = pq = 31 \cdot 61 = 1891$. Let also $\sqrt{56} \equiv x \pmod{1891}$. Now suppose that Alice claims to know x, the square root of 56, but she does not want to reveal it. Bob then wants to prove if Alice really knows x.

[1] Alice chooses $r_1 = 71$ such that $\gcd(71, 1891) = 1$. Then she finds $r_2 \equiv x(1/r_1) \equiv 408(1/71) \equiv 1151 \pmod{1891}$ (because Alice knows $x = 408$).

[2] Alice computes $x_1 \equiv r_1^2 \equiv 71^2 \equiv 1259 \pmod{1891}$, $x_2 \equiv r_2^2 \equiv 1151^2 \equiv 1101 \pmod{1891}$.

[3] Alice sends x_1 and x_2 to Bob.

[4] Bob checks $x_1 x_2 \equiv 1259 \cdot 1101 \equiv 56 \pmod{1891}$.

[5] Bob chooses either $x_1 = 1259$ or $x_2 = 1101$ and asks Alice to provide a square root of either x_1 or x_2.

[6] On request from Bob, Alice sends either $r_1 = 71$ or $r_2 = 1151$ to Bob, since $\sqrt{1259} \equiv 71$ and $\sqrt{1101} = 1151$, i.e., $1259 \equiv 71^2$ and $1101 \equiv 1151^2$.

[7] Bob now is convinced that Alice really knows x, or otherwise, she cannot tell the square root of x_1 or x_2.

Algorithm 4.8.2 (A zero-knowlege identification scheme). Let $n = pq$ be product of two large prime numbers. Let also Alice have the secret numbers $s_1, s_2, \ldots, s_k$ and $v_i \equiv s_i^{-2} \pmod{n}$ with $\gcd(s_i, n) = 1$. The numbers v_i are sent to Bob. Bob tries to verify that Alice knows the numbers $s_1, s_2, \ldots, s_k$. Both Alice and Bob proceed as follows:

[1] Alice first chooses a random numbers r, computes

$$x \equiv r^2 \pmod{n}$$

and sends x to Bob.

[2] Bob chooses numbers $\{b_1, b_2, \ldots, b_k\} \in \{0, 1\}$. He sends these to Alice.

[3] Alice computes
$$y \equiv r s_1^{b_1} s_2^{b_2} \cdots s_k^{b_k} \pmod{n}$$
and sends to Bob.

[4] Bob checks that
$$x \equiv y^2 v_1^{b_1} v_2^{b_2} \cdots v_k^{b_k} \pmod{n}.$$

[5] Repeat Steps [1] to [4] several times (e.g., 20-30 times), each time with a different r.

The zero-knowledge technique is ideally suited for identification of an owner A (who e.g., has a ID number) of a smart card by allowing A to convince a merchant Bob of knowledge S without revealing even a single bit of S [246]. Theoretically, zero-knowledge technique can be based on any computationally intractable problem such as the IFP, DLP, ECDLP KRTP and SQRTP problems. The following is just an example.

Example 4.8.2. In this identification scheme, we assume that there is a smart card owned by e.g., Alice, a card reader machine owned by e.g. a bank, and third party, called the third trust party (TTP).

[1] The TTP first chooses $n = pq$, where p and q are two large primes and $p \equiv q \equiv 3 \pmod{4}$, and computes the PIN number for Alice's smart card such that

$$\text{PIN} \equiv s^2 \pmod{n}. \tag{4.50}$$

(ID is the quadratic residues of both q and q.)

[2] The TTP computes the square root s of ID (he can do so because he knows the prime factorization of n), and stores s in a segment of memory of the smart card that is not accessible from the outside world. The TTP should also made n public, but keep p and q secret. By now the smart card has the information (PIN, n, s), and the card reader has the information n.

[3] The Smart Card or the card holder Alice makes the PIN number to the card reader:

$$\text{Card/Alice} \xrightarrow{\quad \text{PIN} \quad} \text{Card Reader.}$$

[4] Card/Alice generates a random r and compute $t \equiv r^2 \pmod{n}$, and sends t to Bob:

$$\text{Card/Alice} \xrightarrow{\quad t \quad} \text{Card Reader.}$$

[5] The Card Reader selects a random $e \in \{0, 1\}$ and sends to Alice:

$$\text{Card/Alice} \xleftarrow{\quad e \quad} \text{Card Reader.}$$

[6] Card/Alice computes
$$u \equiv r \cdot s^e \pmod{n}$$
and sends it to Card Reader:

$$\text{Card/Alice} \xrightarrow{\quad u \quad} \text{Card Reader.}$$

[7] The Card Reader checks whether or not
$$u^2 \equiv t \cdot \text{PIN}^e \pmod{n}.$$

[8] Repeat the Steps [4]-[7] for different r. If each time,
$$u^2 \equiv t \cdot \text{PIN}^e \pmod{n},$$

then the card is indeed issued by the TTP. That is, the Card Reader has been convinced that the Card has stored s, the square root of PIN modulo n.

Problems for Section 4.8

Problem 4.8.1. Suppose Alice knows:

$$k_1 \equiv \log_{x_1} y_1 \pmod{n},$$
$$k_2 \equiv \log_{x_2} y_2 \pmod{n},$$
$$\alpha = k_1 = k_2,$$

where $n = pq$. Suppose now that Alice wishes to convince Bob that she knows (k_1, k_2, α) as above. Design a zero-knowledge protocol that will convince Bob what Alice claims.

Problem 4.8.2. Suppose Alice knows:

$$M \equiv C^d \pmod{n},$$

where

$$C \equiv M^e \pmod{n},$$
$$ed \equiv 1 \pmod{\phi(n)}),$$
$$n = pq$$
$$(n, e, C) \quad \text{is} \quad \text{public.}$$

Suppose now that Alice wishes to convince Bob that she knows M. Design a zero-knowledge protocol that Bob should be convinced that Alice knows M.

Problem 4.8.3. Let $n = pq$ with p, q primes. Given y, Alice wants to convince Bob that she knows x such that $x^2 \equiv y \pmod{n}$. Design a zero-knowledge protocol that will enable Bob to believe that Alice indeed knows x.

4.9 Deniable Authentication

One of the most important features of authentication is the non-repudiation property, implemented by digital signatures. This marvelous feature of authentication may, however, not be necessary and sometimes even should be avoided whenever possible. In this section, by a combined use of public-key encryption, digital signatures, coding and randomness, a new scheme [272] for deniable/repudiable authentication, suitable for e-voting, is introduced. Two implementations of the scheme are also discussed. The security of the scheme is based on the intractability of the quadratic residuosity problem, which is, in turn, based on the intractability of the integer factorization problem.

In the deniable/repudiable authentication scheme, we first generate a conventional non-deniable/non-repudiable digital signature S using a standard digital signature system/method. Then we convert S into a deniable/repudiable signature S':

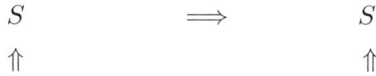

$$S \qquad\qquad \Longrightarrow \qquad\qquad S'$$

$$\Uparrow \qquad\qquad\qquad\qquad\qquad \Uparrow$$

Non-deniable signature   Deniable signature

On receiving the deniable signature, the receiver obviously knows there is a signature attached to the document, say a vote, but he cannot verify who is the author of the vote, as the vote is privacy preserved by the deniable signature, which cannot be verified directly by the sender's public-key. In this section, we propose two implementations for the deniable authentication, both are based on the combined use of public-key encryption, digital signature, coding and randomness. The first implementation works as follows:

[1] *Generate the non-deniable signature* $S = \{s_1, s_2, \ldots\}$. The voter, say, e.g., Bob, uses his private key d to generate his digital signature S on his vote M: $S = M^d \pmod{n}$, where N is a product of two large prime numbers. This signature is ready to verify using Bob's public-key e: $M = S^e \pmod{n}$, where $ed \equiv 1 \pmod{\phi(n)}$.

[2] $S \Longrightarrow (S', B)$: *Obtain the deniable signature* S' *by adding randomness to* S.

  [2-1] $S \Longrightarrow S'$: *Randomize the signature.* Randomly add some extra digits (noises) to the digital signature S to get a corresponding randomized digital signature $S' = \{s'_1, s'_2, \ldots\}$.

  [2-2] $S' \Longrightarrow B$: *Generate a bit string* $B = \{b_1, b_2, \ldots\}$ *for the randomized signature* S'. Generate a binary string B, with each bit in B corresponding to a digit in S', assign 1 to the bit if the corresponding digit appears in both S and S', otherwise, assign 0 to the bit.

  [2-3] $B \Longrightarrow X$: *Generate a random mixed string* X *of squares and pseudo-squares.* Generate a string of integers $X = \{x_1, x_2, \ldots\}$ which are the mixed squares and pseudo-squares, based on the quadratic residuosity problem [268] (it can also even be based on the kth power residuosity problem (kPRP) [267]).

  To get the X string, we choose $N = pq$ with p, q prime. Find a pseudo random square $y \in \mathbb{Z}/n\mathbb{Z}$ such that $y \in \overline{Q}_n$ and $(\frac{y}{n}) = 1$. (n, y) can be made public, but p, q must be kept as a secret. Choose at random the number r_i and compute

$$x_i \equiv \begin{cases} r_i^2 \pmod{n}, & \text{if } m_i = 0 \ (\text{random square}), \\ yr_i^2 \pmod{n}, & \text{if } m_i = 1 \ (\text{pseudo random square}). \end{cases}$$

[2-4] Send $\{(S', X), E\}$ to the Election Centre, where (S', X) is the repudiable digital signature and E the e-vote. The author of the e-vote can deny his authorship of the vote, since S' is a random string of digits which is different from S and which cannot be easily get back to S unless the quadratic residuosity problem can be solved in polynomial-time.

[2-5] $S' \Longrightarrow B$: To verify the signature if needed, the author of the vote who knows the trap-door information (i.e., the prime factors p and q of the composite modulus n) must be presented so that the string B can recovered from X, and hence S can be recovered from S'.

To get the string $B = b_1, b_2, \ldots$ from the string $X = x_1, x_2, \ldots$, one may perform the following operations:

$$b_i \equiv \begin{cases} 0, & \text{if } e_i^p = e_i^q = 1, \\ 1, & \text{otherwise,} \end{cases}$$

where

$$e_i^p = \left(\frac{x_i}{p}\right), \quad e_i^q = \left(\frac{x_i}{q}\right).$$

[2-6] $\{B, S'\} \Longrightarrow S$: Remove the noise from S' to get S according to B.

Example 4.9.1. We demonstrate the above idea in the following simple example for an e-voting system.

[1] *Generate the non-repudiable signature.* The voter, say, e.g., Bob, uses his private key to generate his digital signature S for his vote:

Bob Newman	0215020014052313013	42527067843532368
⇑	⇑	⇑
Name for Signature	Numerical Form (M)	Digital Signature (S)

This digital signature was generated and verified by

$$\begin{aligned} S &\equiv M^d \equiv 215020014052313013^{785558871522137263} \\ &\equiv 42527067843532368 \ (\text{mod } 1832970702926065247) \\ M &\equiv S^e \equiv 42527067843532368^7, \\ &\equiv 215020014052313013 \ (\text{mod } 1832970702926065247), \end{aligned}$$

with

$$ed \equiv 1 \ (\text{mod } \phi(1832970702926065247)).$$

[2] $S \Longrightarrow (S', B)$: *Add repudiable feature to the non-repudiable signature.*

[2-1] $S \implies S'$: *Randomize the signature.* Randomly add some extra digits into the digital signature S to get a corresponding pseudo digital signature S'.

S				4			2			5			2				7	0
⇕ ↕	↕	↕	↕	↕	↕	↕	↕	↕	↕	↕	↕	↕	↕	↕	↕	↕	↕
S'	7	9	1	4	8	5	2	1	4	5	3	2	2	8	9	1	7	0

S	6		7	8		4	3	5					3	2	3	6		8
⇕ ↕	↕	↕	↕	↕	↕	↕	↕	↕	↕	↕	↕	↕	↕	↕	↕	↕	↕
S'	6	3	7	8	9	4	3	5	9	1	3	2	3	2	3	6	6	8

[2-2] $S' \implies B$: *Generate a bit string $B = \{b_1, b_2, \ldots\}$ for the randomized signature S'.* Generate a binary string B with each bit in B corresponding to a digit in S', assign 1 to the bit if the corresponding digit appears in both S and S', otherwise, assign 0 to the bit.

S'	7	9	1	4	8	5	2	1	4	5	3	2	2	8	9	1	7	0
⇕ ↕	↕	↕	↕	↕	↕	↕	↕	↕	↕	↕	↕	↕	↕	↕	↕	↕	↕
B	0	0	0	1	0	0	1	0	0	1	0	0	1	0	0	0	1	1

S'	6	3	7	8	9	4	3	5	9	1	3	2	3	2	3	6	6	8
⇕ ↕	↕	↕	↕	↕	↕	↕	↕	↕	↕	↕	↕	↕	↕	↕	↕	↕	↕
B	1	0	1	1	0	1	1	1	0	0	0	0	1	1	1	1	0	1

[2-3] $B \implies X$: *Generate a random mixed string X of squares and pseudo-squares:* Generate a string of integers $X = \{x_1, x_2, \ldots\}$ which are the mixed squares and pseudo squares. based on the quadratic residuosity problem.

Choose a random pseudo square $y = 1234567 \in \mathbb{Z}/n\mathbb{Z}$ such that $y \in \overline{Q}_n$ and $\left(\frac{y}{n}\right) = 1$. (n, y) can be made public, but p, q must be kept as a secret.

We just calculate the values for x_3 and x_4; all the rest can be calculated in exactly the same way.

Choose at random the number $r_3 = 8194920765$ and $r_4 = 17402983$, and compute (note that $b_3 = 0, b_4 = 1$):

$$x_3 \equiv r_3^2 \equiv 8194920765^2$$

$$\equiv 11697810139289836333 \pmod{18329707027029260265247},$$

$$x_4 \equiv yr_4^2 \equiv 1234567 \cdot 17402983^2$$

$$\equiv 1812621636505510722 \pmod{18329707027029260265247}.$$

[2-4] Send $\{(S', X), E\}$ to the Election Centre, where (S', X) is the deniable/repudiable digital signature and E the e-vote. The author of the e-vote can deny his authorship of the vote, since S' is a random string of digits which is different from S.

[2-5] $S' \Longrightarrow B$: To verify the signature if needed, only the author of the vote who knows the trap-door information (the composite modulus N) can show the Election Centre that he is the author of the vote, anyone else should not be able to verify the authorship of the vote.

We just show how to recover S'_3 and S'_4 by computing b_3 and b_4 from x_3 and x_4 as follows (all the rest are performed in exactly the same way): Since

$$e_3^p = \left(\frac{x_3}{p}\right) = \left(\frac{1169781039289836333}{1353874987}\right) = 1,$$

$$e_3^q = \left(\frac{x_3}{q}\right) = \left(\frac{1169781039289836333}{1353869981}\right) = 1,$$

$$e_4^p = \left(\frac{x_4}{p}\right) = \left(\frac{1812621636505510722}{1353874987}\right) = -1,$$

$$e_4^q = \left(\frac{x_4}{q}\right) = \left(\frac{1812621636505510722}{1353869981}\right) = -1,$$

then

$$b_3 = 0, \text{ as } e_3^p = e_3^q = 1,$$

$$b_4 = 1, \text{ as } e_4^p = e_4^q = -1.$$

[2-6] $\{B, S'\} \Longrightarrow S$: Remove the noise from S' to get S. Since $b_3 = 0$, its corresponding digit 1 in S'_3 is a noise and should be removed from S'. However, as $b_4 = 1$, its corresponding digit 4 should be remained in S'. Clearly, after removing all the noises from S', S' will eventually become S, the true digital signature!

Clearly, anyone who can solve the quadratic residuosity problem (or the kth power residuosity problem) can distinguish the pseudo squares from square (or the pseudo kth powers from the kth powers), and and hence can verify the digital signature and the authorship of the e-vote. But as everybody knows, solving QRP/kPRP is intractable, thus the author can deny his authorship of an e-vote, regardless whether or not he actually did vote.

Problems for Section 4.9

Problem 4.9.1. Give a full computational complexity analysis of the deniable authentication scheme discussed in this section.

Problem 4.9.2. Propose some possible attacks on the deniable authentication scheme.

Problem 4.9.3. Convert the deniable authentication scheme to a practical electronic voting system featured with deniable authorization.

Problem 4.9.4. Develop some other types of deniable authentication scheme, which is not based on the quadratic residuosity problem.

Problem 4.9.5. Develop a better (fast) cryptographic scheme, based on any method such as coding theory and pseudosquares (quadratic residuosity problem), with some added random noise into the ciphertext (the noise should be removed successfully during the decryption process).

4.10 Non-Factoring Based Cryptography

The security of the cryptographic systems and protocols discussed so far are direct or indirect rely on the intractability of the integer factorization, discrete logarithm, and elliptic curve discrete logarithm problems. As can be seen these three problems are essentially one problem, since there is an algorithm, e.g., the xedni calculus, to solve all the three problems, although it is not efficient. More importantly, if a quantum computer becomes available in the market, all the three problems can be solved in polynomial-time. In this section, we shall introduce some cryptographic schemes that are not based on factoring.

Coding-Based Cryptography. We first introduce the most famous code-based cryptosystem, the McEliece system, invented by McEliece in 1978 [141]. One of the most important features of the McEliece system is that it has resistant cryptanalysis to date; it is even quantum computer resisted. The idea of the McEliece system is based on coding theory and its security is based on the fact that decoding an arbitrary linear code is $\mathcal{NP}$-complete.

Algorithm 4.10.1 (McEliece's Coding-Based Cryptography).
Suppose Bob wishes to send an encrypted message to Alice, using Alice's public-key. Alice generates her public-key and the corresponding private key. Bob uses her public-key to encrypt his message and sends it to Alice, Alice uses her own private-key to decrypt Bob's message.

[1] Key Generation: Alice performs:

[1-1]Choose integers k, n, t as common system parameters.

[1-2]Choose a $k \times n$ generator matrix G for a binary (n, k)-linear code which can correct t errors and for which an efficient decoding algorithm exists.

[1-3]Select a random $k \times k$ binary non-singular matrix S.

[1-4]Select a random $k \times k$ permutation matrix P.

[1-5]Compute the $k \times n$ matrix $\widehat{G} = SGP$.

[1-6]Now $(\widehat{G}, t)$ is Alice's public-key whereas (S, G, P) is Alice's private-key.

[2] Encryption: Bob uses Alice's public-key to encrypt his message to Alice. Bob performs:

[2-1]Obtain Alice's authentic public key $(\widehat{G}, t)$.

[2-2]Represent the message in binary string m of length k.

[2-3]Choose a random binary error vector z of length n having at most t 1's.

[2-4]Compute the binary vector $c = m\widehat{G} + z$.

[2-5]Send the ciphertext c to Alice.

[3] Decryption: Alice receives Bob's message m and uses her private-key to recover c from m. Alice performs:

[3-1] Compute $\widehat{c} = cP^{-1}$, where P^{-1} is the inverse of the matrix P.

[3-2] Use the decoding algorithm for the code generated by G to decode $\widehat{c}$ to $\widehat{m}$.

[3-3] Compute $m = \widehat{m}\widehat{S}^{-1}$. This m is thus the original plaintext.

Theorem 4.10.1 (Correctness of McEliece's Cryptosystem). In McEliece's Cryptosystem, m can be correctly recovered from c.

Proof. Since

$$
\begin{aligned}
\widehat{c} &= cP^{-1} \\
&= (m\widehat{G} + z)P^{-1} \\
&= (mSGP + z)P^{-1} \\
&= (mS)G + zP^{-1}, \quad (zP^{-1} \text{ is a vector with at most } t \text{ 1's})
\end{aligned}
$$

the decoding algorithm for the code generated by G corrects $\widehat{c}$ to $\widehat{m} = mS$. Now applying S^{-1} to $\widehat{m}$, we get $mSS^{-1} = m$, the required original plaintext.
□

Remark 4.10.1. The security of McEliece's cryptosystem is based on error-correcting codes, particularly the Goppa [136]; if the Goppa code is replaced by other error-correcting codes, the security will be severely weakened. The McEliece's cryptosystem has two main drawbacks:

(1) the public-key is very large, and

(2) there is a message expansion by a factor of n/k.

It is suggested that the values for the system parameters should be $n = 1024$, $t = 50$, and $k \geq 644$. Thus for these recommended values of system parameters, the public-key has about 2^{19} bits, and the message expansion is about 1.6. For these reasons, McEliece's cryptosystem receives little attention in practice. However, as McEliece's cryptosystem is the first probabilistic encryption and more importantly, it has resisted all cryptanalysis including quantum cryptanalysis, it may be a good candidate to replace RSA in the post-quantum cryptography age.

Lattice-Based Cryptography. Cryptography based on ring properties and particularly lattice reduction is another promising direction for post-quantum cryptography, as lattice reduction is a reasonably well-studied hard problem that is currently not known to be solved in polynomial-time, or even subexponential-time on a quantum computer. There are many types of cryptographic systems based on lattice reduction. In this section, we give a brief account of one if the lattice based cryptographic systems, the NTRU encryption scheme. NTRU is rumored to stand for Nth-degree TRUncated polynomial ring, or Number Theorists eRe Us. Compared with RSA, it is a rather young cryptosystem, developed by Hoffstein, Pipher and Silverman [103] in 1995. We give a brief introduction to NTRU, more information can be found in [103] and [104].

Algorithm 4.10.2 (NTRU Encryption Scheme). The NTRU encryption scheme works as follows:

[1] Key Generation:

    [1-1] Randomly generate polynomials f and g in D_f and D_g, respectively, each of the form:

$$a(x) = a_0 + a_1 x + a_2 x^2 + \cdots + a_{k-2} x^{k-2} + a_{k-1} x^{k-1}.$$

    [1-2] Invert f in $\mathcal{R}_p$ to obtain f_p, and check that g is invertible in f_q.

    [1-3] The public-key is $h \equiv p \cdot g \cdot f_q \pmod{q}$. The private-key is the pair (f, f_p).

[2] Encryption:

    [2-1] Randomly select a small polynomials r in D_r.

    [2-2] Compute the ciphertext $c \equiv r \cdot h + m \pmod{q}$.

[3] Decryption:

    [3-1] Compute $a = \text{center}(f \cdot c)$,

[3-2] Recover m from c by computing $m \equiv f_p \cdot a \pmod{q}$. This is true since

$$a \equiv p \cdot r \equiv +f \cdot m \pmod{q}.$$

In Table 4.2, we present some information comparing NTRU to RSA and McEliece.

	NTRU	RSA	McEliece
Encryption Speed	n^2	$n^2 \approx n^3$	n^2
Decryption Speed	n^2	n^3	n^2
Public-Key	n	n	n^2
Secret-Key	n	n	n^2
Message Expansion	$\log_p q - 1$	$1 - 1$	$1 - 1.6$

Table 4.2. Comparison among NTRU, RSA and McEliece

Quantum Cryptography. It is evident that if a practical quantum computer is available, then all public-key cryptographic systems based on the difficulty of IFP, DLP, and ECDLP will be insecure. However, the cryptographic systems based on quantum mechanics will still be secure even if a quantum computer is available. In this section some basic ideas of quantum cryptography are introduced. More specifically, a quantum analog of the Diffie-Hellman key exchange/distribution system, proposed by Bennett and Brassard in 1984 [17], will be addressed.

First let us define four *polarizations* as follows:

$$\{0°,\ 45°,\ 90°,\ 135°\} \overset{\text{def}}{=} \{\rightarrow,\ \nearrow,\ \uparrow,\ \searrow\}. \tag{4.51}$$

The quantum system consists of a transmitter, a receiver, and a quantum channel through which polarized photons can be sent [19]. By the law of quantum mechanics, the receiver can either distinguish between the *rectilinear polarizations* $\{\rightarrow,\ \uparrow\}$, or reconfigure to discriminate between the diagonal polarizations $\{\nearrow,\ \searrow\}$, but in any case, he cannot distinguish both types. The system works in the following way:

[1] Alice uses the transmitter to send Bob a sequence of photons, each of them should be in one of the four polarizations $\{\rightarrow,\ \nearrow,\ \uparrow,\ \searrow\}$. For instance, Alice could choose, at random, the following photons

$$\uparrow \quad \nearrow \quad \rightarrow \quad \searrow \quad \rightarrow \quad \rightarrow \quad \nearrow \quad \uparrow \quad \uparrow$$

to be sent to Bob.

[2] Bob then uses the receiver to measure the polarizations. For each photon received from Alice, Bob chooses, at random, the following type of measurements $\{+, \times\}$:

[3] Bob records the result of his measurements but keeps it secret:

[4] Bob publicly announces the type of measurements he made, and Alice tells him which measurements were of correct type:

[5] Alice and Bob keep all cases in which Bob measured the correct type. These cases are then translated into bits $\{0, 1\}$ and thereby become the key:

[6] Using this secret key formed by the quantum channel, Bob and Alice can now encrypt and send their ordinary messages via the classic public-key channel.

An eavesdropper is free to try to measure the photons in the quantum channel, but, according to the law of quantum mechanics, he cannot in general do this without disturbing them, and hence, the key formed by the quantum channel is secure.

Problems for Section 4.10

Problem 4.10.1. (Coding-Based Cryptography)

(1) Compare the main parameters (such as encryption and decryption complexity, cryptographic resistance, easy to use, secret-key size, and public-key size, etc) of RSA and McEliece systems.

(2) Show that decoding a general algebraic code is NP-complete.

(3) Write an essay on all possible attacks for the McEliece coding-based cryptosystem.

Problem 4.10.2. (Lattice-Based Cruptography)

(1) Give a critical analysis of the computational complexity of the NTRU cryptosystem.

(2) Show some possible attacks on the Cai-Cusick lattice-based cryptosystem [43].

Problem 4.10.3. (Quantum and DNA Cryptography)

(1) DNA molecular biologic cryptography, e.g., the Rief one-time pad DNA cryptosystem [80], is a new development in cryptography. Give a description of the Rief DNA-based one-time pad.

(2) Write an assay to compare the main features of the classic, the quantum and the DNA cryptography.

4.11 Chapter Notes and Further Reading

Cryptography, particularly public-key cryptography is a very hot topic nowadays in information/network security, since essentially the only practical way to maintain security, privacy and integrity of information is to use public-key cryptography. In this chapter we have introduced some widely used public-key cryptographic schemes and systems, based on the computational intractability of the integer factorization problem, the quadratic residuosity problem, the discrete logarithm problem, and the elliptic curve discrete logarithm problem, etc. There are again many books and papers in this field. The idea of public-key cryptography and digital signatures was first proposed in [70] and [146]. The RSA system was first described in [78] and formally proposed in [201]. The books by Bishop [24], Koblitz ([122] and [123]), Mollin ([153] and [162]), Pieprzyk, Hardjono and Seberry [172] Smart [238], Stinson [242], Trappe and Washington [247], and Wagstaff [251] are strongly recommended. For a quick survey on cryptography, see, e.g., the papers by Koblitz ([124] and [125]).

Again, many other related books and papers in this field have also been listed in the Bibliography section at the end of this book for further reading, include: Bauer [16], BlumG [26], Diffie and Hellamn [71], ElGamal [74], Goldwasser and Micali [84], Kate and Lindell [116], Koblitz [120], McC [140], McEliece [141], Menezes and Vanstone [144], A. Menezes, van Oorschot and Vanstone, [145], MeyerM [148], Miller [150], Miller [174], Pomerance [182], Rothe [207], Schneier [215], Schnorr [216], Shamir [221], Singh [237], Tilborg [246], Wiener [255], and Yan [268].

Readers who are interested in cryptanalysis may consult the following two research monographs for more information: Stamp and Low [240] and Yan [270], and also the following two forthcoming books in 2009: Hinek [102] and Joux [115].

Bibliography

1. L. M. Adleman, "A Subexponential Algorithmic for the Discrete Logarithm Problem with Applications to Cryptography", *Proceedings of the 20th Annual IEEE Symposium on Foundations of Computer Science*, IEEE Press, 1979, pp 55–60.

2. L. M. Adleman, "Algorithmic Number Theory – The Complexity Contribution", *Proceedings of the 35th Annual IEEE Symposium on Foundations of Computer Science*, IEEE Press, 1994, pp 88–113.

3. L. M. Adleman, C. Pomerance, and R. S. Rumely, "On Distinguishing Prime Numbers from Composite Numbers", *Annals of Mathematics*, **117**, (1983), pp 173–206.

4. L. M. Adleman and M. D. A. Huang, *Primality Testing and Abelian Varieties over Finite Fields*, Lecture Notes in Mathematics **1512**, Springer, 1992.

5. M. Agrawal, N. Kayal and N. Saxena, "Primes is in P", *Annals of Mathematics*, **160**, 2(2004), pp 781–793.

6. W. Alford, G. Granville and C. Pomerance, "There Are Infinitely Many Carmichael Numbers", *Annals of Mathematics*, **140**, (1994), pp 703–722.

7. J. A. Anderson and J. M. Bell, *Number Theory with Applications*, Prentice-Hall, 1997.

8. G. E. Andrews, *Number Theory*, W. B. Sayders Company, 1971. Also Dover Publications, 1994.

9. T. M. Apostol, *Introduction to Analytic Number Theory*, Corrected 5th Printing, Undergraduate Texts in Mathematics, Springer, 1998.

10. A. O. L. Atkin and F. Morain, "Elliptic Curves and Primality Proving", *Mathematics of Computation*, **61**, (1993), pp 29–68.

11. E. Bach, M. Giesbrecht and J. McInnes, *The Complexity of Number Theoretical Algorithms*, Technical Report 247/91, Department of Computer Science, University of Toronto, 1991.

12. E. Bach and J. Shallit, *Algorithmic Number Theory I – Efficient Algorithms*, MIT Press, 1996.

13. A. Baker, *A Concise Introduction to the Theory of Numbers*, Cambridge University Press, 1984.

14. R. C. Baker and G. Harman, "The Brun-Tichmarsh Theorem on Average", In: *Proceedings of a Conference in Honor of Heini Halberstam*, Volume 1, 1996, pp 39–103.

15. R. J. Baillie and S. S. Wagstaff, Jr., "Lucas Pseudoprimes", *Mathematics of Computation*, **35**, (1980), pp 1391–1417.

16. F. L. Bauer, *Decrypted Secrets – Methods and Maxims of Cryptology*, 2nd Edition, Springer, 2000.
   smallskip

17. C. H. Bennett and G. Brassard, "Quantum Cryptography: Public Key Distribution and Coin Tossing", *Proccedings of the IEEE International Conference on Computersm Systems and Singal Processing*, IEEE Press, 1984, pp 175–179.

18. C. H. Bennett, "Quantum Information and Computation", *Physics Today*, October 1995, pp 24–30.

19. C. H. Bennett, G. Brassard and A. K. Ekert, "Quantum Cryptography", *Scientific American*, October 1992, pp 26–33.

20. C. H. Bennett, "Strengths and Weakness of Quantum Computing", *SIAM Journal on Computing*, **26**, 5(1997), pp 1510–1523.

21. E. Bernstein and U. Vazirani, "Quantum Complexity Theory", *SIAM Journal on Computing*, **26**, 5(1997), pp 1411–1473.

22. D. J. Bernstein, *Proving Primality After Agrawal-Kayal-Saxena*, Dept of Mathematics, Statistics and Computer Science, The University of Illinois at Chicago, 25 Jan 2003.

23. R. Bhattacharjee and P. Pandey, *Primality Testing*, Dept of Computer Science & Engineering, Indian Institute of Technology Kanpur, India, 2001.

24. M. Bishop, *Introduction to Computer Security*, Addison-Wesley, 2003.

25. I. F. Blake, G. Seroussi and N. P. Smart, *Advanced in Elliptic Curve Cryptography*, Cambridge University Press, 2006.

26. M. Blum and S. Goldwasser, "An Efficient Probabilistic Public-key Encryption Scheme that Hides all Partial Information", *Advances in Cryptography – CRYPTO '84*, Lecture Notes in Computer Science **196**, Springer, 1985, pp 289–302.

27. E. Bombieri, "The Riemann Hypothesis", *The Millennium Prize Problems*, Clay Mathematical Institute and American Mathematical Institute, 2006, pp 107–124.

28. D. Boneh, "Twenty Years of Attacks on the RSA Cryptosystem", *Notices of the AMS*, **46**, 2(1999), pp 203–213.

29. G. Brassard, "A Quantum Jump in Computer Science", *Computer Science Today – Recent Trends and Development*, Lecture Notes in Computer Science **1000**, Springer, 1995, pp 1–14.

30. R. P. Brent, "Irregularities in the Distribution of Primes and Twin Primes", *Mathematics of Computation*, **29**, (1975), pp 43–56.

31. R. P. Brent, "An Improved Monte Carlo Factorization Algorithm", *BIT*, **20**, (1980), pp 176–184.

32. R. P. Brent, "Some Integer Factorization Algorithms using Elliptic Curves", *Australian Computer Science Communications*, **8**, (1986), pp 149–163.

33. R. P. Brent, "Primality Testing and Integer Factorization", *Proceedings of Australian Academy of Science Annual General Meeting Symposium on the Role of Mathematics in Science*, Canberra, 1991, 14–26.

34. R. P. Brent, "Uses of Randomness in Computation", Report TR-CS-94-06, Computer Sciences Laboratory, Australian National University, 1994.

35. R. P. Brent, "Some Parallel Algorithms for Integer Factorisation", *Proc. Fifth International Euro-Par Conference (Toulouse, France, 1-3 Sept 1999)*, Lecture Notes in Computer Science **1685**, Springer, 1999, pp 1-22.

36. R. P. Brent, "Uncertainty Can Be Better Than certainty: Some Algorithms for Primality Testing", Mathematical Institute, Australian National University, 2006.

37. R. P. Brent, "Recent Progress and Prospects for Integer Factorisation Algorithms", *Proc. COCOON 2000 (Sydney, July 2000)*, Lecture Notes in Computer Science **1858**, Springer, 2000, pp 3-22.

38. R. P. Brent, G. L. Cohen and H. J. J. te Riele, "Improved Techniques for Lower Bounds for Odd Perfect Numbers", *Mathematics of Computation*, **57**, (1991), pp 857–868.

39. D. M. Bressoud, *Factorization and Primality Testing*, Undergraduate Texts in Mathematics, Springer, 1989.

40. E. F. Brickell, D. M. Gordon and K. S. McCurley, "Fast Exponentiation with Precomputation" (Extended Abstract), *Advances in Cryptography*, EURO-CRYPT '92, Proceedings, Lecture Notes in Computer Science **658**, Springer, 1992, pp 200–207.

41. J. P. Buhler (editor), *Algorithmic Number Theory*, Third International Symposium, ANTS-III, Proceedings, Lecture Notes in Computer Science **1423**, Springer, 1998.

42. S. A. Burr (editor), *The Unreasonable Effectiveness of Number Theory*, Proceedings of Symposia in Applied Mathematics **46**, American Mathematical Society, 1992.

43. J. Y. Cai and T. W. Cusick, "A Lattice-Based Public-Key Cryptosystem", *Information and Computation*, **151**, 1-2(1999), pp 17–31.

44. J. R. Chen, "On the Representation of a Large Even Integer as the Sum of a Prime and a Product of at Most Two Primes" (in Chinese), Kexue Tongbao (Science Buletin), **17**, (1966), pp 385–386.

45. J. R. Chen, "On the Representation of a Large Even Integer as the Sum of a Prime and a Product of at Most Two Primes", *Scientia Sinica*, **16**, (1973), pp 157–176.

46. J. R. Chen and T. Wang, "On the odd Goldbach problem," *Acta Mathematica Sinica*, **32**, (1989) pp 702–718.

47. L. Childs, *A Concrete Introduction to Higher Algebra*, 2nd Edition, Springer, 2000.

48. S. Chowla, "There Exists an Infinity of 3 – Combinations of Primes in A. P.", *Proceedings of Lahore Philosiphic Society*, **6**, (1944), pp 15–16.

49. H. Cohen, *Advanced Number Theory*, Dover Publications, 1980.

50. H. Cohen, *A Course in Computational Algebraic Number Theory*, Graduate Texts in Mathematics **138**, Springer, 1993.

51. H. Cohen and H. W. Lenstra, Jr., "Primality Testing and Jacobis", *Mathematics of Computation*, **42**, 165(1984), pp 297–330.

52. H. Cohen and G. Frey, *Handbook of Elliptic and Hyperelliptic Curve Cryptography*, Chapman & Hall/CRC Press, 2006.

53. S. Cook, "The P versus NP Problem", *The Millennium Prize Problems*, Clay Mathematical Institute and American Mathematical Institute, 2006, pp 87–104.

54. J. W. Cooley and J. W. Tukey, "An Algorithm for the Machine Calculation of Complex Fourier Series", *Mathematics of Computation*, **19**, (1965), pp 297–301.

55. D. Coppersmith, "Small Solutions to Polynomial Equations, and Low Exponent RSA Vulnerability", *Journal of Cryptology*, **10**, (1997), pp 233–260.

56. D. Coppersmith, "Finding a Small Root of a Bivariate Integer Equation; Factoring with High Bits Known", *Cryptography and Lattices – CaLC 2001*, Lecture Notes in Computer Science **2146**, Springer, 2001, pp 20–31.

57. D. Coppersmith, "Solving Low Degree Polynomials", *Presentation at Asiacrypt*, Taipei, Taiwan, 1 December 2003.

58. D. Coppersmith, M. Franklin, J. Patarin and R. Reiter, "Low-Exponent RSA with Related Messages", *Advances in Cryptology - Eurocrypt'96*, Lecture Notes in Computer Science **1070**, Springer, 1996, pp 1–9.

59. T. H. Cormen, C. E. Ceiserson and R. L. Rivest, *Introduction to Algorithms*, MIT Press, 1990.

60. J. S. Coron and A. May, "Deterministic Polynomial-Time Equivalence of Computing the RSA Secret Key and Factoring", *Journal of Cryptology*, **20**, 1(2007), pp 39–50.

61. D. A. Cox, *Primes of the Form $x^2 + ny^2$*, Wiley, 1989.

62. R. Crandall and C. Pomerance, *Prime Numbers – A Computational Perspective*, Second Edition, Springer, 2005.

63. E. Croot, A. Granville, R. Pemantle and P. Tetali, "Running Time Predictions for Factoring Algorithms" *Algorithmic Number Theory*, Lecture Notes in Computer Science **5011**, Springer, 2008, pp 1–36.

64. H. Davenport, *The Higher Arithmetic*, 7th Edition, Cambridge University Press, 1999.

65. J. H. Davenport, *Primality Testing Revisited*, School of Mathematical Sciences, University of Bath, UK, 1992.

66. J. M. Deshouillers, H. J. J. te Riele and Y. Saouter, "New Experimental Results Concerning the Goldbach Conjecture", *Algorithmic number theory*, Lecture Notes in Computer Science **1423**, Springer, 1998, pp 204–215.

67. J. M. Deshouillers, G. Effinger, H. J. J. te Riele, D. Zinoviev, "A Complete Vinogradov 3-Primes Theorem under the Riemann Hypothesis". *Electronic Research Announcements of the American Mathematical Society*, **3**, (1997), pp 99–104.

68. D. Deutsch, "Quantum Theory, the Church–Turing Principle and the Universal Quantum Computer", Proceedings of the Royal Society of London, Series **A**, **400**, (1985), pp 96–117.

69. L. E. Dickson, *History of the Theory of Numbers I – Divisibility and Primality*, G. E. Stechert & Co., New York, 1934.

70. W. Diffie and E. Hellman, "New Directions in Cryptography", *IEEE Transactions on Information Theory*, **22**, 5(1976), pp 644–654.

71. W. Diffie and E. Hellman, "Privacy and Authentication: An Introduction to Cryptography", *Proceedings of the IEEE*, **67**, 3(1979), pp 393–427.

72. P. G. L. Dirichlet, *Lecturers on Number Theory*, Supplements by R. Dedekind, American Mathematics Society and London Mathematics Society, 1999.

73. J. D. Dixon, "Factorization and Primality tests", *The American Mathematical Monthly*, June-July 1984, pp 333–352.

74. T. ElGamal, "A Public Key Cryptosystem and a Signature Scheme based on Discrete Logarithms", *IEEE Transactions on Information Theory*, **31**, (1985), pp 496–472.

75. Euclid, *The Thirteen Books of Euclid's Elements*, Translated by T. L. Heath, *Great Books of the Western World* **11**, edited by R. M. Hutchins, William Benton Publishers, 1952.

76. Euclid, *The Thirteen Books of Euclid's Elements*, Second Edition, Translated by Thomas L. Heath, Dover Publications, 1956.

77. E. Fouvry, "Theorème de Brun-Titchmarsh: Application au Theorème de Fermat", *ventiones Mathematicae*, **79**, (1985), pp 383–407.

78. M. Gardner, "Mathematical Games – A New Kind of Cipher that Would Take Millions of Years to Break", *Scientific American*, **237**, 2(1977), pp 120–124.

79. M. R. Garey and D. S. Johnson, *Computers and Intractability – A Guide to the Theory of NP-Completeness*, W. H. Freeman and Company, 1979.

80. A. Gehani, T. H. LaBean and J. H. Rief, "DNA-Based Cryptography", *Aspects of Molecular Computing*, Lecture Notes in Computer Science **2950**, 2004, pp 167–188.

81. C. F. Gauss, *Disquisitiones Arithmeticae*, G. Fleischer, Leipzig, 1801. English translation by A. A. Clarke, Yale University Press, 1966. Revised English translation by W. C. Waterhouse, Springer, 1975.

82. S. Goldwasser and J. Kilian, "Almost All Primes Can be Quickly Certified", *Proceedings of the 18th ACM Symposium on Theory of Computing*, Berkeley, 1986, pp 316–329.

83. S. Goldwasser and J. Kilian, "Primality Testing Using Elliptic Curves", *Journal of ACM*, **46**, 4(1999), pp 450–472.

84. S. Goldwasser and S. Micali, "Probabilistic Encryption", *Journal of Computer and System Sciences*, **28**, (1984), pp 270–299.

85. D. M. Gordon and K. S. McCurley, "Massively Parallel Computation of Discrete Logarithms", *Advances in Cryptography*, Crypto '92, Proceedings, Lecture Notes in Computer Science **740**, Springer, 1992, pp 312–323.

86. D. M. Gordon, "Discrete Logarithms in $GF(p)$ using the Number Field Sieve", *SIAM Journal on Discrete Mathematics*, **6**, 1(1993), pp 124–138.

87. D. M. Gordon, "Strong RSA Keys", *Electronic Letters*, **20**, 12(1984), pp 514–516.

88. D. M. Gordon, "Strong Primes are Easy to Find", *Advances in Cryptography – Eurocrypt 84*, Lecture Notes in Computer Science **209**, 1985, pp 49–89.

89. A. Granville, "Primality Testing and Carmichael Numbers", *Notice of the American Mathematical Society*, **Vol 39**, (1992), pp 696–700.

90. A. Granville, "It is Easy to Determine Whethere a Given Integer is Prime", *Bulletin (New Series) of the American Mathematical Society*, **Vol 42**, 1(2004), pp 3–38.

91. A. Granville, "Smooth Numbers: Computational Number Theory and Beyond", *Algorithmic Number Theory*, Edited by J. P. Buhler and P. Stevenhagen, Cambridge University Press, 2008, pp 267–323.

92. B. Green and T. Tao, "The Primes Contain Arbitrarily Long Arithmetic Progressions", *Annal of Mathematics*, **167**, (2008), pp 481–547.

93. R. B. Griffiths and C. S. Niu, "Semiclassical Fourier Transform for Quantum Computation", *Physical Review Letters*, **76**, (1996), pp 3228–3231.

94. B. H. Gross, An Elliptic Curve Test for Mersenne Primes, *Journal of Number Theory*, **110**, 2005, pp 114-119.

95. F. Guterl, "Suddenly, Number Theory Makes Sense to Industry", *International Business Week*, 20 June 1994, pp 62–64.

96. G. H. Hardy, *A Mathematician's Apology*, Cambridge University Press, 1979.

97. G. H. Hardy and E. M. Wright, *An Introduction to Theory of Numbers*, 5th Edition, Oxford University Press, 1979.

98. G. H. Hardy and J. E. Littlewood, "Some Problems of 'Partitio Numerorum' III: on the Expression of a Number as a Sum of Primes", *Acta Mathemtica*, **44** (1923) pp 1–70. Reprinted in *Collected Papers of G. H. Hardy*, Volumn I, pp 561–630, Clarendon Press, Oxford, 1966

99. I. N. Herstein, *Topics in Algebra*, Second Edition, Wiley, 1975.

100. I. N. Herstein, *Abstract Algebra*, Third Edition, Wiley, 1999.

101. M. Hellman, "Private Communications", 2001–2003.

102. M. J. Hinek, *Cryptanalysis of RSA and Its Variants*, Chapman & Hall/CRC Press, 2009.

103. J. Hoffstein, J. Pipher and J. H. Silverman, "A Ring-Based Public-Key Cryptosystem", *Algorithmic Number Theory ANTS-III*, Lecture Notes in Computer Science **1423**, Springer, 1998, pp 267–288.

104. J. Hoffstein, N. Howgrave-Graham, J. Pipher, J. H. Silverman and W. Whyte, "NTRUEncrypt and NTRUSign: Efficient Public Key Algorithmd for a Post-Quantum World", *Proceedings of the International Workshop on Post-Quantum Cryptography (PQCrypto 2006)*, 23-26 May 2006, pp 71–77.

105. L. Hua, *Introduction to Number Theory*, English Translation from Chinese by Peter Shiu, Springer, 1980.

106. R. J. Hughes, "Cryptography, Quantum Computation and Trapped Ions", *Philosophic Transactions of the Royal Society London*, Series **A**, **356**, (1998), pp 1853–1868.

107. R. M. Huizing, "An Implementation of the Number Field Sieve", *Experiemnetal Mathematics*, **5** 3(1996), pp 231–253.

108. D. Husemöller, *Elliptic Curves*, Graduate Texts in Mathematics **111**, Springer, 1987.

109. K. Ireland and M. Rosen, *A Classical Introduction to Modern Number Theory*, 2nd Edition, Graduate Texts in Mathematics **84**, Springer, 1990.

110. T. H. Jackson, *From Number Theory to Secret Codes*, A Computer Illustrated Text, Adam Hilger, Bristol, 1987.

111. T. H. Jackson, *From Polynomials to Sum of Squares*, A Computer Illustrated Text, Adam Hilger, Bristol, 1996.

112. M. J. Jacobson, N. Koblitz, J. H. Silverman, A. Stein, E. Teske, "Analysis of the Xedni Calculus Attack", *Designs, Codes and Cryptography*, **20**, 2000, pp 41-64.

113. D. Johnson, A. Menezes and S. Vanstone, "The Elliptic Curve Digital Signatures Algorithm (ECDSA)", *International Journal of Information Security*, **1**, 1(2001), pp 36–63.

114. M. F. Jones, M. Lal and W. J. Blundon, "Statistics on certain large primes", *Mathematics of Computation*, **21** (1967), pp 103–107.

115. A. Joux, *Algorithmic Cryptanalysis*, Chapman & Hall/CRC Press, 2009.

116. J Kate and Y. Lindell, *Introduction to Modern Cryptography*, Chapman & Hall/CRC, 2008.

117. K. Kato, N. Kurokawa and T. Saito, *Number Theory 1: Fermat's Dream*, AMS, 2000.

118. J. Kilian, *Uses of Randomness in Algorithms and Protocols*, MIT Press, 1990.

119. D. E. Knuth, *The Art of Computer Programming II: Seminumerical Algorithms*, 3rd Edition, Addison-Wesley, 1998.

120. N. Koblitz, "Elliptic Curve Cryptography", *Mathematics of Computation*, **48**, (1987), pp 203–209.

121. N. Koblitz, *Introduction to Elliptic Curves and Modular Forms*, 2nd Edition, Graduate Texts in Mathematics **97**, Springer, 1993.

122. N. Koblitz, *A Course in Number Theory and Cryptography*, 2nd Edition, Graduate Texts in Mathematics **114**, Springer, 1994.

123. N. Koblitz, *Algebraic Aspects of Cryptography*, Algorithms and Computation in Mathematics **3**, Springer, 1998.

124. N. Koblitz, "A Survey of Number Theory and Cryptography", in: *Number Theory*, Edited by . P. Bambah, V. C. Dumir and R. J. Hans-Gill, Birkhäser, 2000, pp 217–239.

125. N. Koblitz, "Cryptography", in: *Mathematics Unlimited – 2001 and Beyond*, Edited by B. Enguist and W. Schmid, Springer, 2001, pp 749–769.

126. E. Kranakis, *Primality and Cryptography*, John Wiley & Sons, 1986.

127. L. J. Lander and T. R. Parkin, "Consecutive primes in arithmetic progression", *Mathematics of Computation*, **21**, (1967), page 489.

128. S. Lang, *Elliptic Functions*, 2nd Edition, Springer, 1987.

129. D. H. Lehmer, On Lucas's Test for the Primality of Mersenne's Numbers, Journal of London Mathematical Society, **10**, 1935, pp 162–165.

130. H. W. Lenstra, Jr., *Primality Testing Algorithms*, Séminaire N. Bourbaki 1980-1981, Number 561, pp 243-257. Lecture Notes in Mathematics **901**, Springer, 1981.

131. H. W. Lenstra, Jr., "Factoring Integers with Elliptic Curves", *Annals of Mathematics*, **126**, (1987), pp 649–673.

132. A. K. Lenstra and H. W. Lenstra, Jr. (editors), *The Development of the Number Field Sieve*, Lecture Notes in Mathematics **1554**, Springer, 1993.

133. A. K. Lenstra, H. W. Lenstra, Jr., and L. Lovász, "Factoring Polynomials with Rational Coefficients", *Mathematische Annalen*, **261**, (1982), pp 515–534.

134. H. W. Lenstra, Jr., and C. Pomerance, "Primality Testing with Gaussian Periods", Reprint, 2008.

135. E. Lucas, Noveaux Théorèmes d'Arithmétique Supèrieure, C. R. Acad. Sci. Paris, **83**, 1876, pp 1286–1288.

136. F. J. MacWilliams and N. J. A. Sloana, *The Theory of Error Correcting Codes*, North-Holland, 2001.

137. U. Maurer, "Fast Generation of Prime Numbers and Secure Public-Key Cryptographic Parameters", *Journal of Cryptology*, **8**, 3(1995), pp 123–155,

138. J. H. McClellan and C. M. Rader, *Number Theory in Digital Signal Processing*, Prentice-Hall, 1979.

139. K. S. McCurley, "The Discrete Logarithm Problem", in: *Cryptology and Computational Number Theory*, edited by C. Pomerance, Proceedings of Symposia in Applied Mathematics **42**, American Mathematics Society, 1990, pp 49–74.

140. K. S. McCurley, "Odds and Ends from Cryptology and Computational Number Theory", in: *Cryptology and Computational Number Theory*, edited by C. Pomerance, Proceedings of Symposia in Applied Mathematics **42**, American Mathematics Society, 1990, pp 49–74.

141. R. J. McEliece, *A Public-Key Cryptosystem based on Algebraic Coding Theory*, JPL DSN Progress Report 42-44, 1978, pp 583–584.

142. H. McKean and V. Moll, *Elliptic Curves – Function Theory, Geometry, Arithmetic*, Cambridge University Press, 1997.

143. A. R. Meijer, "Groups, Factoring, and Cryptography" *Mathematics Magazine*, **69**, 2(1996), pp 103–109.

144. A. Menezes and S. A. Vanstone, "Elliptic curve cryptosystems and their implementation", *Journal of Cryptology*, **6**, (1993), pp 209–224.

145. A. Menezes, P. C. van Oorschot and S. A. Vanstone, *Handbook of Applied Cryptosystems*, CRC Press, 1996.

146. R. C. Merkle, "Secure Communications over Insecure Channels" *Communications of the ACM*, **21**, (1978), pp 294–299. (Submitted in 1975.)

147. J. F. Mestre, "Formules Explicites et Minoration de Conducteurs de Variétés algébriques" *Compositio Mathematica*, **58**, (1986), pp 209–232.

148. B. Meyer and and V. Müller, "A Public Key Cryptosystem Based on Elliptic Curves over $\mathbb{Z}/n\mathbb{Z}$ Equivalent to Factoring", *Advances in Cryptology*, EUROCRYPT '96, Proceedings, Lecture Notes in Computer Science **1070**, Springer, 1996, pp 49–59.

149. G. Miller, "Riemann's Hypothesis and Tests for Primality", *Journal of Systems and Computer Science*, **13**, (1976), pp 300–317.

150. V. Miller, "Uses of Elliptic Curves in Cryptography", *Advances in Cryptology*, CRYPTO '85, Proceedings, Lecture Notes in Computer Science **218**, Springer, 1986, pp 417–426.

151. R. A. Mollin, *Fundamental Number Theory with Applications*, Chapman & Hall/CRC Press, 1998.

152. R. A. Mollin, *Algebraic Number Theory*, Chapman & Hall/CRC Press, 1999.

153. R. A. Mollin, *An Introduction to Cryptography*, Chapman & Hall/CRC, 2001.

154. R. A. Mollin, *RSA and Public-Key Cryptography*, Chapman & Hall/CRC Press, 2003.

155. P. L. Montgomery, "Speeding Pollard's and Elliptic Curve Methods of Factorization", *Mathematics of Computation*, **48**, (1987), pp 243–264.

156. F. Morain, *Courbes Elliptiques et Tests de Primalité*, Université Claude Bernard, Lyon I, 1990.

157. F. Morain, "Primality Proving Using Elliptic Curves: An Update", *Algorithmic Number Theory*, Lecture Notes in Computer Science **1423**, Springer, 1998, pp 111–127.

158. F. Morain, "Impementing the Asymptotically Fast Version of Elliptic Curve Primality Proving Algorithm", *Mathematics of Computation*, **76**, 257(2007), 493–505.

159. P. L. Montgomery, "A Survey of Modern Integer Factorization Algorithms", *CWI Quarterly*, **7**, 4(1994), pp 337–394.

160. M. A. Morrison and J. Brillhart, "A Method of Factoring and the Factorization of F_7", *Mathematics of Computation*, **29**, (1975), pp 183–205.

161. R. Motwani and P. Raghavan, *Randomized Algorithms*, Cambridge University Press, 1995.

162. R. A. Mollin, *RSA and Public-Key Cryptography*, Chapman & Hall/CRC Press, 2003.

163. M. B. Nathanson, *Elementary Methods in Number Theory*, Springer, 2000.

164. M. A. Nielson and I. L. Chuang, *Quantum Computation and Quantum Information*, Cambridge University Press, 2000.

165. I. Niven, H. S. Zuckerman and H. L. Montgomery, *An Introduction to the Theory of Numbers*, 5th Edition, John Wiley & Sons, 1991.

166. S. Pohlig and M. Hellman, "An Improved Algorithm for Computing Logarithms over GF(p) and its Cryptographic Significance", *IEEE Transactions on Information Theory*, **24**, (1978), pp 106–110.

167. J. O'Connor and E. Robertson, *The MacTutor History of Mathematics Archive*, http://www.groups.dcs.st-and.ac.uk/~history/Mathematicians.

168. A. M. Odlyzko, "Discrete Logarithms in Finite Fields and their Cryptographic Significance", *Advances in Cryptography*, EUROCRYPT '84, Proceedings, Lecture Notes in Computer Science **209**, Springer, 1984, pp 225–314.

169. O. Ore, *Number Theory and its History*, Dover Publications, 1988.

170. A. Papanikolaou and S. Y. Yan, "Prime Number Generation Based on Pocklington's Theorem", *International Journal of Computer Mathematics*, **79**, (2002), pp 1049–1056.

171. C. P. Pfleeger, *Security in Computing*, Prentice-Hall, 1997.

172. J. Pieprzyk, T. Hardjono and J. Seberry, *Fundamentals of Computer Security*, Springer, 2003.

173. R. G. E. Pinch, "Some Primality Testing Algorithms", *Notices of the American Mathematical Society*, **40**, 9(1993), pp 1203–1210.

174. R. G. E. Pinch, *Mathematics for Cryptography*, Queen's College, University of Cambridge, 1997.

175. S. C. Pohlig and M. Hellman, "An Improved Algorithm for Computing Logarithms over GF(p) and its Cryptographic Significance", *IEEE Transactions on Information Theory*, **24**, (1978), pp 106–110.

176. J. M. Pollard, "A Monte Carlo Method for Factorization", *BIT*, **15**, (1975), pp 331–332.

177. J. M. Pollard, "Monte Carlo Methods for Index Computation (mod p)", *Mathematics of Computation*, **32**, (1980), pp 918–924.

178. C. Pomerance, "Analysis and Comparison of some Integer Factoring Algorithms", in H. W. Lenstra, Jr. and R. Tijdeman, eds., *Computational Methods in Number Theory*, No **154/155**, Mathematical Centrum, Amsterdam, 1982, pp 89–139.

179. C. Pomerance, "The Quadratic Sieve Factoring Algorithm", *Proceedings of Eurocrypt 84*, Lecture Notes in Computer Science **209**, Springer, 1985, pp 169–182.

180. C. Pomerance, "Very Short Primality Proofs", *Mathematics of Computation*, **48**, (1987), pp 315–322.

181. C. Pomerance (editor), *Cryptology and Computational Number Theory*, Proceedings of Symposia in Applied Mathematics **42**, American Mathematical Society, 1990.

182. C. Pomerance, "Cryptology and Computational Number Theory – An Introduction", *Cryptology and Computational Number Theory*, edited by C. Pomerance, Proceedings of Symposia in Applied Mathematics **42**, American Mathematical Society, 1990, pp 1–12.

183. C. Pomerance, "On the Role of Smooth Numbers in Number Theoretic Algorithms", *Proceedings of the Intenational Congress of Mathematicians*, Zurich, Switzerland 1994, Birkhauser Verlag, Basel, 1995, pp 411–422.

184. C. Pomerance, "The Number Field Sieve", *Mathematics of Computation, 1943-1993, Fifty Years of Computational Mathematics*, edited by W. Gautschi, Proceedings of Symposium in Applied Mathematics **48**, American Mathematical Society, Providence, 1994, pp 465–480.

185. C. Pomerance, "A Tale of Two Sieves", *Notice of the AMS*, **43**, 12(1996), pp 1473–1485.

186. C. Pomerance, "Smooth Numbers and the Quadratic Sieve", *Algorithmic Number Theory*, Edited by J. P. Buhler and P. Stevenhagen, Cambirdge University Press, 2008, pp 69–81.

187. C. Pomerance, "Elementary Thoughts on Discrete Logarithms", *Algorithmic Number Theory*, Edited by J. P. Buhler and P. Stevenhagen, Cambirdge University Press, 2008, pp 385–395.

188. C. Pomerance, *Primality Testing: Variations on a Theme of Lucas*, Dept of Mathematics, Dartmouth College, Hanover, NH 03874, USA.

189. C. Pomerance, J. L. Selfridge and S. S. Wagstaff, Jr., "The Pseudoprimes to $25 \cdot 10^9$", *Mathematics of Computation*, **35**, (1980), pp 1003–1026.

190. V. R. Pratt, "Every Prime Has a Succinct Certificate", *SIAM Journal on Computing*, **4**, (1975), pp 214–220.

191. M. O. Rabin, "Probabilistic Algorithms for Testing Primality", *Journal of Number Theory*, **12**, (1980), pp 128–138.

192. D. Redmond, *Number Theory: An Introduction*, Marcel Dekker, New York, 1996.

193. P. Ribenboim, *The Little Book on Big Primes*, Springer, 1991.

194. P. Ribenboim, "Selling Primes", *Mathematics Magazine*, **68**, 3(1995), pp 175–182.

195. P. Ribenboim, *The New Book of Prime Number Records*, Springer, 1996.

196. J. Richstein, "Goldbach's Conjecture up to $4 \cdot 10^{14}$", *Mathematics of Computation*, **70**, (2001), pp 1745-1749.

197. H. J. J. te Riele, "Factorization of RSA-140 using the Number Field Sieve", http://www.crypto-world.com/announcements/RSA140.txt, 4 February 1999.

198. H. J. J. te Riele, "Factorization of a 512-bits RSA Key using the Number Field Sieve", http://www.crypto-world.com/announcements/RSA155.txt, 26 August 1999.

199. H. J. J. te Riele, W. Lioen and D. Winter, "Factoring with the Quadrtaic Sieve on Large Vector Computers", *Journal of Computational and Applied Mathematics*, **27**, (1989), 267–278.

200. H. Riesel, *Prime Numbers and Computer Methods for Factorization*, Birkhäuser, 1990.

201. R. L. Rivest, A. Shamir and L. Adleman, A Method for Obtaining Digital Signatures and Public Key Cryptosystems, *Communications of the ACM*, **21**, 2(1978), pp 120–126.

202. R. L. Rivest and R. D. Silverman, Are 'Strong' Primes Needed for RSA, *Communications of the ACM*, **21**, 2(1978), pp 120–126.

203. R. L. Rivest and B. Kaliski, RSA Problem, In: *Encyclopedia of Cryptography and Security*, Edited by H. C. A. van Tilborg, Springer, 2005.

204. H. E. Rose, *A Course in Number Theory*, 2nd Edition, Oxford University Press, 1994.

205. M. Rosen, A Proof of the Lucase-Lehmer Test, *American Mathematics Monthly*, **95**, 1988, pp 855–856.

206. K. Rosen, *Elementary Number Theory and its Applications*, 4th Edition, Addison-Wesley, 2000.

207. J. Rothe, *Complexity Theory and Cryptography*, Springer, 2005.

208. J. J. Rotman, *An Introduction to the Theory of Groups*, Springer, 1994.

209. J. J. Rotman, *Advanced Modern Algebra*, Prentice-Hall, 2002.

210. J. J. Rotman, *A First Course in Abstract Algebra*, 3rd Edition, Prentice-Hall, 2008.

211. R. Rumely, "Recent advances in primality testing", *Notices of the AMS*, **30**, 8(1983), pp 475–477.

212. A. Salomaa, *Public-Key Cryptography*, 2nd Edition, Springer, 1996.

213. Y. Saouter, *Vinogradov's Theorem is True up to 10^{20}*, Publication Interne No. 977, IRISA, 1995.

214. J. F. Schneiderman, M. E. Stanley and P. K. Aravind, "A Pseudo-Simulation of Shor's Quantum Factoring Algorithm", *arXiv:quant-ph/0206101v1*, 16 Jun 2002.

215. B. Schneier, *Applied Cryptography – Protocols, Algorithms, and Source Code in C*, 2nd Edition, John Wiley & Sons, 1996.

216. C. P. Schnorr, "Efficient Identification and Signatures for Smart Cards", *Advances in Cryptography*, CRYPTO '89, Proceedings, Lecture Notes in Computer Science **435**, Springer, 1990, pp 239–252.

217. R. Schoof, "Elliptic Curves over Finite Fields and the Computation of Square Roots mod p", *Mathematics of Computation*, **44**, (1985), pp 101–126.

218. R. Schoof, "Four Primality Testing Algorithms", *Algorithmic Number Theory*, Edited by J. P. Buhler and P. Stevenhagen, Cambridge University Press, 2008, pp 69–81.

219. M. R. Schroeder, *Number Theory in Science and Communication*, 3rd Edition, Springer Series in Information Sciences **7**, Springer, 1997.

220. A. Shamir, "Factoring Numbers in $\mathcal{O}(\log n)$ Arithmetic Steps", *Information Processing Letters*, **8**, 1(1979), pp 28–31.

221. A. Shamir, "How to Share a Secret", *Communications of the ACM*, **22**, 11(1979), pp 612–613.

222. D. Shank, "Class Numbers, a Theory of Factorization and Genera", *The 1969 Number Theory Institute, Stony Brook, NY, Proceedings of Symposia in Pure Mathematics*, **20**, 1969, pp 415–440.

223. P. Shor, "Algorithms for Quantum Computation: Discrete Logarithms and Factoring", *Proceedings of 35th Annual Symposium on Foundations of Computer Science*, IEEE Computer Society Press, 1994, pp 124–134.

224. P. Shor, "Polynomial-Time Algorithms for Prime Factorization and Discrete Logarithms on a Quantum Computer", *SIAM Journal on Computing*, **26**, 5(1997), 1484–1509.

225. V. Shoup, "Searching for Primitive Roots in Finite Fields", *Mathematics of Computation*, **58**, 197(1992), pp 369–380.

226. J. H. Silverman and J. Tate, *Rational Points on Elliptic Curves*, Undergraduate Texts in Mathematics, Springer, 1992.

227. J. H. Silverman, *The Arithmetic of Elliptic Curves*, Graduate Texts in Mathematics **106**, Springer, 1994.

228. J. H. Silverman, *A Friendly Introduction to Number Theory*, Second Edition, Prentice-Hall, 2001.

229. J. H. Silverman, "The Xedni Calculus and the Elliptic Curve Discrete Logarithm Problem", *Designs, Codes and Cryptography*, **20**, 2000, pp 5-40.

230. J. H. Silverman, "The Xedni Calculus and the Elliptic Curve Discrete Logarithm Problem", *Designs, Codes and Cryptography*, **20**, 2000, pp 5-40.

231. J. H. Silverman and J. Suzuki, "Elliptic Curve Discrete Logarithms and the Index Calculus", *Advances in Cryptology – ASIACRYPT '98*, Springer Lecture Notes in Computer Science **1514**, 1998, pp 110–125.

232. R. D. Silverman, 'The Multiple Polynomial Quadratic Sieve", *Mathematics of Computation*, **48**, 177(1987), 329–339.

233. R. D. Silverman, "A Perspective on Computational Number Theory", *Notices of the American Mathematical Society*, **38**, 6(1991), pp 562–568.

234. R. D. Silverman, "Massively Distributed Computing and Factoring Large Integers", *Communications of the ACM*, **34**, 11(1991), pp 95–103.

235. D. R. Simon, "On the Power of Quantum Computation", *Proceedings of the 35th Annual IEEE Symposium on Foundations of Computer Science*, IEEE Press, 1994, pp 116–123.

236. S. Singh, *The Code Book – The Science of Secrecy from Ancient Egypt to Quantum Cryptography*, Fourth Estate, London, 1999.

237. S. Singh, *The Science of Secrecy – The Histroy of Codes and Codebreaking*, Fourth Estate, London, 2000.

238. N. Smart, *Cryptography: An Introduction*, McGraw-Hill, 2003.

239. R. Solovay and V. Strassen, "A Fast Monte-Carlo Test for Primality", *SIAM Journal on Computing*, **6**, 1(1977), 84–85. "Erratum: A Fast Monte-Carlo Test for Primality", *SIAM Journal on Computing*, **7**, 1(1978), page 118.

240. M. Stamp and R. M. Low, *Applied Cryptanalysis*, Wiley, 2007.

241. J. Stillwell, *Elements of Number Theory*, Springer, 2000.

242. D. R. Stinson, *Cryptography: Theory and Practice*, 2nd Edition, Chapman & Hall/CRC Press, 2002.

243. N. S. Szabo and R. I. Tanaka, *Residue Arithmetic and its Applications to Computer Technology*, McGraw-Hill, 1967.

244. US Department of Commerce/National Institute of Standards and Technology, "Digital Signature Standard (DSS)Proposed by NIST", *Federal Information Processing Standards Publication FIPS-PUBH 186-2*, 2000.

245. L. M. K. Vandersypen, M. Steffen, G. Breyta, C. S. Tannoni, M. H. Sherwood, and I. L. Chuang, "Experiemental Realization of Shor's Quantum Factoring Algorithm Uisng Nuclear Magnetic Resonance", *Nature*, **414**, 20/27 December 2001, pp 883–887.

246. H. C. A. van Tilborg, *Fundamentals of Cryptography*, Kluwer Academic Publishers, 1999.

247. W. Trappe and L. Washington, *Introduction to Cryptography with Coding Theory*, 2nd Edition, Prentice-Hall, 2006.

248. I. M. Vinogradov, "Representation of an odd number as the sum of three primes" (in Russian), *Dokl. Akad. Nauk USSR*, **16**, (1937) pp 179–195.

249. S. Wagon, "Primality Testing", *The Mathematical Intelligencer*, **8**, 3(1986), pp 58–61.

250. S. S. Wagstaff, Jr., "Prime Numbers with a Fixed Number of One Bits or Zero Bits in Their Binary Representation", *Experimental Mathematics*, **10**, 2(2001), pp 267–273.

251. S. S. Wagstaff, Jr., *Cryptanalysis of Number Theoretic Ciphers*, Chapman & Hall/CRC Press, 2002.

252. S. S. Wagstaff, Jr., "Is there a shortage of primes for cryptography?", *International Journal of Network Security*, **3**, (2006), pp 296–299.

253. Y. Wang, "On the Least Positive Primitive Root", *The Chinese Journal of Mathematics*, **9**, 4(1959), pp 432–441.

254. L. C. Washington, *Elliptic Curves: Number Theory and Cryptography*, Chapman & Hall/CRC Press, 2008.

255. H. Wiener, "Cryptanalysis of Short RSA Secret Exponents", *IEEE Transactions on Information Theory*, **36**, 3(1990), pp 553–558.

256. A. Wiles, "Modular Elliptic Curves and Fermat's Last Theorem", *Annals of Mathematics*, **141**, (1995), pp 443–551.

257. A. Wiles, "The Birch and Swinnerton-Dyer Conjecture", *The Millennium Prize Problems*, Clay Mathematical Institute and American Mathematical Institute, 2006, pp 31–41.

258. H. S. Wilf, *Algorithms and Complexity*, 2nd Edition, A. K. Peters, 2002.

259. H. C. Williams, "A Modification of the RSA Public-Key Encryption Procedure", *IEEE Transactions on Information Theory*, **26**, (1980), pp 726–729.

260. H. C. Williams, "The Influence of Computers in the Development of Number Theory", *Computers & Mathematics with Applications*, **8**, 2(1982), pp 75–93.

261. H. C. Williams, "A $p+1$ Method of Factoring", *Mathematics of Computation*, **39**, (1982), pp 225–234.

262. H. C. Williams, "Factoring on a Computer", *Mathematical Intelligencer*, **6**, 3(1984), pp 29–36.

263. H. C. Williams, "An M^3 Public-Key Encryption Scheme", *Advances in Cryptology - CRYPTO '85*, Lecture Notes in Computer Science **218**, Springer, 1986, pp 358–368.

264. H. C. Williams, *Édouard Lucas and Primality Testing*, John Wiley & Sons, 1998.

265. C. P. Williams and S. H. Clearwater, *Explorations in Quantum Computation*, The Electronic Library of Science (TELOS), Springer, 1998.

266. S. Y. Yan, "Primality Testing of Large Numbers in Maple", *Computers & Mathematics with Applications*, **29**, 12(1995), pp 1–8.

267. S. Y. Yan, A New Cryptographic Scheme based on the kth Power Residuosity Problem, 15th British Colloquium for Theoretical Computer Science (BCTCS15) , Keele University, 14-16 April 1999.

268. S. Y. Yan, *Number Theory for Computing*, 2nd Edition, Springer, 2002.

269. S. Y. Yan, "Computing Prime Factorization and Discrete Logarithms: From Index Calculus to Xedni Calculus", *International Journal of Computer Mathematics*, **80**, 5(2003), pp 573–590.

270. S. Y. Yan, *Cryptanalytic Attacks on RSA*, Springer, 2008.

271. S. Y. Yan and G. James, "Testing Mersenne Primes with Elliptic Curves", *Computer Algebra in Scientific Computing*, Lecture Notes in Computer Science **4194**, Springer, 2006, pp 303–312.

272. S. Y. Yan, C. Maple and G. James, "A New Scheme for Deniable/Repudiable Authentication", *Computer Algebra in Scientific Computing*, Lecture Notes in Computer Science **4770**, Springer, 2007, pp 424–432.

Index

B-smooth number, 238
$\lambda(n)$, 58
$\mu(n)$, 58
$\phi(n)$, 56
$\pi(x)$, 139
$\sigma(n)$, 53
$\tau(n)$, 53
$\zeta(s)$, 1
b-sequence, 162
kth (higher) power non-residue, 109
kth (higher) power residue, 109
kth Power Residuosity Problem (kPRP), 272
kth Root Problem (kRTP), 271
kth power non-residue, 87
kth power residue, 87
$n-1$ primality test, 151

Primality test by trial divisions, 150

additive group, 14
additive identity, 17
additive inverse, 17
algebraic computation law, 121
algebraic equation, 46
algebraic integer, 22, 240
algebraic number, 21, 240
almostfield, 187
APR-test, 177
arithmetic function, 50
arithmetic progression of consecutive primes, 9, 11
arithmetic progression of primes, 9
associativity, 13
asymmetric key cryptosystem, 290
authentication, 288

base-2 pseudoprimality test, 160
binary Goldbach conjecture, 3

Carmichael number, 159

Carmichael's λ-function, 58, 79
Carmichael's theorem, 79
CFRAC factoring algorithm, 232
CFRAC method, 230
Chinese Remainder Theorem (CRT), 81
Chinese test, 161
closure, 13
coding-based cryptography, 346
combined test, 170
common modulus attack, 308
common multiple, 29
commutative group, 13
commutative ring, 16
commutativity, 14
complete system of residues, 67
completely multiplicative function, 51
complex zeros, 141
composite number, 24
compositeness tests, 157
congruence classes, 65
congruent, 63
conic, 114
consecutive pairs of quadratic residues, 88
consecutive triples of quadratic residues, 89
Continued FRACtion (CFRAC) method, 210
continued fraction algorithm, 45
convergent, 39
convergents, 47
Converse of the Fermat little theorem, 78
Converse of Wilson's theorem, 80
cryptanalysis, 287
cryptography, 287
cryptology, 287
cubic Diophantine equation, 114
cubic integer, 240

cyclic group, 14

degree of polynomial, 19
deterministic encryption, 320
Diffie-Hellman-Merkle key-exchange,
    326
Diophantine geometry, 113
Dirichlet L-functions, 145
Dirichlet characters, 145
Dirichlet series, 145
discrete logarithm, 108
discrete logarithm problem, 257
dividend, 24
division algorithm, 24
division ring, 16
divisor, 23
domain, 50

ECPP (Elliptic Curve Primality
    Proving), 174
ECPP Algorithm, 175
ElGamal cryptosystem, 329
elliptic curve, 114
elliptic curve analogue of Diffie–
    Hellman, 333
elliptic curve analogue of ElGamal, 334
elliptic curve analogue of Massey–
    Omura, 334
elliptic curve analogue of RSA, 335
elliptic curve cryptography (ECC), 331
Elliptic Curve Digital Signature
    Algorithm (ECDSA), 337
elliptic function, 117
elliptic integral, 117
embedding messages on elliptic curves,
    332
ENIGMA code, 288
equivalence classes, 65
equivalence relation, 65
Euclid, 25
Euclid's algorithm, 35
Euclid's *Elements*, 37
Euclidean prime, 178
Euler probable prime, 166
Euler pseudoprime, 166
Euler's (totient) ϕ-function, 56
Euler's criterion, 90
Euler's pseudoprimality test, 166
Euler's theorem, 78
even number, 24
extended Euclid's algorithm, 74

factor, 23
factoring by trial divisions, 212

fast group operations, 135
fast modular exponentiations, 132
fast point additions, 135
Fermat probable prime, 159
Fermat pseudoprime, 159
Fermat's factoring algorithm, 213
Fermat's little theorem, 77
Fibonacci numbers, 168
field, 16
finite fields, 18
finite group, 14
finite order of a point on an elliptic
    curve, 119
finite simple continued fraction, 40
fixed-point attack, 308
Fundamental Theorem of Arithmetic,
    27

Galois field, 18
Gauss sum primality test, 179
Gauss's lemma, 92
Gaussian integer, 22
Gaussian prime, 22
Generalized (Extended) Riemann
    Hypothesis, 3
Generalized Riemann Hypothesis, 147
generating function, 145
geometric composition law, 117
Goldbach Conjecture, 3
greatest common divisor (gcd), 28
group, 13
group laws on elliptic curves, 117

height, 122
high-order congruence, 85
hybrid cryptosystem, 291

identity, 13
incongruent, 64
index calculus method, 265
index of a to the base g, 108
index of an integer modulo n, 108
infinite fields, 18
infinite group, 14
infinite order of a point on an elliptic
    curve, 119
infinite simple continued fraction, 42
integer factorization problem, 209
integral domain, 16
inverse, 13
inverse of RSA function, 301
irrational numbers, 42
irreducible polynomial, 21

Jacobi sum test, 180
Jacobi sums, 180
Jacobi symbol, 98

large random prime generation, 202
lattice-based cryptography, 348
least common multiple (lcm), 29
least non-negative residue, 64
least residue, 92
Legendre symbol, 90
Legendre's congruence, 226
Legendre, A. M., 90
Lehman's method, 209
Lenstra's Elliptic Curve Method
    (ECM), 210, 222
linear congruence, 74
linear Diophantine equation, 46
Lucas numbers, 168
Lucas probable prime, 169
Lucas pseudoprimality test, 170
Lucas pseudoprime, 169
Lucas sequences, 168
Lucas test, 169
Lucas theorem, 169

Möbius μ-function, 58
Möbius inversion formula, 60
Massey–Omura cryptosystem, 329
Mersenne Prime Conjecture, 8
Mersenne primes, 8
Miller-Rabin test, 163
minimal polynomial, 22
mock residue symbol, 177
modular arithmetic in $\mathbb{Z}/n\mathbb{Z}$, 69
modular inverse, 72
modulus, 64
monic, 19
multiple, 23
Multiple Polynomial Quadratic Sieve
    (MPQS), 210
multiple polynomial quadratic sieve
    (MPQS), 235
multiplicative function, 51
multiplicative group, 14
multiplicative identity, 17
multiplicative inverse, 17, 72

non-secret encryption, 289
non-singular curve, 115
non-singular elliptic curve, 115
non-witness, 165
non-zero field element, 17
nontrivial divisor, 24
nontrivial square root of 1, 161

nontrivial zeros, 141
norm, 22
NTRU, 348
Number Field Sieve (NFS), 210, 239,
    268

odd number, 24
odd perfect number, 9
one-way trap-door function, 301
order of a modulo n, 103
order of a field, 18
order of a point on an elliptic curve,
    119

P vs NP Problem, 129
partial quotients, 39
perfect number, 8, 9
perfect square, 32
period, 44
periodic simple continued fraction, 44
Pocklington's theorem, 173, 202
Pohlig-Hellman cryptosystem, 337
point at infinity, 116
polarization, 349
Pollard's ρ factoring algorithm, 218
Pollard's ρ-method, 210, 214
Pollard's "$p - 1$" factoring algorithm,
    219
Pollard's "$p - 1$" method, 219
polynomial, 19
polynomial congruence, 85
polynomial congruential equation, 85
polynomial security, 320
powerful number, 34
Pratt's primality proving, 154
Primality test based on order of
    integers, 152
Primality test based on primitive roots,
    151
prime counting function, 139
prime factor, 26
prime field, 18
prime number, 24
Prime Number Theorem, 139
prime power, 18
primitive root of n, 104
privacy, 288
private key, 291
probabilistic encryption, 320, 322
probable prime, 159, 204
probable safe prime, 204
probable strong prime, 205
proper divisor, 23

pseudofield, 188
pseudoprime, 159
public-key, 291
public-key cryptography, 289
public-key cryptosystem, 291
purely periodic simple continued
    fraction, 44

quadratic congruence, 86
quadratic integer, 240
quadratic irrational, 44
quadratic non-residue, 87
Quadratic reciprocity law, 95
quadratic residue, 87
Quadratic Residuosity Problem (QRP),
    271, 321
Quadratic Sieve (QS), 234
quantum algorithm for discrete
    logarithms, 268
quantum algorithm for integer
    factorization, 253
quantum cryptography, 349
quantum register, 253
quotient, 24

Rabin's M^2 encryption, 314
randomized encryption, 320
rank of an elliptic curve, 121
rational integer, 22
rational integers, 241
rational line, 114
rational number, 114
rational numbers, 40
rational point, 114
rational prime, 22
real base logarithm, 108
real number, 44
real zeros, 141
real-valued function, 50
rectilinear polarization, 349
reduced system of residues modulo n,
    68
reflexive, 65
relatively prime, 28
remainder, 24
repeated doubling and addition, 136
repeated doubling method, 333
repeated squaring and multiplication,
    132
residue, 64
residue class, 65
residue classes, 65
residue of x modulo n, 65

Riemann ζ-function, 1
Riemann Hypothesis (RH), 142
Riemann's Hypothesis, 1
ring, 15
ring with identity, 16
root of polynomial, 19
RSA Assumption, 293
RSA assumption, 302
RSA conjecture, 302
RSA cryptosystem, 292
RSA function, 301
RSA problem, 301

safe prime, 204
safe prime generation, 204
secret-key, 291
secret-key cryptosystem, 291
semantic security, 320
Shanks' baby-step giant-step method
    for discrete logarithms, 258
Shanks' class group method, 210
Shanks' SQUFOF method, 210
Sieve of Eratosthenes, 25
Silver–Pohlig–Hellman algorithm, 260
simple continued fraction, 39
singular curve, 115
size of point on elliptic curve, 122
smooth number, 238
Solovay-Strassen test, 166
Sophie Germain Primes, 7
square number, 32
square root method, 260
SQuare RooT Problem (SQRTP), 271
strong prime, 204
strong probable prime, 163
strong pseudoprimality test, 161, 163
strong pseudoprime, 163
subgroup, 14
symmetric, 65

ternary Goldbach conjecture, 3
torsion subgroup, 121
transitive, 65
trial division, 210
trivial divisor, 24
trivial zeros, 141
Twin Prime Conjecture, 5

uccinct primality certification, 154

Wiener's Diophantine attack, 305
Williams' M^2 encryption, 317
Williams' M^3 encryption, 319
Wilson's primality test, 154

Wilson's theorem, 79
witness, 165

zero of polynomial, 19

About the Author

SONG Y. YAN majored in both Computer Science and Mathematics, and obtained a Doctorate in Mathematics (Number Theory) from the Department of Mathematics at the University of York, England, UK. His current research interests include Number Theory, Complexity Theory, Cryptography, and Information Security. His other publications include:

[1] *Cryptanalytic Attacks on RSA*, Springer, 2008.

[2] *Number Theory for Computing*, Second Edition, Springer, 2002. (Polish Translation, PWN, Warsaw, 2006; Chinese Translation, Tsinghua University Press, Beijing, 2008).

[3] *Perfect, Amicable and Sociable Numbers: A Computational Approach*, World Scientific, 1996.

Song is currently Visiting Professor in the Department of Mathematics at Harvard University and the Department of Mathematics at the Massachusetts Institute of Technology, USA, supported by a Global Research Award from the Royal Academy of Engineeering, London, UK. He can be contacted by syan@math.harvard.edu, syan@math.mit.edu, or songyuanyan@hotmail.com.

Printed in the United States of America